WORKING WITH PAPER

WORKING WITH PAPER

GENDERED PRACTICES IN THE HISTORY OF KNOWLEDGE

EDITED BY
Carla Bittel
Elaine Leong
Christine von Oertzen

University of Pittsburgh Press

Published by the University of Pittsburgh Press, Pittsburgh, Pa., 15260

Manufactured in the United States of America
Printed on acid-free paper
10 9 8 7 6 5 4 3 2 1

Cataloging-in-Publication data is available from the Library of Congress

ISBN 13: 978-0-8229-4559-8
ISBN 10: 0-8229-4559-2

Cover art: Page taken from the recipe book of Margaret Boscawen (d. 1688) and Bridget Fortescue (1666–1708). Devon Heritage Centre, 1262M/0/FC6.
Cover design: Joel W. Coggins

CONTENTS

ACKNOWLEDGMENTS

This project began with three colleagues having casual conversations about note-books, and resulted in a working group dedicated to collectively unpacking the *trias* of paper, gender, and the making of knowledge. *Working with Paper: Gendered Practices in the History of Knowledge* is the product of this group, which came together at the Max Planck Institute for the History of Science (MPIWG) in Berlin. After an initial conference in January 2016, that autumn we moved on to intensive work and two authors' workshops. Our gratitude goes to the participants of the opening conference and workshops, most of whom contributed essays to this book. From September to December 2016, the authors in this volume spent four months at the MPIWG as visiting scholars, taking time out of their teaching schedules to attend reading groups and colloquia, work on their own contributions, read each other's work, and develop our common framework through continued conversations. Department II of the MPIWG served as a congenial and stimulating environment for our collaboration and we thank its director, Lorraine Daston, for her continued support for this project.

Over the past several years we have all benefited from discussions on paper, gender, and knowledge in a number of seminars, workshops, and institutional settings around the globe. We thank the organizers and the audiences of these events for their constructive feedback and comments. We are particularly grateful to audiences at the History of Science Society meetings and the Department II colloquium at the MPIWG. During 2016, many scholars visiting the MPIWG were drawn into long deliberations about paper, gender, and knowledge. Some suggested additional readings, others helped us clarify our arguments—thank you to you all! A number of scholars and friends deserve our special thanks: Francesca Bray, Lucia Dacome, Vera Keller, Sally Gregory Kohlstedt, Nina Lerman, James Secord, Sally Shuttleworth, and Elizabeth Siegel Watkins. In Berlin,

Gangolf Ulbricht introduced us to hands-on papermaking in his workshop at the Künstlerhaus Bethanien.

The work to bring an edited volume with thirteen essays and more than fifty images to publication is no small feat. Here, we are grateful to Abby Collier at the University of Pittsburgh Press for her enthusiastic support for this project and to her team for their professionalism and efficiency in taking our manuscript to publication. At the MPIWG, our work was made much easier by our wonderful academic support staff. We are hugely indebted to our student assistant extraordinaire Laura Selle and to our colleagues in the library, especially Urte Brauckmann.

Finally, we are grateful to the many libraries and archives, referenced in this volume, who preserve precious materials and permit us to work with their paper.

INTRODUCTION

Paper, Gender, and the History of Knowledge

Carla Bittel, Elaine Leong, & Christine von Oertzen

*Consider what it took to craft a message at Queen Elizabeth's behest in sixteenth-*century England. Only the finest quality of Venetian writing paper would do. The style of the handwriting and the layout of the page had to be carefully matched with the content of the note. Once the text was written, the letter was folded in a particular form, locked, and artfully sealed with brightly colored silk floss in ways that would significantly speak to Her Majesty's intent.[1] To make the work of many seem like the work of one, the process required refined skills of both male and female members of the queen's household. The resulting letter emerged out of a set of social interactions that transformed paper from a material to a carrier of information, to an object of communication. It yielded complex meaning in which the inscription on the page fused with the material it carried and how it was packaged. Making and understanding such a letter thus required knowledge in all three categories in which paper operated: as a material, a writing substrate, and an object.

In *Working with Paper,* we demonstrate that the making and keeping of knowledge was never just about words or writing, but inseparable from material and social practices embedded in daily life. Paper has mattered in a myriad of ways beyond the seemingly transparent surface of the page. Intrinsically interwoven with reading and writing practices of the learned, the material was also utilized for a more varied spectrum of functions and uses. We argue that paper, so ubiquitous that it was often taken for granted, was enmeshed with social constructions of femininity and masculinity, similarly so pervasive as to be sometimes imperceptible. This volume sets out to denaturalize the material and the social. By making paper and gender visible in practices of knowledge production, we bridge the history of quotidian and scholarly practices, forging a history of knowledge that uniquely enfolds material histories with the gendered worlds that made them.

Working with Paper follows a diverse range of papers to demonstrate the vital role of the material and the social in practices of knowledge production. The abovementioned translucent Venetian writing paper, folded and stitched together with brightly colored silk floss to form a precious letter, is just one example of the kinds of materials and objects, and their respective meanings, scrutinized in this volume. Others include brown paper plasters to ease headaches, French paper twists to curl hair, and semitransparent tracing paper to sketch postural outlines. We see how millions of mass-produced, heavy-duty, card-size questionnaires, produced for Prussia's 1871 census, were shipped to and worked on by middle-class housewives in their own parlors, and observe how hefty eighteenth-century accounting books and small, resilient parchment slips attached to infants' waists were used to monitor the heavy traffic of children within and beyond Madrid's foundling house. We learn how the thin strips of papier-mâché, held together by a secret paste, were molded into human and animal anatomical models that educated not just medical students but also the "peasant" makers of the models themselves and the Egyptian students French educators sought to "civilize." In each case, we highlight the ways in which different kinds of paper—as material, meaning, and object—shaped epistemic practices, the work, and the people engaging with it.

To examine these uncharted waters, each contribution follows a "sociomaterial" paper trail, tracking the substance as it formed, functioned, and moved within and between a wide range of epistemic communities. By examining the "constitutive entanglement(s)" of the social and the material, we reveal new connections between sites of knowledge making, sets of knowers, and different kinds of epistemic work.[2] Focusing on material worlds, we bring the manipulations and uses of paper to the fore and shine light on the women and men, the families and communities of readers, writers, makers, workers, and consumers who performed such operations. We showcase their social and work environments, relationships, and networks as central to the historical narrative of knowledge practices.

We unlock these multilayered social and material entanglements by utilizing gender as a category of historical analysis. As a method developed to analyze the concrete and symbolic organization of social life through sets of relationships of power and difference, gender analysis enables us to dissect paper practices as technological "transactions" within the everyday power structures of a particular time and place.[3] By following paper—and the tools made from it—into the quotidian practices and workflows of knowledge production, we take inspiration from recent works in the history of gender and technology that redefine technology itself to encompass a wide spectrum of artifacts, knowledge, and practices, or in other words, "people's ways of making and doing things."[4] With this definition, scholars have helped shift our attention from masculine models of invention and creation to gendered ways of knowing and using, highlighting the complex ways in which knowledge, gender, and materials are "mutually constitutive."[5] In our case studies, the object of study (paper) and the methodology (gender analysis) form the grid through which we see knowledge practices unfold in multiple and continual negotiations of power. This new perspective emphasizes how knowledge production is inextricable from the power structures of the quotidian in which all epistemic work is carried out. Thus, via paper and gender, we contend that the history of knowledge practices is one tangled in the order and disorder of everyday life.[6]

United by their focus on paper and paper use, the essays in this volume cover case studies in Europe and North America. Spanning five hundred years from the early modern period to the mid-twentieth century, they collectively weave together a long-view history of working and knowing with paper. Through the afterword, we extend the discussions begun in the volume by reflecting on them from the East Asian context. Three main themes run through our essays and are reflected in the organization of the book and this introduction. The first part, "Beyond the Page: The Sociomaterial History of Paper," introduces readers to an expansive, rich world of papers and paper use. The second, "Transcending Boundaries: Tools and Technologies," showcases how this new focus on the sociomaterial offers us opportunities to extend current understandings of "paper tools" and "paper technologies" in the history of science. The final part, "Knowledge, Power, and the Everyday," demonstrates how attentiveness to the sociomateriality of paper and the use of gender analysis as a methodology prompt us to reconsider knowledge practices as dynamic negotiations of power and difference deeply rooted in daily life. References to particular essays weave in and out throughout this introduction, regardless of their position within the book, for these themes permeate all of our essays.

THE SOCIOMATERIAL HISTORY OF PAPER

Over the expansive period covered by this volume, paper became central to daily life and thus the larger history of gender, working, and knowing in the West. Beginning in the late medieval period, a wave of economic and consumer activity, nourished by the spoils of empire, brought many types of paper into English and European marketplaces. An expansion of literacy among both men and women accompanied a growing trade in books and papers. While work roles and learning became increasingly bifurcated and prescribed along gendered lines—and those divisions were powerful—there was a great deal of negotiation, as the sociomaterial history of paper demonstrates. Epistemic dynamics did not always abide by dualistic boundaries and barriers around bodies, sites, and materials, particularly around paper. Our case studies are connected by their attentiveness to paper as a material and the ways in which its various properties (portability, foldability, malleability, etc.) frame the production of knowledge and the construction of gender norms. Taken together, they make clear that paper, as a material, was far from stable and monolithic. Paper's various meanings shifted across boundaries of time and space, as did its "affordances," the possible or perceived actions taken with it.[7] In fact, it is perhaps more accurate to speak of our object(s) of study as "papers" or "paperstuff," as they encompass such a wide variety of types and uses. The images in this volume pay witness to this dazzling array.

Paper and its material properties have long been a topic of fascination for scholars. As Jacob Eyferth and others have shown us, paper was used in Asia for about one thousand years before it spread to Europe. In China, Japan, and beyond, it was cheaper, more common, and even more ubiquitously used than in the West, mostly because papermakers from very early on perfected methods to make paper directly from plant fibers.[8] For the Euro-American context addressed in this volume, paper has been studied in a number of fields. Professional papermakers and conservators such as Timothy Barrett have conducted detailed studies on the techniques of papermaking and analyzed the composition of various types of handmade paper.[9] As a crucial resource in postmedieval Europe, paper—and the industry that produces it—has long come under the scrutiny of economic and business historians.[10] Historians of the book and bibliographers have charted the central role played by paper in the business of book production and reception.[11] Scholars working on early modern print have argued that paper costs contributed to a great portion of the total cost of book production during this period, enabling them to analyze intended audiences and potential consumers through observations of the physical size of the book and use of paratextual apparatus such as blank spaces and decorative patterns.[12] Likewise, historians of science have emphasized key milestones in the production of paper, claiming that the advent of wood pulp in the beginning of the nineteenth century and the mass production of cheap paper revolutionized the market for books and journals as well as reading habits.[13] These studies make clear that *how* paper was made played a crucial role in the history of knowledge.

Two dominant papermaking techniques were in use in Europe and America from the mid-sixteenth century to the mid-twentieth century, and the long-view history presented by our collective essays straddles a major technological change in paper production. Until circa 1800, paper was largely handmade from recycled linen rags that were sorted, fermented or retted, washed, stamped or beaten, and then formed into thin sheets that were consequently pressed, dried, and sized.[14] By the eighteenth century, increased paper use had created a linen shortage and, across Europe, different groups of intellectuals and bureaucrats sought to solve this problem by finding alternate materials for papermaking. In our volume, Gabriella Szalay tells this story through the paper trials of the Protestant pastor Jacob Christian Schäffer.[15] Schäffer strove to find alternatives for paper production, experimenting with material ranging from poplar wool to wasps' nests to sawdust. These trials involved rich negotiations between bureaucrats, scholars, and papermakers. In Szalay's narrative, governance, natural history, experimentations, and artisanal expertise coalesced in a potent mix, in which multiple notions of masculinity were equally at stake.

After 1800 the papermaking process changed significantly. It became increasingly industrialized and mechanized, and wood pulp replaced the oft scarce linen rags as the main base for paper production. The use of chlorine bleach, rosin, and alum altered the color, appearance, and feel of paper and the new process produced long sheets of paper, ready to be cut to size, freeing papermakers from the size constraints imposed by the form.[16] The changeover from handmade, linen-based paper to factory-produced, wood pulp paper was gradual and not necessarily an easy feat, due to the deep affordances given to paper as a material and the firmly rooted practices of paper use. Makers and users of paper alike were attuned to the material differences in linen- and wood-based papers, and complex schemes were designed to assess and define various grades of paper, from color to smoothness to foldability. Not only were measures taken to differentiate the two kinds of paper as separate products destined for particular purposes but efforts were also made to create a wood-based paper that possessed many of the material properties associated with handmade linen-based papers. Given that the shift from handmade to industrialized paper prompted myriad sets of negotiations conducted by multiple actors, long-view studies offer opportunities to untangle changing sets of affordances given to paper.

Our volume shifts our current understandings of paper and its place in the history of knowledge in a number of ways by constructing a continuous history of paper across the handmade and industrialized periods of papermaking. First, we demonstrate that across the different geographical regions and time periods examined in this book, variations in environmental conditions, techniques, and ingredients produced, materially, a wide range of different papers. Paper producers and consumers, whether living in the sixteenth or twentieth century, were well aware of vast arrays of paper, and a vibrant paper market flourished across Europe and America. For the early modern context, studies on the "material text" have particularly brought this to our collective attention in recent years.[17] Paying attention to the materiality of paper such as watermarks, for example, has enabled scholars such as Nancy Pollard Brown to reconstruct communities of readers and writers (in this case underground networks of Elizabethan recusants) connected through social relationships framed by patronage and clientage.[18] It has also prompted others to bring to light the various materials other than linen extant in early modern papers, to reconstruct the history of printed damask paper, and to recover the history of erasable notebooks.[19]

In this volume, Elaine Leong, Gabriella Szalay, Heather Wolfe, and Simon Werrett join these revisionist studies in painting a rich material history of paper. They point to the wide variety of papers, from brown to cap to gray to fine white, and push back against traditional views that early modern paper was expensive and scarce, opening up a world where consumers were offered different kinds of paper at diverse price points.[20] They show that historical actors viewed this colorful world of paper through sophisticated and nuanced lenses, selecting papers with specific material qualities for their tasks at hand. These choices and uses of paper were profoundly shaped by social, class, and cultural circumstances. By electing to use a particular kind of paper in a certain way, actors were coproducing affordances attached to materials and objects.[21] As Werrett argues in his essay on waste and scrap paper in early modern England, "the social and material are inseparable: there is a mutual interaction between people and things in which the identities of both are transformed." Taking his cue from the organization theorist Wanda Orlikowski, Werrett writes of the sociomaterial nature of paper use, which, "often linked to particular sites or occasions, gestured toward particular negotiations of gender."[22]

Notions of sociomateriality run through our case studies, regardless of time period. Wolfe's examination of letter-writing paper in sixteenth- and seventeenth-century England argues that letter writers and recipients alike were attuned to the nuanced social messages encoded in the choice of letter paper. Bad paper, or "rude paper"—that is, paper whose quality (in size, whiteness, or smoothness) did not quite match the social status of the recipient—required explanation and apologies.[23] While dealing with mass-produced papers, contributors writing on post-1800 case studies also open conversations about the materiality of modern paper, foregrounding the multiple sorts of paper products available in markets in Europe and America.[24] Christine von Oertzen's study of the Prussian census compilation, for example, shows how the census bureau took inspiration from contemporary playing cards to create a sturdy and robust census card suited for the repeated handling required by the multiple rounds of sorting and counting.

A second theme weaving through these case studies is a strong sense that, across different geographical and temporal contexts, paper was used in a multitude of ways—hence the earlier mention of paperstuff as a descriptor of these kinds of objects. It is our contention that while the current historiography tends to emphasize paper's role within bookish and, largely, learned male practices, this represents merely one strand of its place in the history of knowledge. Here, we aim to highlight two kinds of activities—work with paper objects and ephemeral paper work. Both of these practices come to the fore in Anna Maerker's study of nineteenth-century papier-mâché anatomical models where figures such as Louis Thomas Jérôme Auzoux utilized the material to both fashion cheap, transportable, "dissectable" life-sized models and create elaborate labeling systems as well as an array of printed textbooks, pamphlets, and posters to accompany the anatomical objects. Workers, in Auzoux's factories, thus engaged in paper work in a number of ways, from mixing paper paste to molding shapes to making labels to reading and learning from printed books. Paper-based objects, traditional paper technologies, and paper work merged together in an entangled system of learning, making, and knowing.

A similar web of activities is outlined in Leong's and Werrett's essays. Leong explores a range of early modern everyday technologies using paper—from plasters and bandages to filter papers to pie dish liners—and connects these practices with the very paper technologies—loose paper slips and notebooks—on which the know-how is inscribed. Werrett's discussion of eighteenth-century hairdressing places how-to guides, printed on paper, for methods and styles of hairdressing alongside the use of twisted paper to produce "French curls" or the brown paper used to dry curled hairs in wig making or to cover one's face while powder was puffed onto the wig to whiten it. These cases demonstrate how working with paper encompasses a wide range of knowledge activities across a range of epistemic spaces, and that in many cases, knowing and doing or making were intimately connected by the very materials used.

Our essays also join recent conversations in exploring another world of print beyond books—the murky space where print and manuscript cultures collide. Here, we refer to printed blank forms such as questionnaires, census cards, and phrenological charts, where the order of knowledge was structured and predetermined by the questions and prompts in print but where, oftentimes, users and consumers engage with these structures in surprising and unexpected ways, forging new paths through seemingly fixed routes.[25] In this volume, Dan Bouk's study of the questionnaires used to gather data for population research shows that the objects were, concurrently, an instructional and a reference text for interviewers, a site for calculation for researchers such as Pascal K. Whelpton, and, subsequently, a part of the Scripps Foundation archive. In this case, the portable and easy-to-file-and-refile paper cards enabled researchers to bridge the spaces between the homes and parlors visited during their fieldwork and the "laboratory" space at the research institute.

The meanings afforded to paperstuff by these different historical actors shifted across the life span of the object. In a number of these cases, the paperstuff is at once ephemeral and permanent, dependent on the layers of meaning and value attached by producers and users alike. As a material, paper is susceptible to the elements and prone to decay and disintegration. The survival of papers and paperstuff is thus highly dependent on the social and epistemic values afforded to the objects by their users and producers. Papers and paperstuff have to be saved, stored, and archived.[26] While institutional and scientific concerns might have led to the archiving of the questionnaires and cards in Bouk's studies, in other contexts the preservation of paperwork was closely tied to acts of sociability. An example is provided by Carla Bittel's essay on phrenological charts; in the initial medical encounter they served as a crucial interface between client and practitioner, yet once at home they became objects of sociability, as family and friends collectively reviewed and discussed readings in parlors and sitting rooms. Practitioners and clients likely afforded different sets of meanings to the paper chart. For the practitioner it was a tool of communication, a means of guiding the encounter and, undoubtedly, a way to advertise their practice and services to others. For the client, the chart was a treasured souvenir of their experience, offering crucial information about themselves that prompted continual study, conversation, and even negotiation. Paper, open to printed and handwritten notes, transportable, shareable, and easy to store, was an ideal material for these various functions and, crucially, afforded with multiple layers of meaning.[27] The ways in which paper was perceived, as a material and as a tool and technology, is thus coproduced with design and use.

TOOLS AND TECHNOLOGIES

By paying heed to the sociomaterial aspects of paper and paper work, the authors in this volume take paper tools and technologies in a literal, three-dimensional sense. The paper in question never fades into the background as a simple, transparent bearer of marks and signs.[28] Rather, papers and paperstuff take center stage as material tools and technologies, and through them our authors uncover a host of places, skills, and epistemologies that have been bypassed in many material histories of science. As we explore the gendered context of working on, with, and around paper, we provide new evidence that it is neither possible nor fruitful to disentangle knowledge making from its ambient cultures.[29] In this sense, our volume pushes our field further toward a history of knowledge in which science is an important part.[30]

Our sociomaterial account of knowing with paper builds on the expansive historiographical interest in the materials of science, as well as the ongoing examination of the fundamental role of paper in the production of knowledge.[31] Historians of science and medicine have explored the means of knowledge production by examining the tools of observation, data collection, and knowledge transfer via paper to gain a deeper understanding of its role in epistemic practices. They have argued for the epistemic function of paper tools and paper technologies and over the last few decades what began as a rich and dynamic conversation has grown into a lively subfield.[32] In his influential ethnography of scientific practice, Bruno Latour deployed the notion of "paperwork," framing it as processes and practices of representation to systemize and simplify scientific objects, thereby making them manageable, visible, and negotiable.[33] Building on Latour's work on paper and his concept of "chains of inscriptions," Ursula Klein coined the notion of *paper tools* to analyze the sign system of chemical formulas invented by the Swedish chemist Jacob Berzelius.[34] Arguing that "paper tools were fully comparable to physical laboratory tools or instruments," she showcased paper as a central notation device external to the mental process.[35]

Beyond experimental practices, as scholars examined information management strategies and knowledge production processes, they came to talk of *paper technologies,* a term introduced by Anke te Heesen in connection with the humble notebook.[36] Following this, Ann Blair, James Delbourgo, Fabian Krämer, Richard Yeo, and others have explored the scribal techniques to note, keep track of, organize, memorize, and condense written observations.[37] These studies have uncovered how philosophers and naturalists used and experimented with different formats of writing paper—from commonplace books to loose paper slips—to best cope with the overload of information.[38] However, note-taking and other techniques of keeping and collating facts, such as list making or clipping and pasting notes, have mostly been considered as textual practices, or "writing technologies" performed with paper.[39] In this sense, Andrew Mendelsohn and Volker Hess talk about paper technologies as a term of art in which "humanist textual methods . . . came to bear on medicine," as generalizations from case histories were achieved through techniques of "collecting, formatting, selecting, reducing, comparing, [and] sorting."[40]

Taking inspiration from historiographies that set out to investigate paper and the people engaging with it, we bring a new focus on sociomateriality to the rich existing conversation on paper tools and technologies. First and foremost, in the context of bureaucracies social historians, anthropologists, media, and legal historians have understood documents as objects rather than carriers of inscription.[41] In doing so, they have brought to light the hard work of anonymous scribes in the early modern period to master the growing amounts of official paperwork.[42] Studies committed to the archaeology of media have explored the performative potential of what Ben Kafka aptly termed the "raw materials of power,"[43] such as legal paperwork put onstage by the traveling court, or files stacked up high on a clerk's desk in an antechamber to demarcate boundaries of authority.[44] Additionally, they have drawn attention to technological innovations of the eighteenth and nineteenth century such as the library catalog, the file cabinet, and the typewriter, which in turn helped to establish extensive formal communication systems, and, consequently, ever more paper to store and to handle.[45] Latour, in his thick ethnographic description of the legal clerk work within the French supreme court, concluded that the workings of the court are best understood by following the dossiers of different ages and colors from clerk to clerk as the dossiers grow like fruit that ripens.[46] Meanwhile, historians of imperial governance have unearthed the "micropractices" of writing elaborated through bureaucratic forms and the recruitment of native scribes, making colonial rule a pervasive "paper reality," in which written documents represent the world.[47]

The essays of this volume offer a fresh perspective on the notion of paper tools and technologies within the history of science, as they engage with the approaches outlined above.[48] Bouk questions Latour's concept of a document as an immutable and combinable mobile, and argues that paper documents were useful precisely because of the multiple purposes they could serve as they moved not only through space but through time. Elena Serrano explores te Heesen's observation that "the notebook collects people" further to highlight that accounting books were purposely written for others to look at with inquisitive eyes—a fact reassured in Madrid's foundling house by formal revision and signing ceremonies.[49]

Collectively, our authors call attention to ways in which gender and paper-stuff served to contest and restructure the very making of knowledge. In Serrano's study, aristocratic women drew on feminine virtues of housekeeping to reform the accounting system of the institution they took over from the church. Through various forms of paper work, such as changing the hefty ledger books and attaching parchment slips inscribed with names and other crucial information around the waists of babies, they created a panoptical paper system designed to remediate the institution's horrific death rates, and with it, their status, credibility, and reputation.

If paper tools served as mechanisms for ordering, the essays in this volume thus point to the cracks in the order.[50] For example, the material properties of paper shine in Beth Linker's essay, as she examines how translucent tracing paper, intended for mapmaking and mechanical engineering, became a medium for "physiological feminism." Adopted by the emerging posture sciences of twentieth-century America, an inscription device known as the "schematograph" used drafting paper to swiftly draw an outline of an examinee's human form. Linker situates the material, which was first used as a tool distinctly designed to record female patients, at the center of a debate on the role of the observer, the privacy of the observed, and of discussions on whether women's bodies were just as physically able and capable as men's.

Despite the wealth of studies on paper tools and technologies, this literature is one in which gender is mostly not accounted for.[51] Not only do the activities of learned men preoccupy the bulk of existing studies but the impact of material paper tools and technologies on the construction of gendered relationships and hierarchies in society has not been fully explored. Through explorations of the sociomateriality of paper, we address this lacuna in the current literature. In so doing, we untangle the multilayered conjunctions of paper, doing full justice to the riches that a history of working and knowing with paper has to offer.

GENDER, POWER, AND THE EVERYDAY

With our case studies, we find ourselves at the doorstep of daily life, entering households and offices, kitchens and laboratories, private rooms and public institutions. We observe paper moving through the hands of historical actors as well as in and out of communities of people, producing knowledge as they relate to one another in ways deemed feminine or masculine. In this volume, paper emerges as an essential "everyday technology," folded and filed into gender structures and daily activities in multiple contexts.[52] By simultaneously unpacking the material history of paper and interrogating the historical meanings assigned to gender, our authors challenge conjoined bedrock assumptions about the static nature of materials and dualistic gender. Instead, they illustrate material fluidity and gender negotiations, and open new histories of epistemology.

As scholars of gender, we build on the rich literature in the history of gender and science that has traced and untangled the gendering of knowledge, expertise, and authority. We are equally inspired by recent work that has unpacked the multiple masculinities and femininities historically operating in the realm of science.[53] To challenge the primacy of modern disciplines and sites of "Western science"—that is, the laboratory, the academy, or the scientific society—we join historians who more recently have looked beyond the academy—and in doing so, stretched the definitional boundaries of scientific work itself.[54] We question dichotomies of popular versus orthodox science, along with binary spheres of masculine and feminine, to find unrecognized sites and forms of science and knowledge making through the practices of paper use.[55]

This book extends these perspectives by drawing upon methodologies and theories developed by historians of gender and technology, particularly Francesca Bray's historical investigations of domestic spaces and agricultural technologies in late imperial China, and likewise to Jacob Eyferth's ethnographic analysis of a twentieth-century Chinese community of artisanal bamboo papermakers. Bray's notion of "gendered technics" and Eyferth's concept of socially embedded skill, which he develops in his afterword to this volume, figure prominently in this collection of Euro-American case studies, as they help reveal the intricate ways in which knowledge, gender, and materials are coproduced.[56] Gendered "technics," or gendered "technological attributes and practices," contends Bray, are mutually shaping in the construction of social order. She describes technologies as "a network of objects, activities, knowledge, and meaning" and emphasizes that "skills, judgment, and choice are involved at every step—and that bricolage, negotiation, creative compromise and distributed cognition are at the core of all technical activities."[57]

Historians working on Anglo-American contexts have demonstrated how skills and technological systems have become aligned with gendered labor and gender role expectations, from laundry to tailoring to cigar making.[58] They have shown how technologies, from the cupboard to the brassiere to the automobile, acquired feminine and masculine associations and characteristics.[59] Scholars working on textiles have identified clothing as a potent signifier of sex difference, especially with the coming of the industrial era, when it was used to discipline sexed bodies, serve gender performances, and symbolize gender transgressions at the same time.[60] Applying these approaches to paper, our authors uncover epistemic practices as everyday technologies and skills entrenched with gendered meanings. In so doing, they open up new avenues for the history of science as they demonstrate how the very making of knowledge is infused with quotidian negotiations of what counts as masculine or feminine. Throughout this book, knowledge practices and gender relationships materialize through the making, using, and handling of paper.

Highlighting the reciprocity between users, paper technologies, and different forms of labor and skill, we show that in the processes of knowledge making, gender was materially transacted on, with, and around paper in a multitude of ways. Matthew Eddy demonstrates how scribal lessons, individual comportment, and family dynamics played out in the education of Margaret Monro (1727–1802), who learned to use pen and paper as tools to achieve reasoned enlightenment. For both Monro and her father, a professor of medicine at the University of Edinburgh, the handcrafted notebook forged a collaborative learning project, as he instructed Monro to copy his advice on unfolded paper with large margins and to bind these together, leaving every other page blank to "allow Corrections that you and I make in [it]." Exploring the letters flowing in and out of the household of the English naturalist John Ray, Elizabeth Yale reveals the inseparability of learned and household work, mirrored in everyday paper use. Considering the roles that John Ray, his wife, Margaret Ray, and their daughters played in the naturalist's correspondence, Yale makes clear the stakes in representing the pursuit of natural knowledge as much less a masculine, and more a domestic, enterprise. This aspiration encompassed the naturalist's papers' afterlives, enabling Margaret to order those materials available for archival preservation and posthumous publication and to negotiate the value of the library she had to sell.

Gender transactions around paper reveal power relationships, many of them unequal. Nina Lerman has argued that "categories of difference," often aligned with "categories of power," have enabled some to engage with particular technologies and limited others.[61] To this end, we recognize how intersectional elements—gender along with sexuality, race, class, ability/disability—impact our historical actors and translate into hierarchies.[62] In the European and North American contexts covered in this volume, we find countless examples of cultures that are cordoned off technologically, barring particular groups or peoples, either overtly or implicitly. Within cultures of writing, reading, and paper activity, paper and its practices have been demarcated as accessible or inaccessible, teachable and learnable, or not, often depending on one's gender, racial, or class identity. While paper may have been available to those across economic spectra, its uses and practices served as indicators of class. In nineteenth-century America, members of the middle classes performed their status through certain types of paper work. Phrenological practitioners and consumers, examined by Bittel in this volume, used paper to record skull measurements that assessed degrees of femininity and masculinity. Those paper tools also did the work of race, as practitioners and middle-class consumers used charts to assign value to so-called Caucasian features. Whether in the contexts of slavery and colonialism or in the metropoles of empires, possessing and using paper tools often indicated status, required specific skill sets, and enabled belonging to networks of knowledge.

We find our actors operating in contexts governed by "categories of difference," and within certain gender, racial, and class structures that worked to the inclusion of some and exclusion of others. For several chapters in this volume, paper practices serve as markers of cultural belonging or exclusion, as some claimed scribal and material practices as a part of their "civilization." As Maerker shows, in Egypt, French medical educators, seeking to expand medical education, promoted the use of paper anatomical models to train local practitioners, who were reluctant to examine human corpses. Auzoux's paper models were intended to transform slaves into healers, and entrench Western medicine there. Paper, pasted and molded into human forms, thus became a centerpiece of claims to European superiority.

While categories of difference produce power differentials, paper's material malleability also reveals how actors negotiate gender and knowledge, and thus, power. These negotiations prove that much can be gained when we interrogate paper and gender relations together, categories naturalized in most accounts of knowledge production. By moving from analyzing the written to examining the material, and discovering the sociomaterial, we see how paper makes gender more visible and tangible, and in turn how gender relationships bring to light the tactile journeys and life cycles of material things. As the following essays demonstrate, in order to fully understand histories of knowledge, we must uncover the intricate negotiations of power and difference underlying epistemic practices—from the quotidian to the most elaborate alike.

PART I

Beyond the Page

The Sociomaterial History of Paper

ONE

Letter Writing and Paper Connoisseurship in Elite Households in Early Modern England

Heather Wolfe

Writing a letter requires a series of small but consequential decisions based on two baseline questions: What is my relationship to the recipient of my letter? What do I hope to achieve by writing it? The answers to these questions dictate the style, content, and materiality of a letter. In the early modern period in particular, material aspects such as the quality and size of paper, the style of handwriting, and the way in which a letter was folded and sealed were significant indicators of the sender's intent, reflecting a complex and unwritten set of sociomaterial practices communicated by example rather than by rule.

A letter from one of Queen Elizabeth I's favorites, Robert Cecil, Earl of Salisbury, to clerk of the signet Thomas Windebank, offers a window into the technological difficulties one might face in preparing a letter to be sent. Cecil describes an exquisitely written letter from Queen Elizabeth to King James VI of Scotland, which he is tasked with sealing and sending, as being written "in the style with those of the House of Florence, as Buchanan's scholar will blush when he reads it." He continues in his praise: "But it was not for nothing, that the son of Catherine de Medicis was so far gone with Sir John Smith's *muy linda e poderosa Reina.*"[1] The queen's letter is untraced, but from this brief description we learn that it is both eloquent and powerful, diplomatically and personally. Cecil would have known that the queen wished to have it sealed in the same fashion as her other personal letters to James VI, but it was past eleven at night, he was "very near his pillow-bere [pillowcase]," and his options were limited. Thus, he includes an important postscript: "Tell her Majesty that I will try how finely I can seal up her letter, but my house is such a Friary now, without any women, as If she send not my Lady Scudamore to me with a Needle and a thread, I must go to some of my Neighbours either at Essex house or Russell house for some of those utensils. If you find that the Queen will be content to be put out of her, *Passer solitarius in tecto*, tell her some of these toys of mine."[2]

In this extraordinary passage, Cecil reveals his quandary: he is surrounded only by men at his lodgings at the Savoy, none of whom possesses the skills to seal the queen's letter according to her specifications. He asks Windebank to ask the queen to send him one of her most trusted confidantes and sometime bedfellow, Mary Scudamore, Lady Scudamore, with a needle and thread. In exchange for sending out Lady Scudamore, as "a sparrow on a roof," he jokingly suggests that Windebank share some of Cecil's "toys"—that is, some of his jokes or puns—with the queen. If Windebank did not succeed in having the queen dispatch Lady Scudamore to Cecil's house, Cecil would be forced to seek out the ladies of Essex House or Russell House for the proper "utensils." These aristocratic women were not far from his lodgings—they were also located south of the Strand between Ivybridge Lane and Middle Temple—while the queen and her gentlewomen of the bedchamber were at Richmond Palace, more than ten miles down the river, in Surrey.

Why would Cecil need a woman to seal a letter from the queen, and why, specifically, Lady Scudamore? What feminine skills and "utensils" were needed?[3] Why wouldn't the queen seal it herself? After all, the technology of sealing is fairly straightforward: expose a stick of hard wax to a heat source until it softens, let it drip onto the closed letter, and then swiftly apply the seal matrix with firm pressure and a slight rocking motion, either directly on the wax or on a piece of paper covering the wax. This task was completed thousands of times a day, by thousands of men and women across Europe.

In this pre-envelope era, letters were written on the front side of either a bifolium (a sheet of paper folded in half, creating four pages) or on a single half sheet.[4] The letter was then folded into a rectangular packet for sending. The sender would write the address on the outside of the packet, providing more or less detail depending on whether the letter was being sent by a common carrier or a trusted servant or friend. The recipient's first encounter with the letter was a folded and sealed packet, addressed in a way that conveyed the letter's purpose, whether it be official business, confidential information, household updates, gratitude, apology, a favor, love.[5] Another important contextual clue to a letter's purpose was the mode of delivery: common carrier, servant, personal bearer, secretary, confidante.[6] In many cases, the person delivering the letter conveyed the most important messages orally, the text of the letter itself serving as a credential authorizing the bearer to serve as the letter writer's surrogate.

In sixteenth- and seventeenth-century England, there were a handful of dominant techniques for folding and sealing a letter, with many variations.[7] These variations could be localized to a particular person, a particular pair of correspondents, or a particular coterie, or exist within larger more porous social networks. The same person could use different packeting styles for different purposes. The technologies behind the folds appear to be entirely undocumented by contemporaries. While there are no mentions in the conduct literature or letter-writing manuals of the period, the style of closure clearly played a role in the reception of a letter and the technology must have been transferred person-to-person.

The most common method for creating a packet was the simple tuck-and-seal method; men and women of all social classes used this format for everyday communications. It was simple because no cutting was required; all that was needed was sealing wax, a seal matrix, and a heat source. One side of the folded letter is tucked into the other side, and the seal is applied to the seam. When the seal is broken or removed upon opening, a small fragment of paper is often torn off with it (see fig. 1.1). Other, more complicated and secure techniques, seen in correspondence between political and aristocratic elites, required much more time and expertise. For example, a triangle- or sword-shaped paper "lock" could be cut from a blank part of the paper on which the letter itself is written, and then carefully tapered and pushed through a slit in the multilayered folds of the packet that was either perpendicular or parallel to the fore edge. The small flap that emerged from the slit was flattened and melted wax was placed on it. The other end of the strip was then wrapped around the fore edge onto the wax and pressed into place with a seal matrix. The superscription, or address, was written partially on the paper lock and partially on the packet. Since the recipient had to tear or cut the paper lock in order to read the letter, the letter was essentially tamper-proof, since it could not be resealed using the original triangle strip, with its tell-tale matching laid lines, chain lines, and handwriting. Although the paper lock was usually discarded upon receipt, remnants on hundreds of surviving letters, as well as surviving slits and partially obscured superscriptions (because some of the text was written on the now-missing paper lock) on thousands of others, allow us to recover the letterlocking techniques used, and as a result, better understand the intentions of senders and the expectations of recipients.

To call these styles of closure "locks" is perhaps anachronistic, implying that all letters were sealed to avoid interception or prying eyes. In some cases this

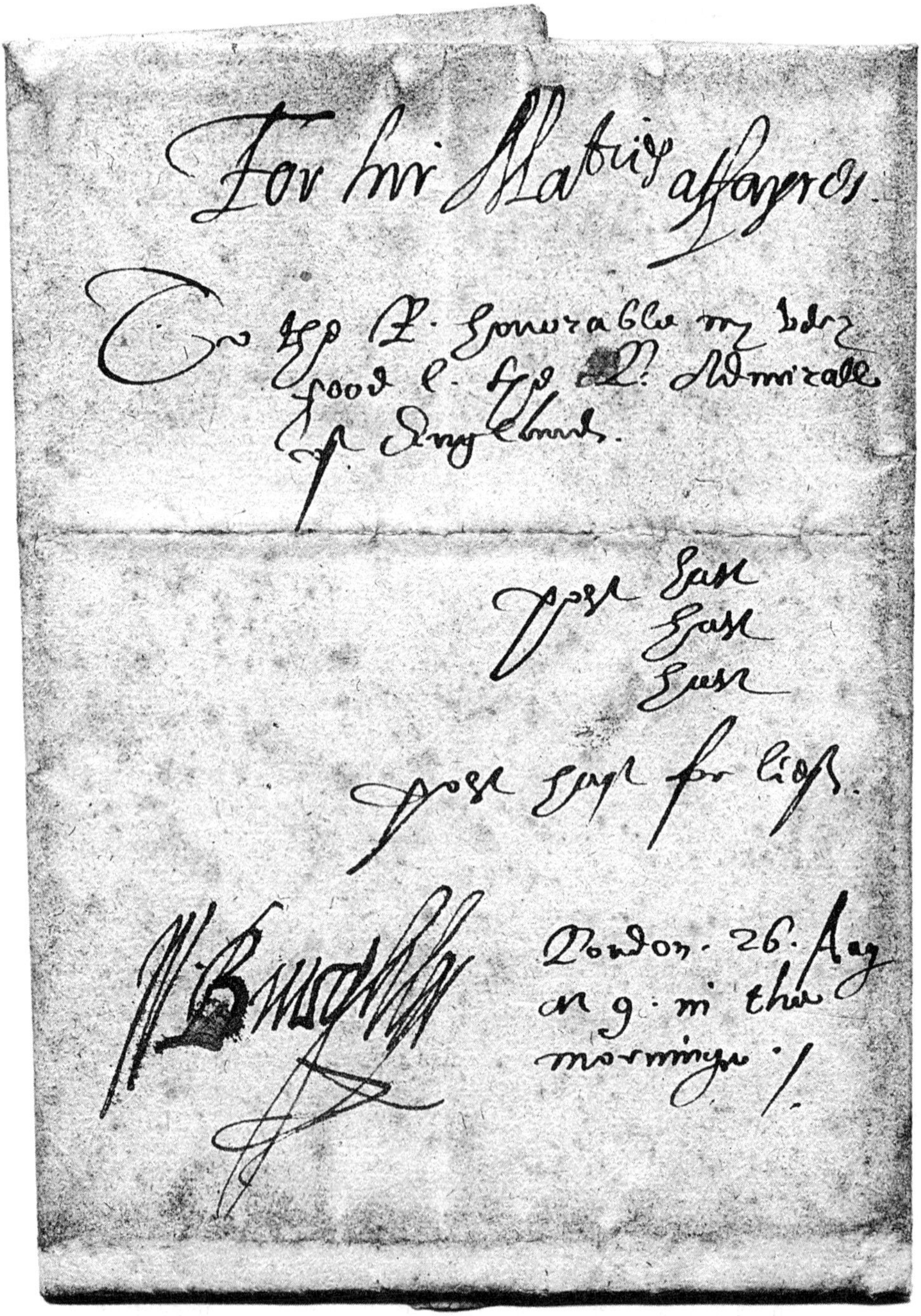

Figure 1.1.a. Front of a letter packet opened and then refolded using the simple tuck and seal method, with a tear and wax residue remaining at the location of the removed seal. Letter from William Cecil, Lord Burghley, to Charles Howard, Lord Admiral, August 26, 1588, MS X.d.494, Folger Shakespeare Library, Washington, DC.

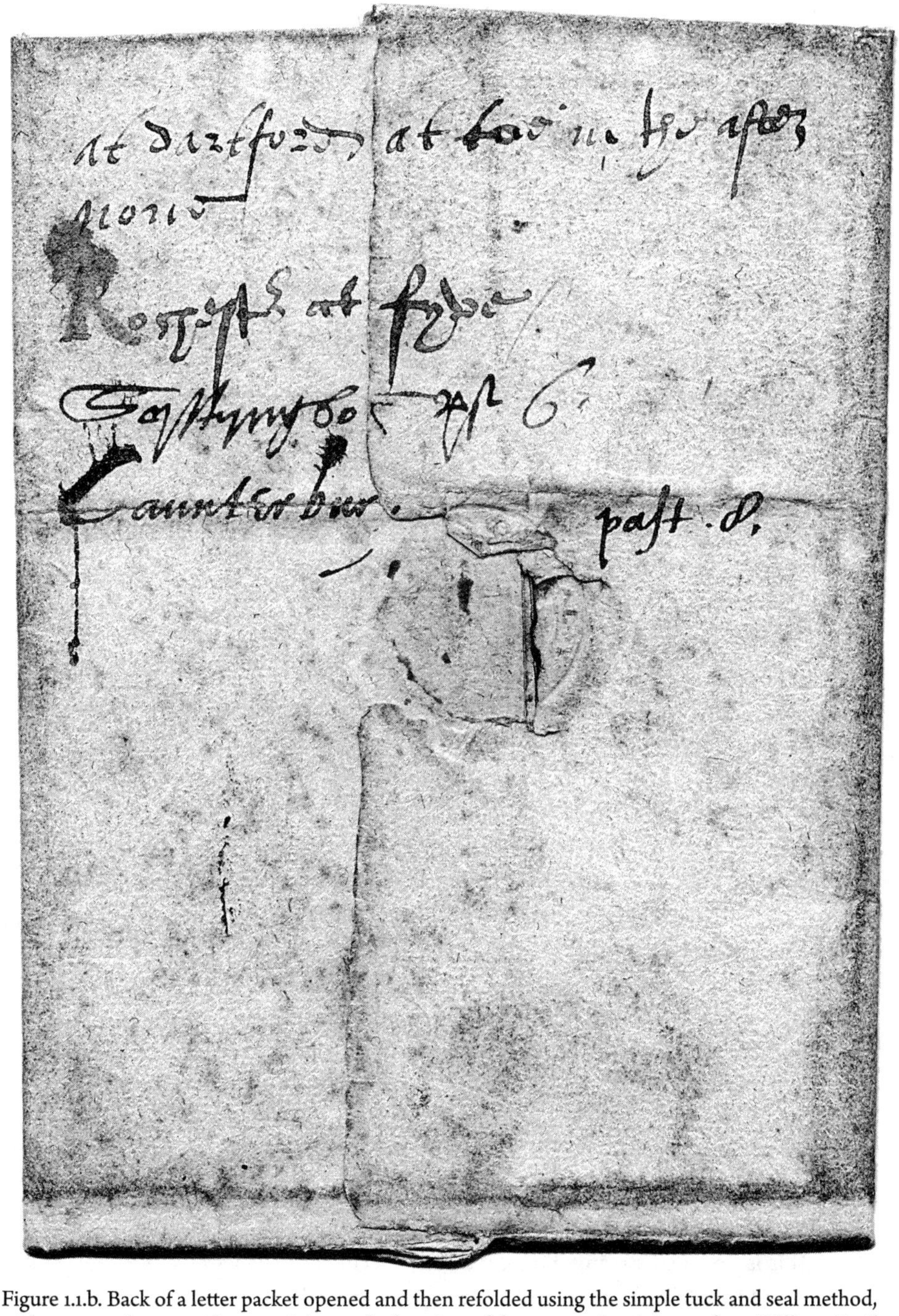

Figure 1.1.b. Back of a letter packet opened and then refolded using the simple tuck and seal method, with a tear and wax residue remaining at the location of the removed seal. Letter from William Cecil, Lord Burghley, to Charles Howard, Lord Admiral, August 26, 1588, MS X.d.494, Folger Shakespeare Library, Washington, DC.

may be true, but in the majority of surviving letters, the type of closure reflected a form of sociability, hospitality, networking, and identity formation. The extent to which someone individualized a form of closure reflected the extent to which that person wanted the closure to be noticed as part of the letter's meaning.

Going back to Cecil's "friary" problem: the fact that Cecil was in need of a woman to provide the needle, thread, and sewing expertise to seal the queen's letter indicates that the letter required a different sealing technique altogether. Material evidence on hundreds of surviving letters indicates that the queen wanted her letter to be pleated and closed with silk embroidery floss and wax seals.[8] First, a letter written on a fine, thin piece of paper (usually Venetian) was folded in half from top to bottom three or four times to create six or eight pleats, or long, narrow panels. This packet was then folded so that the two edges joined, in order to create a small rectangular packet roughly one inch wide and three inches long. Silk floss was then either wrapped around the fore edge of the folded packet multiple times and sealed in place on both sides of the packet or woven through a hole or holes in the packet, if extra security was desired. In some surviving examples, the floss was wrapped around *both* sides of the packet before being secured with small seals on both sides, resembling a modern gift-wrapped package, albeit with a seal, rather than a bow (see plates 1 and 2).

Cecil had no way to access silk floss in his "friary"—it was kept in women's spaces, such as the closet, the bedchamber, or other social gathering spots for women in aristocratic households, where they sewed, read, played music, and wrote letters.[9] Even if he could locate it among their coffers, chests, and desks, his early humanist education was very different from the education of aristocratic girls in the interconnected arts of sewing, reading, and writing. The gatekeepers for access to needles and silk floss in his household were absent, and only another woman, preferably Lady Scudamore, would have the knowledge to find the right utensils and use them appropriately. The women of Essex House and Russell House, though his near neighbors and kinswomen, were politically toxic because Robert Devereux, the Earl of Essex, was under house arrest at York House after a spectacularly unsuccessful campaign against the Irish rebels. Essex had recently entered a truce with the Irish chieftains and then deserted his post in Ireland and returned to England against the wishes of the queen, and his future was currently being weighed by the queen (he was executed twelve months later). At the time of his "friary" letter to Windebank, Cecil was closely involved in these deliberations.

This is not the first time that the queen specified how a letter she had written should be folded and sealed. Four years earlier, in 1596, Sir Thomas Lake informed Cecil that the queen wished to retrieve from Cecil an as-yet unsent letter she had written so that it could be reopened "and folded in a small plight [pleat] like those of her own hand, and so sealed."[10] The letter was from Elizabeth to Catherine de Bourbon, the King of France's sister.[11] This request reveals Elizabeth's concern that the letter's method of folding and closure match the intimate and reassuring style of the letter's textual message. It needed to be "pleated" into a small packet and closed with silk floss.

Indeed, an examination of several hundred examples of silk-flossed letters reveals a clear pattern of physical features: most are written on fine Italian paper, in an italic rather than a secretary script, and in the handwriting of the sender rather than by a secretary. The address on the superscription indicates that the letter is to be conveyed by a trusted bearer, dispensing with full names or descriptive locations in favor of more affectionate addresses. The content, too, of letters secured with silk floss, follows a pattern that seeks to solidify or restore preestablished social bonds: sending a gift, expressing gratitude and respect, requesting a favor, or offering one's service. The tone is often less formal than usual because of the familiarity between correspondents, reflecting or aspiring to a relationship of trust, confidence, and intimacy. This bond is further enforced by the time taken to apply the silk floss, and by the choice of colors used, which could convey symbolism as well: black when the sender was in mourning or pink for fidelity, for example.

When these attributes are combined, the small packet, conveyed into the recipient's hands by a trusted bearer, becomes something greater than a simple letter: an intimate token of love and affection even if the text itself is not a love letter. Alison Wiggins describes the silk-flossed letters received by Bess of Hardwick "as the products of a culture of epistolary politeness, and as gifts in themselves," shiny, silky, colorful, pretty tokens rather than substantive texts.[12] Similarly, Rayne Allinson writes that letters written in Queen Elizabeth's own handwriting "were a particularly personal kind of gift, since they embodied a trace of the monarch-author's metaphysical presence" while at the same time carrying the expectation of reciprocity.[13] If the letter not only was in the queen's autograph but followed the other conventions of the pleated silk-flossed letters as well—the folds, the paper, the script, the tone, the silk embroidery floss—the effect upon the recipient was multiplied (see plate 2).

The earliest silk-flossed letters in England appear in Elizabeth's correspondence with the Duke of Anjou during their marriage negotiations in 1579–1581. Like many fashions, it was initially adopted by her most intimate associates, the Earl of Leicester followed by the Earl of Essex, before radiating outward to their wives, daughters, and sisters, who used the technique sparingly so as not to dilute its exclusivity. The majority of surviving silk-flossed letters from the Elizabethan and early Jacobean periods emanate from a few interrelated circles of influential and well-educated aristocratic men and women associated mainly with the Sidney, Russell, Hastings, and Essex households. Receiving or sending a silk-flossed letter was an invitation to this exclusive club.

While a pleated letter secured with silk floss was ostensibly a private and intimate form of communication, as Cecil's "friary" problem indicates, it would most likely not have been created in isolation. The sociomaterial practices that went into creating a letter in an early modern elite household or secretariat, from selecting the paper, preparing the ink and quill, writing, subscribing, signing, folding, wrapping in embroidery floss or creating a form of paper lock, sealing, and superscribing, were embedded in complementary technologies and spaces, and dependent on a range of variables and people.[14] Cecil's predicament points to Ruth Oldenziel's observation that certain modes of production are "embedded in the family economy, in which family members shared their talent, capital and time."[15] This integrated production method falls apart when there is domestic imbalance. But to suggest that silk-flossing was a gendered activity does not point to the complexity of the letter-writing environment. While it is probably safe to suggest that male secretaries tended to use paper locks and women were the suppliers of silk floss and needles for creating pleated silk-flossed letters, we simply cannot know the full extent to which the skill sets were interchangeable. We rely on passing allusions and the material objects themselves to create a narrative that makes sense. Cecil, as most likely did the other leading men in Elizabeth's court, deferred to female knowledge and skills in order to execute the queen's wishes: that her letter be flossed and sealed in order to convey the right message to its recipient. The final result is a seemingly private communication that has in fact been written by the queen (or drafted first by Cecil for the queen's approval), edited, copied out by the queen in her own handwriting, sent by Windebank back to Cecil for sending, delayed by Cecil until he found someone who could close it "finely" with needle and thread, and then given to a bearer to present to King James in Edinburgh.

Wiggins observes that "letters with ribbon and floss remind us that letters are things themselves."[16] Telling the story of these things through the example of a moment of failure—Cecil's inability to finely seal a letter without access to any women—helps us to focus on the skills and tools behind the production of early modern letters, and to recognize the connection between message and materiality forged by the collaborative work of the queen's extended household. It takes the work of many to make it seem like the work of one.

UNDERSTANDING FINE PAPER

The second part of this essay explores what happens when you move an inanimate agent—in this case, early modern writing paper—from the background into the foreground. Paper is the foundation for written communication, yet overlooked by historians as a category of analysis. It was used by men and women alike to reinforce and establish social distinctions, hierarchical relationships, and alliances. The quality, size, whiteness, smoothness, and origin of paper have the power to signify as much as the text itself. Indeed, the intentions of early modern letter writers can in part be gleaned not only from the way they folded and sealed their letters but also from their paper choices. Paper was a nuanced and flexible substrate, embedded with codes of social behavior hidden in plain sight.

The paper historian Ruby Reid Thompson writes that "a significant element of personal taste and preference determines the choice of a certain quality and format of paper by a particular individual."[17] Stationers, haberdashers, and other vendors sold many different varieties of papers, in a range of quantities, sizes, prices, and qualities. Individuals generally purchased paper by the quire (twenty-five sheets), but occasionally by the ream (twenty quires, or five hundred sheets). Manuscript account books and receipts suggest a wide range of availability. For example, Ordinary, Fine, Best, Best (second size), Demy, Royal, Royal Italian, Royal Best, Imperial, Imperial Best, Dutch, Dutch Fine, Dutch Royal, Frankfurt, Troyes Demy, Venice, Venice Demy, Ruled, White, Writing, Pot, Gilt, and Gilt Large all appear in account books from 1570 to 1640. The terminology was not fully standardized, with names referring to quality, size, type of watermark, place of production, or some combination thereof.[18] Writing paper was sized with animal gelatin to prevent the iron gall ink from sinking into paper, and it needed a smooth and receptive surface in order for the nib of the quill pen to do its best work. Different qualities of paper required different nib-cutting and blotting techniques, and a writer needed to understand a sheet of paper's specific material attributes before selecting a pen.

Helen Smith rightly observes that the convenient literary trope of paper as a passive, virginal, and often gendered surface largely contradicted reality.[19] Some writers recognized an alternative trope: instead of a sheet of paper as a meaningless and innocent blank space in need of inky activation, a sheet of paper needed to be interpreted so that the ink could perform accordingly and effectively. For example, Nehemiah Rogers compared the relationship between a preacher and his audience to that of a penman and his paper. A writer must cut his nib according to the type of paper he writes upon: paper with a "hard and crosse graine" will damage a "Tender Penn," paper with a "fine, and tender graine" works best with a "smaller penn" and "Ordinarie Paper," and "Pot-paper of a middle nature" requires a nib that is "neither too soft, nor too hard."[20] Presumably this metaphor resonated with his readers because of their firsthand experience of mismatching nib to paper.

The size of the paper conveyed meaning as well. An etiquette manual published in 1675 encourages letter writers to "make use of large Paper rather than small, and a whole sheet (though we write but six lines in the first Page) rather than half a one [a quarto bifolium], is no inconsiderable piece of ceremony, one shewing reverence and esteem, the other familiarity or indifference."[21] Thomas Docksie included an apologetic postscript in an update to his master, Sir Walter Bagot, in 1612: "I must crave pardon for writing on so bad paper for I have lost the key of my desk and can come to no other."[22] The smaller size of the paper, less than half of a sheet, suggests that it was cut from part of the address leaf of a bifolium letter (a standard letter was a bifolium, or a sheet folded in half). It has blots and stains that might have been on the paper before he wrote his letter, since the bottom blot appears to be cut in half. He did not want to offend his master by using paper inappropriate to the message, and so drew attention to the locked writing desk to avoid impropriety.

It is also possible that the presence of deckle edges on a sheet of paper, or unevenly trimmed edges, could have an ill effect on the sender's intentions. The poet Philip Sidney opens a formal letter to the queen with: "This rude piece of paper shall presume because of your Majesty's commandment."[23] Sidney's letter is written on a bifolium, or folded sheet of paper, with an ordinary pot watermark and unevenly trimmed edges: perhaps not the sort of paper fit for a queen. Similarly, one of the Duke of Anjou's love letters to the queen begins by asking forgiveness since the letter is not written on such fine paper nor with such delicacy and dignity as he could wish, as he is in an inconvenient place.[24] The paper is cut from a much larger sheet, and has unevenly trimmed edges. In a Latin letter from Dr. Pierre Beutterich to D. Rogers in November 1578, Beutterich laments the poor reception in England of a letter of his to Queen Elizabeth, explaining: "If the writing was not very elegant nor the paper very fine, that was because I wrote it in my tent with my knees for a table."[25] Passing comments like these force us to look at the substrate more carefully, and reconsider what we might be missing. In these examples, the writers attempt to create a boundary between the paper's rudeness or badness and their own good intentions, suggesting that without explicit intervention, the poor quality of paper could somehow reflect the moral integrity of the writer. In fact, paper could be a tool for inclusion and identity formation, defining relationships and networks.

We know that some types of writing paper were better than other types because of their names and prices in account books. For example, gilt, fine, best, royal, large, and imperial all indicate expensive and high-quality paper. Bureaucratic letters occasionally include requests for fine paper, such as William Herle's request for a quire of fine paper and some hard wax for sealing letters from the Bishop of Ross in 1571, or, in the same year, the Bishop of Ross's receipt of a ream of fine paper from Mr. George Hackert, or, in 1640, Sir William Galley's request to Richard Harvey for a ream of "good fine writing paper."[26] The most striking example of an elite person acquiring multiple qualities of paper is Bess of Hardwick, the wealthiest and most powerful woman in England after the queen. On June 25, 1592, she purchased 1,600 sheets of writing paper in London, including 2 reams of pot paper (1,000 sheets), 6 quires of Frankfurt paper (300 sheets), and 6 quires of Venice paper (300 sheets).[27] On the same day and from the same shopkeeper, one Mr. Prescott, she also purchased hard and soft wax for sealing letters, and many yards of silver cloth, black satin, and black velvet. Bess displayed her wealth and power in monumental ways, building Chatsworth, a grand country house, commissioning massive embroidery projects, and consolidating her wealth through her own four marriages and the marriages of her children and stepchildren. She was a busy correspondent and accountant who needed different types of paper for different purposes, in addition to materials for making clothing and other luxury goods. It would have been obvious to her and the recipients of her letters that the pot paper (produced in Normandy, with a pot watermark), at 4 pence a quire, would have been used for day-to-day business and family correspondence and notes, while the higher-quality Frankfurt paper and Venice paper, at 10 pence a quire and just over 7 pence a quire, respectively, were reserved for more important letters that needed to communicate power and prestige.[28]

Perhaps the easiest-to-spot examples of fine paper in the wild are letters written on Venice paper, noticeably thinner, smoother, and lighter in color than pot paper, with a double pennant on a flagstaff watermark (fig. 1.2). Venice paper, typically slightly smaller than pot paper, was often used for letters circulating in elite circles, conveying respect, humility, and gratitude and reinforcing social bonds between women, between men, and between men and women. For example, the poet John Donne's eight groveling letters in early 1602 to his father-in-law, Sir George More and employer, Lord Keeper Thomas Egerton, after his clandestine marriage to Anne More, are written on Venice paper. The rhetoric and the location of Donne's signature on these letters suggest humility and poverty, while the paper itself speaks power and luxury despite his troubled circumstances.[29] Not surprisingly, letters sent from Elizabeth's palaces in the 1590s and by Cecil's secretariat also tended to be written on Venice paper.[30]

A letter's meaning and intent could be further strengthened if the fine paper had gilt edges. Thomas Massinger's manual *The Secretary in Fashion* (1654) recommends that letters should be written in a fair hand, "upon fine perfumed and gilted paper (if you please) and with large Margins." We may never be able to detect four-hundred-year-old perfume on a letter, and on paper so thin the presence of gilt is almost imperceptible to the naked eye; in fact, Robert Hooke reported to the Royal Society in July 1665 that he had made a microscopic observation of the "very smooth and even gilt edge of fine Venice-paper" as well as of a moth wing and "fine lawn" (a kind of linen) in order to reveal the intricacy of the

Figure 1.2. An example of a watermark on Venice paper, on a letter from the Earl of Middlesex to Sir Robert Pye, February 12, 1623/24, MS X.d.134, Folger Shakespeare Library, Washington, DC.

substrates.[31] The intimacy of discovering the presence of a gilt edge upon opening a letter undoubtedly would add to the impact of the letter's nontextual gesturing. Donne wrote to his brother-in-law Sir Robert More and to his friend Sir Robert Harley, among others, on gilt-edged Venice paper in 1612, 1613, and 1614 (see plate 3).[32] In 1610 Jane Skipwith wrote three affectionate letters to her cousin Lewis

Bagot and an apology letter to her uncle on gilt-edged Venice paper, enclosing the pleated packets with silk floss (see plate 1a for one of these letters).[33] Frances Howard, recently widowed by Henry Pranell and known to be in love with the Earl of Southampton, wrote a letter to the famous astrologer Simon Forman on gilt-edged Venice paper on March 2, 1600, seeking his judgment on whom she might next marry.[34] Her paper choice reflects her class, but also her desire to influence his decision through a display of wealth and power. Other evidence for the use of gilt paper for letter writing can be found in account books: the antiquary Edward Dering bought gilt paper for his wedding in 1619 and again in 1627; Sir Arthur Capell purchased ten quires of gilt paper and ten quires of "ordinary paper" in 1629, shortly after a failed courtship; and Sir Francis Windebank requisitioned one ream of large gilt paper in 1639 for official use.[35] Nadine Akkerman has shown that Elizabeth of Bohemia, after the deaths of her husband and father, began using stationery edged in black.[36] Another letter from Donne to his brother-in-law is edged in blue.[37] Documented examples are scarce because the presence of colored edges is not usually recorded as a physical attribute in bibliographic descriptions of early modern letters. However, careful visual inspection of the edges of letters between early modern elite men and women would most likely reveal many more instances and add to our knowledge of when and why gilt-edged paper would have been used.

While gilt edges were a subtle signal of status and intent, decorated borders was anything but subtle. The borders of a letter from Aletheia Talbot, Countess of Arundel, to her grandmother, Bess of Hardwick, illustrated in a 1974 Sotheby's catalog (current location unknown) are highly ornamented with gold leaf patterns and monogram letters.[38] Written by the countess while in her early twenties, it is a short and formulaic letter apologizing for not visiting her grandmother, and promising her sincere and affectionate desire to serve Bess. The text is filler: the point of the letter is the uniqueness of the border, a beautiful token that delights and surprises, substituting for her physical absence. The time and care taken in creating it (whether it was the work of the countess or someone else), and its echoing of the embroidered tapestries created by Bess, will hopefully save her from Bess's disapproval or disappointment.

A simple letter of recommendation from Catherine de Bourbon, sister to King Henri IV of France, to Queen Elizabeth from the late sixteenth century has colorful decorated borders adorned with ships, monograms, and cupids (see plate 5). This makes the letter itself a token of love and respect. Catherine opens and closes the letter with the French *fermesse* symbol (resembling an italicized dollar sign with the bar tilted to connect the two ends of the *S*) as a further sign of her affection.[39] She indicates her humility and respect by leaving significant space between the end of the letter, her subscription, and her signature (Wiggins refers to this as a split subscription).[40] It is endorsed in a contemporary hand: "A letter of the King's sister [to] *the* Queen, more conceited for the paper than respective for the manner of writing." In this sense, "conceited" means "elaborately or ingeniously wrought" (a definition that is now obsolete) rather than our modern sense of the word.[41] Another letter from Catherine de Bourbon to Elizabeth, written on thin, milky translucent paper, has exquisitely cut ornamental margins. This one, lengthier than the first, assures the queen of her Protestant convictions, continued love and friendship, and willingness to serve her in any way she can (see plate 4).

A letter from Francis Gainsford to the Earl of Essex in 1598 is written in a predictably ornamental style, matching the border in its excessive use of praise to frame a simple request for a job captaining a company in Ireland.[42] It differs from the other three examples in that the sender is not familiar to the recipient, and so Gainsford must remind Essex of his devoted military service to him and his undying respect. In this respect, the letter signals itself as an extraordinarily important letter, a matter of life and death, compared to the letters sent between royalty or aristocrats. Gainsford's use of the fermesse in the corners reveals a surprising intimacy: the symbol typically signifies fidelity in love when it appears in letters, but can also signify friendship and political or spiritual fidelity. Through the fermesse, he seems to be suggesting that their duty to each other is both personal and political.

These three surviving examples with decorated borders are unique, but their similarity in motifs suggests a connection to the decorative arts in domestic spaces of large country houses—patterns and symbols on wallpaper, tapestries, ceiling beams, mantels, embroidered pillows and furniture coverings, rugs, and the samplers made by girls as their first exposure to the arts of sewing, reading, and writing.[43] At the same time, it harks back to the borders of illuminated personal devotional manuscripts or books of hours. The stationery's visual and textual messages work in harmony to create a gift deserving of the recipient's appreciation and favor.

The style of letter locking and the quality of writing paper in the early modern period was something that could probably have been detected by just about anyone, and is something that we should learn to observe when interpreting the correspondence of the political and aristocratic elite. Some forms of closure were practical and utilitarian, and some required skills and supplies that the letter writer did not possess. The many and varied ways in which surviving letters were folded and sealed suggest that the recipient of the letter was not meant to read the text alone but to read the *entire* material object, beginning with the closure style, the seal, the quality and size of paper, the physical layout of the text, the amount of blank space, and finally, the text itself. Like the paper itself, the form of closure could signal love, secrecy, urgency, or, more commonly, that it was just another letter.

Today we know immediately when we encounter a piece of paper if it is cheap, expensive, or unusual. We know from the paper itself, from the text and its physical layout on the paper, and from the paper preparation (the way it is folded, shaped, or cut), each working in harmony or disharmony to send an overall message. Checks are sent in opaque patterned security envelopes. Greeting cards arrive in colored envelopes and the cards themselves are printed on card stock rather than paper. We would never expect to receive an electricity bill on expensive Fabriano paper, or a wedding invitation scribbled on a Post-it note. We don't need written instructions to tell us how to interpret the relationship between substrate and text, between quality and expense, and neither did early modern consumers of writing paper. By learning how to distinguish between early modern paper types and reimagining the three-dimensionality of now-flattened correspondence by using the surviving fold lines to reverse engineer the process of making a letter packet, we can make visible the shared work and intentionality that went into the creation of a seemingly familiar everyday object: the handwritten letter.

By unpacking the steps and players involved in the creation of a seemingly very simple object—the letter—we are able to uncover previously hidden dependencies and collaborations, and to view the substrate itself as part of the social bond or contract between the sender and recipient, animated by the intervention of human agents. We are also able to further untangle the complexity of gender in the household. Prescriptive literature, scripture, and cultural norms dictated that the husband was head of the household, but, as Amanda Flather notes, "the economic and practical demands of running a complex multifunctional household . . . rendered doctrines of male dominance irrelevant," and "the organisation of domestic space was highly dynamic and at any given moment it could be used to assert authority or to signify commonality, depending upon the context."[44] Although gender roles were well defined in aristocratic households, domestic and work-related tasks and spaces frequently overlapped, collapsing the male/public, female/private dichotomy reflected in the conduct literature. Nowhere was this more evident than through the sociomateriality of letter writing.

TWO

Papering the Household

Paper, Recipes, and Everyday Technologies in Early Modern England

Elaine Leong

Across early modern Europe, the household—as a domestic space and as a collective of historical actors—played a crucial role in a range of medical activities. In many cases of ill health, it was the first port of call.[1] Household collectives often worked together to diagnose sickness and ailments, nurse and care for the sick, source and produce drugs, and promote general health maintenance through diet and exercise.[2] To accomplish this work, they utilized a broad knowledge base, a range of material objects, and a number of complex processes. Many recorded their practical knowledge in the form of recipes—short tidbits of instructions detailing how to make a range of medicines and foodstuffs. Hundreds of handwritten recipe books now survive in library and archives. Attesting to the ubiquity of the practice and the eagerness with which householders formulated, recorded, and exchanged recipe knowledge, these books now serve as ideal entries into explorations of medical knowledge and practices in early modern households.[3]

Many of the diverse tasks under the umbrella of household health activities involve the production and use of everyday technologies.[4] Alembics and furnaces were used to produce distilled medicines and fortifying spirits. Pots, pans, and ovens were used to prepare healthful meals and preserve foodstuffs to consume throughout the year. Drugs, to be ingested or applied to the surface of the body, eased ailments and balanced the humors. Printed books and blank notebooks carried and recorded medical knowledge. The modern counterparts of many of these technologies are well studied by historians of medicine.[5] And recent work on medical technologies such as trusses, electricity, paper anatomical models, and medical trade catalogs have demonstrated the rewards of taking these studies into premodern contexts.[6] As Christelle Rabier argues, "technology proves a useful concept to explore the multifaceted nature of medicine beyond the traditional world of doctors and their theories."[7]

With acknowledgment that this is not an actors' category, the term "everyday technologies" is used here to signal the analytical framework underpinning this study, drawn from recent studies in the history and anthropology of technology.[8] Like many in this volume, this essay adopts Francesca Bray's view of technology "as a network of objects, activities, knowledge and meaning" and "repertoires of skilled material practices and artefacts that serve to produce food, shelter and clothing, to communicate, to control, to worship, to fight or to trade."[9] Within this view, technology is a fundamental way through which gender is expressed in societies and an integral part of everyday life. This focus on the quotidian pushes us to investigate a history of "technology in use" and to bring out the users and consumers' experiences and interventions, helping us understand the formation and shaping of gender dynamics and relationships.[10] Just as within the larger umbrella of other domestic labor, gender played a crucial role in shaping medical knowledge and material practices. For example, while the master of the house was often involved in domestic activities and in representing the interests of the household within wider medical economies, many of the everyday health-related tasks fell to the women of the house, be they the mistress, the housekeeper, or the maid.[11]

One material—paper—appears and reappears across the wide spectrum of home-based medical activities. Notebooks and paper slips were used to organize, store, and manage medical knowledge. Paper plasters and bandages were used to apply ointments and salves to the body. Other kinds of papers enabled householders to separate and preserve medicaments and foodstuffs. In other words, paper filtered, contained, and bound knowledge, objects, and the human body. At first sight, paper might seem to be an odd entryway into the entangled web of household medical activities. After all, early modern paper is often presented as scarce and expensive. Contemporary writers are said to have filled pages to the brim with notes and reused and repurposed paper because it was costly and hard to come by.[12] Within histories of the book trade, paper is frequently touted as the single highest expense in the book production process.[13] Concurrently, standard narratives about paper use in the early modern period tend to emphasize its role in reading and writing practices and knowledge transmission. For example, recent studies on "paper technologies," as termed by Anke te Heesen, have sharpened our focus on the epistemic role of items such as notebooks, slips, cards, and clippings.[14] Examining what is inscribed on the paper, researchers have, for the most part, highlighted the work of (often scholarly) men. By widening our scope to explore other kinds of paper use, we uncover an alternate yet interconnected world of knowledge practices conducted in spaces such as kitchens, stillrooms, and parlors, and by a wide range of knowers.[15]

Framed within the intersection of paper, gender, technology, and histories of knowledge, I follow a trail of paper through a number of connected practices in early modern households in three main areas: surgery (especially the application of medicines to the surface of the body), medicine and food production, and practices of reading and writing health-related knowledge. In other words, this chapter explores both technologies using paper and paper technologies, bringing together the activities of making and doing with those of reading and writing. Emerging is a story in which paper, far from being scarce and expensive, functioned as the key material in a range of everyday tasks, some designed to be ephemeral and others permanent.[16] These tasks all played on the material properties of paper—absorbency, malleability, foldability—and took advantage of the ease with which one can cut, paste together, and in some cases, burn paper. By attending to the material history of paper and paper use, I take a fresh look at the various processes used by householders to produce quotidian goods and open new conversations about the place of everyday technologies in histories of premodern medicine and gendered labor in early modern England.

PAPER PLASTERS

Recent studies have highlighted a range of surgical-related healing work occurring across diverse spaces in early modern Europe.[17] Within the various rooms of the early modern home, householders dressed wounds, alleviated burns, and concocted assorted waters and syrups to beautify and perfume the body. Paper played a central role in many of these practices. Made into a plaster, paper was one of the most common materials used to apply medicines to the exterior of the human body. Such plasters addressed a range of ailments, from those of the surface of the body, such as burns and sores, to internal ailments, such as headaches.

The ubiquity of paper in household surgical activities is evident from its frequent appearances in contemporary household medical recipe books. One example is a mid-seventeenth-century recipe book associated with a "Mrs Carlyon" that includes a recipe for a headache remedy. This involved putting a mixture of powdered cumin and egg between two folds of brown paper and, with a handkerchief, securing this paper plaster on the sufferer's head for twenty-four hours.[18] Another example can be found in the book of Lady Johanna St. John (1631–1705). Intended for "a knubb or hardness in a breast," the remedy involves spreading honey on a brandy-soaked piece of brown paper, applying it to the sore, and as the plaster dried, replacing it with freshly made ones.[19]

The specificity of brown paper in many of these recipes is telling. In the early modern period, the English imported a wide range of papers from countries such as France, Italy, and the Netherlands.[20] Paper sold at various price points included a smooth, high-end white writing paper; a cheap, rough brown or gray paper used for wrapping; a blue paper designated as the wrapper for sugar; and many more. A 1695 paper auction at a coffeehouse in Corn Hill, for example, offered "Several sorts of Writing Papers, Small and large, brown and whited brown, Bisket Papers, large white and whited brown Paper for Paste boards for pressing Papers, Drafts, Maps, all sorts of Past Boards for the same use."[21] As Heather Wolfe outlines in her chapter in this volume, consumers not only were aware of the different kinds and grades of paper available but also shopped accordingly.

Contemporary household account books reveal that many stocked their cupboards with several kinds of papers at the same time. The accounts of an elderly single gentlewoman living in Herefordshire, Joyce Jeffreys (ca. 1570–1650), include a number of references to paper purchases.[22] In 1638, for example, Jeffreys bought five separate lots of paper. In April she obtained a quire of paper for five shillings; in May she spent three shillings and sixpence on ten quires of white paper; in July she purchased brown paper for fourpence; and in October she bought cap paper for a penny and "a quier of browne paper to putt under pies" for twopence.[23] These kinds of purchases recur throughout the ten-year run of the accounts. Aside from white, brown, and cap paper, Jeffreys also purchased "lardge fine writing paper," "white writing paper," "cours writing paper," and "brown cap paper."[24] Of all the different kinds of paper listed in Jeffreys's account, brown paper makes the most frequent appearances and was purchased on twenty-one separate occasions, often in the amount of two or three quires.[25] Brown paper was, in fact, the workhorse of the paper family within the household. A relatively cheap material, it was used to wrap objects and line cake tins, and acted as a carrier to deliver medicines continually to the surface of the body.

Leather and cloth, of course, also shared some of the material qualities of paper and were sometimes used interchangeably. For instance, Rebekah Winche's mid-seventeenth-century recipe book contains instructions to make an ointment for a burn or scald; this involved boiling herbs in an olive oil, white wine, and wax mixture and "clapping" it onto the burn with either paper or cloth.[26] There were advantages to using (or perhaps reusing) brown wrapping paper over linens or leather. A recipe "for a cancer" attributed to a Lady Honeywood in the St. John recipe book involves mixing the juice of four herbs with "beanflower" and honey, spreading it on the softest brown paper, and applying it to or close to the sore. The resulting plaster was to be changed twice a day and the discarded paper taken away and burned.[27] As is often the case, the recipe creators offered no explanation for why one might need to burn the soiled paper. However, this step quickly and safely disposed of corrupt bodily liquids and perhaps offered a symbolic effect of destroying the sore alongside the paper. The recipe recalls practices common in magical and devotional medicine in the period. During the time of the 1656 plague in Naples, for example, infected people pressed printed devotional images of the Jesuit Saint Francis Xavier against the area of infection and, afterward, discarded the paper. The action was believed to expel the contaminated fluids from the sores, which were thought to be absorbed by the paper.[28] In Protestant England, the makers and users of Honeywood's remedy likely afforded different meanings to burning the paper plaster. Nonetheless, we should not understate the potential symbolism of such an act.

By the seventeenth century, when the recipe books described above were created, paper plasters had long been in use in English medicine. As Orietta Da Rold has shown, recipes for paper plasters began to appear in late medieval medical manuscripts not long after paper arrived in England. For example, a fourteenth-century translation of Lanfranc of Milan's surgical tract included a remedy using paper to apply a tooth-whitening powder. Used in a similar way as the headache remedy described above, the recipe involves filling a folded piece of paper with a mixture of flour, *sal ana*, and honey.[29] Following in the tradition of Lanfranc, seventeenth-century surgeons also adopted the practice of using paper plasters. In his 1617 *The Surgeons Mate*, John Woodall (1570–1643) listed gray and white paper alongside other surgical instruments as necessary to keep in a "Surgeons Chest."[30] He also offered a number of paper plaster recipes, including instructions for *unguentum tripharmacon* and *mel saponis*, and advised young surgeons, while at sea, to use paper to treat gunshot wounds in order to spare linen.[31]

Following paper plasters through the archive highlights the continuities of paper use in surgical practices across different medical sites, from households to naval ships. This focus on the commonalities and intersections between the surgical work carried out by female householders and well-known male practitioners prompts us to question conventional boundaries between domestic and "professional" practices.[32] This theme of shared practices continues in the realm of drug production where paper was used to preserve, to filter, and to contain.

MAKING MEDICINES AND FOODSTUFFS

"We use Paper to filter several Liquors, to cover bottles, pots, and to wrap up several Medicines. In short, we make use of a thousand things, and a thousand inventions."[33] So writes the French apothecary Moise Charas (1619–1698) in *The Royal Pharmacopœa, Galenical and Chymical.* Here paper is used to apply medicines, to filter liquids, to dry substances, to shape plasters, and to preserve drugs. Unsurprisingly, paper also served very similar functions in bustling kitchens and stillrooms. Wrapping paper, notated as gray brown or cap paper, was often used to line vessels to strain liquids. For example, a recipe to make a syrup with steel filings, appearing in a number of different collections, required the boiling of the filings in mace-spiked white wine with subsequent filtering through a paper-lined glass funnel.[34] Treated paper, such as oiled or greased paper, was used to shape and preserve pills and ointments. Recipes for the popular "Paracelsian plaster" often ended with instructions to keep the medicine in oiled papers, changing the paper once a year.[35] One version of this recipe claimed that doing so will allow the medicine to keep its virtue for fifty years.[36] The Paracelsian plaster was both expensive and labor- and time-intensive to produce, accounting for the attention paid to ensuring the preservation of the medicine.

Paper was also used to preserve in another context. Both apothecaries like Charas and householders utilized the absorbent qualities of paper to dry and preserve plant materials. In Charas's recipe for *hiera picra,* for example, the maker is instructed to dry the saffron on paper before the fire.[37] Susanna Packe, a mid-seventeenth-century gentlewoman, uses a similar method to dry cherries.[38] The absorbent qualities of paper could also be manipulated to protect foodstuffs and materia medica in another sense. In a recipe to make the syrup of lemon, the maker is instructed to wrap a lemon in wet paper and to lay it in the warm ashes to roast. This method ensures that the lemon will cook without burning, as "if it be the least burnt it is good for nothing."[39] In another similar usage, the makers of twice-baked "frate" biscuits are told to do the second baking in a "cofen" made of paper.[40] In both these instances, paper serves as a protective layer against the heat of the oven.

In fact, baking was one of the main areas where paper was used as an everyday technology. Early modern cooks were often told to bake cakes, biscuits, and other goods on floured baking papers.[41] Recalling the entry for "a quier of browne paper to putt under pies" in the Jeffreys account book, we can assume that at least some of these cases involved purposely made and bought brown paper.[42] One recipe for a "Speciall Cake bread" invited the baker to "take a browne paper & drye it very well & strowe it with flower & lay it under ye Cake."[43] The additional drying step in this recipe suggests that damp, perhaps washed, brown paper was common in the household, hinting at the practices of reuse as outlined by Simon Werrett in his contribution to this volume. The recipe archive also yields one final fascinating use of paper, aside from easing the removal of foodstuffs from kitchen equipment. In a recipe for biscuits, the baker is told that "the oven must be soe hot as to turne a peece of white paper browne."[44] In times before the oven thermometer, this was a simple and easy way to ensure that the oven was at the right heat point.

Thus far, paper has mainly functioned as tool or technology to enable householders to perform particular tasks.[45] However, there is also evidence that paper served as a materia medica—actual "paper cures," as it were. A recipe for a nosebleed recommended grinding up an oven-dried, vinegar-soaked piece of paper and stuffing it up the patient's nose "often, as well when it does not as when it bleeds."[46] A cure for a wart advised the use of "oil of brown paper" and a recipe for "an admirable beauty water for the face commonly called Pig water by som Puppie water" requires "Venice paper cutt into small bitts" to be used alongside puppies, pigs, snails, bread, and wine.[47] For the makers of these recipes, paper acted as a medicinal ingredient actively encouraging bodily changes.

The exploration of paper use in the production of medicine and foodstuffs confirms earlier claims about continuities across home-based medicine and practices offered in the wider medical markets.[48] The dual focus on medicine and food production, moreover, takes us further into the space of everyday household technologies—the processes of cooking, medicine making, and food preserving. In doing so, it firmly locates medicine and healthcare activities within a range of everyday tasks. As a three-dimensional tool, paper also played a key role elsewhere in early modern homes: the study and library. As we will see, as a material and an everyday technology, paper connects various kinds of epistemic work from making and doing to reading and writing.

MANAGING INFORMATION IN THE HOUSEHOLD: PAPER TECHNOLOGIES

In recent years, the reading, note-taking, and information management strategies of male scholars have come under the scrutiny of historians and literary scholars alike.[49] How readers engaged with, appropriated, sorted, filtered, and stored information with paper technologies had profound impact on both knowledge making and knowledge transmission.[50] In this last section of this essay, I take a look at the notebooks within which the recipes earlier described are stored. Householders, male or female, adopted many of the strategies we associate with scholars to manage their collections of practical knowledge. Their work with paper—in the form of homemade notebooks, loose paper slips, and knowledge codification systems—form the final paper-based technology in this essay.

Early modern English householders were bombarded with health-related information.[51] Contemporary booksellers' shelves were crammed full of medical textbooks, pharmacopoeias, herbals, surgical handbooks, expositions on new medical theories, regimen guides, and more. Medical consultations, whether face-to-face or via correspondence, with physicians and other medical practitioners inevitably yielded new medical know-how. Around dinner tables, householders exchanged tips on how to make secret remedies or make the most of one's herb garden, advised on where to obtain the best spices and debated about the almost miraculous properties of exotic materia medica. With their family's health maintenance and potential sicknesses never far from their minds, householders eagerly recorded and treasured this kind of information. As a result, many collectively compiled large notebooks are filled to the brim with medical knowledge. With new, seemingly unique and important information flooding in from so many avenues, householders often felt the sense of the information overload experienced by Renaissance scholars, and like scholars, they used a range of sophisticated information management (or note-taking) technologies to shape, control, and create their treasuries of health.[52] Here, we will look at two different paper technologies: alphabetization as an organizational scheme and the use of multiple paper notebooks and paper slips to sort and categorize knowledge.

Alphabetizing the Body

The development of paper technologies as information management systems went hand in hand with the explosion of books and knowledge in the early modern period. Scholars used a range of different systems as memory aids and to ease further readings through their own notes.[53] Recipe collectors, faced with similar problems, also adopted a wide variety of organizational systems. Some, following long-standing medical genres such as the *receptaria*, listed remedies in a head-to-toe arrangement; others, following the pharmacopoeia and *antidotaria* traditions, grouped the medicines by kind; for example, putting together instructions to make waters, balms, and pills. Tables of contents or indices, compiled and updated as new recipes arrived into the compilers' hands, were also common methods of organization.[54] One particular method, alphabetization, became popular with recipe collectors in the seventeenth century and showcases the complex decisions required of many paper technologies.[55]

Alphabetization came to the fore in the sixteenth century as an information management strategy and was widely used in a range of knowledge fields from natural history to medicine.[56] By the seventeenth century it was a popular and common scheme for organizing all kinds of knowledge, including medical recipes. In his 1631 publication *An Alphabetical Book of Physicall Secrets*, the medical practitioner Owen Wood drew attention to his method of organization, claiming that "Remedies may easily be found out, the Diseases being set down Alphabetically."[57] Householders, compiling their own book of recipes, also embraced this fashion to alphabetize. The collection of Lady Johanna St. John, already known to us by her paper plaster recipes, is particularly illustrative of these issues.

The St. John recipe book is filled with medical remedies addressing all sorts of ailments and sicknesses from agues to coughs to fevers, collected from a wide range of sources from household members such as Mrs. Patrick to the fashionable physician Dr. Thomas Willis.[58] In that sense, within the book, recipes from university-trained physicians jostled next to practical expertise from friends and family members in much the same way as they might on the early modern medical marketplace.[59] The book is alphabetically organized, beginning with a section on ague and aches and ending with a section on wounds, women's health, and worms. That is, in her alphabetizing, St. John was also categorizing and defining ailments and expressing her view of the body and ideas of disease.

Yet the assignment of recipes to each letter section is not as simple as it might first appear. Early modern medicaments tended to concurrently target multiple bodily complaints or ailments and thus this system of categorization required St. John to identify and pick out the central ailment/symptom addressed by each recipe. The section headed "W" presents an interesting example. Alongside recipes to cure wounds and warts is one titled "Dr Wilsons Broth to prevent miscarrying." This is followed by a recipe "For a weak Back in a woman with child & to prevent miscarrying" and others such as "To Make a woman have easy Labor," "Mrs Tassels cawdle to bring away the Afterberth or any part left behind," and "To clens the Mother and cause Conception Dr Row."[60] Undoubtedly, the instructions to encourage conception, to avoid miscarrying, to hasten a painless and easy labor, or, in the worst-case scenario, to bring away a dead child are all collected under the "W"' section, as they were seen by St. John to be concerned with the womb, or perhaps women's secrets or ailments.[61]

Or this was her opinion at one point in time. On subsequent readings, St. John recategorized and reassigned recipes to new lettered sections through the use of marginal annotations. For example, a letter "A" is written next to "Mrs Tassels cawdle to bring away the Afterberth or any part left behind"; the letter "M" next to "To Prevent miscarrying after any fright or hurt or disorder"; and "C" next to "Dr Willis for one that had a Falce Conception" (fig. 2.1).[62] In reassigning these recipes, St. John deconstructed the overarching category of women's ailments and returned the different components into their own separate distinct entries. In all of these reassignments, there is a general shift toward specificity. Where St. John was content to group remedies addressing particular areas of the body together in her initial categorization process, her second pass at the process worked to emphasize ailments and symptoms. Her play with the alphabet and information management reveals the flexibility with which early modern householders considered medical categories and their continued experimentation with categorizing and ordering knowledge. St. John's paper-based practices of alphabetization

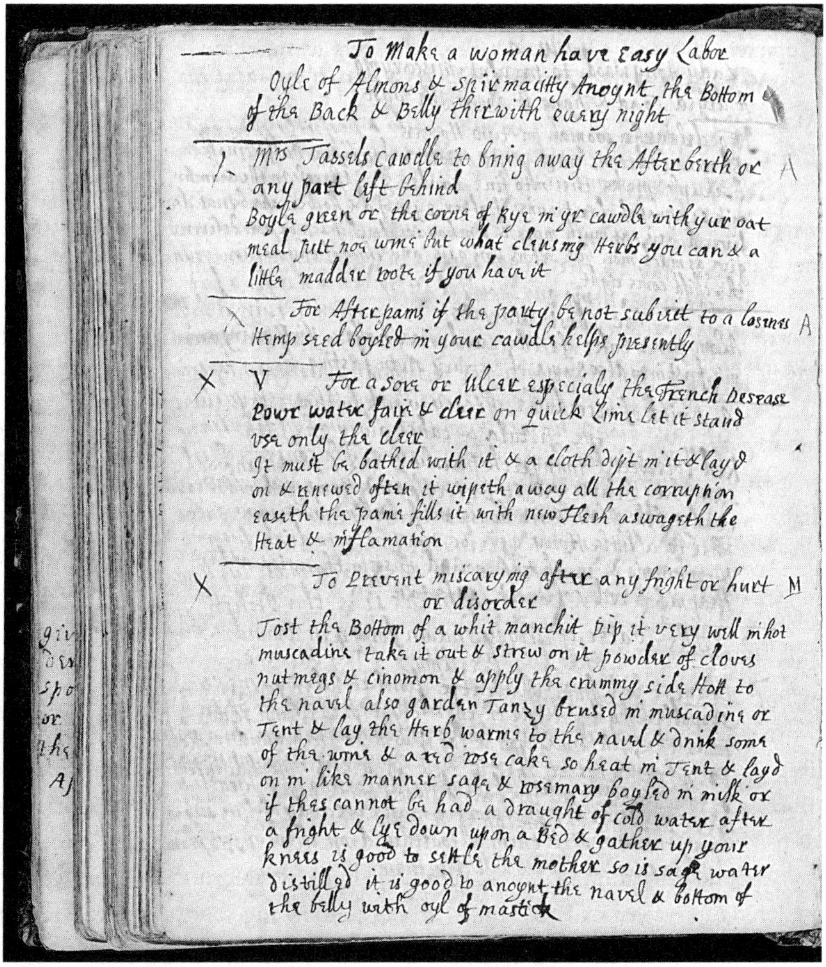

To Make a woman have easy Labor
Oyle of Almons & Spirmacitty Anoynt the Bottom of the Back & Belly therwith every night
Mrs Tassels cawdle to bring away the After birth or any part left behind A
Boyle green or the corne of Rye in yr cawdle with yur oat meal putt noe wine but what cleansing Herbs you can & a little madder roote if you have it
For After pains if the party be not subject to a loosness A
Hemp seed boyled in your cawdle helps presently
X V For a Sore or Ulcer especialy the French Desease
Powr water fair & clear on quick Lime let it stand use only the clear
It must be bathed with it & a cloth dipt in it & layd on & renewed often it wipeth away all the corruption easeth the pain fills it with new Flesh aswageth the Heat & inflamation
X To Prevent miscaryinge after any fright or hurt or disorder M
Tost the Bottom of a white manchit dip it very well in hot muscadine take it out & strew on it powder of cloves nutmegs & cinomon & apply the crummy side Hott to the navel also garden Tanzy brused in muscadine or Tent & lay the Herb warme to the navel & drink some of the wine & a red rose cake so heat in Tent & layd on in like manner sage & rosemary boyled in milk or if thes cannot be had a draught of cold water after a fright & lye down upon a Bed & gather up your knees is good to settle the mother so is sage water distilled it is good to anoynt the navel & bottom of the belly with oyl of mastick

Figure 2.1. A page taken out of the household recipe book of Lady Johanna St. John, showing how she used alphabetization as a way to categorize and recategorize medical knowledge. Western MS 4338, fol. 205v, Wellcome Collection, London.

and knowledge categorization, much like those of learned men and physicians, was an epistemic activity.

Managing Health and Household with Paper

Aside from adopting strategies inscribed on paper to manage information, early modern note-takers also used multiple notebooks and loose paper slips to sort, separate, and define epistemic spaces. For example, Angus Vine has written of Francis Bacon's use of waste and neat books and Fabian Krämer has shown in detail how Ulisse Aldrovandi used a system of paper slips, linen bags, and notebooks to organize his *Pandechion*.[63] Though perhaps not comparable in size and scope to Aldrovandi's *Pandechion*, householders also adopted complex paper systems of knowledge organization. An excellent example is the medical archive of Margaret Boscawen (d. 1688), a mid-seventeenth-century gentlewoman living in Cornwall.

During her lifetime Boscawen filled multiple notebooks and a flurry of loose paper slips with medical recipes and other information. Inheriting these papers, her daughter Bridget Fortescue continued to add and extend the collection.[64] Paper shines in this archive in a number of ways. It is folded, cut to size, bound, glued, clipped, or sewn together to shift, collate, and systematize the gathered medical knowledge. A set of loose paper slips filled with recipes that had once been sent in letters serves as a good entry point to the archive.[65] The loose slips are now preserved in small bundles. Some bundles, like the group of recipes for plague and general cordial waters, suggest a loose categorization by ailment, perhaps made by Boscawen, Fortescue, a later family member, or a modern archivist.[66] Each slip is endorsed, sometimes multiple times, on the reverse, indicating that they were not accidentally saved but constituted a system of information storage. For example, a slip with an untitled recipe is endorsed on the back with two labels: "for the eyes:" and "From ye the gentlewoman yt came heare wth wilmat Hoskins for ye cannings in her eyes." Another slip with two recipes for worms has written on the back: "For ye water that is to be maid in ye sun," "a Receite for wormes," and in a different ink "and for ye Rickets," suggesting that the Boscawen/Fortescue women considered these recipes under two categories—as production tasks and as remedies for different yet specific ailments.

Another indication that these slips constituted an information management system is Boscawen or Fortescue's habit of pinning together related recipes. In some cases, the pinned-together recipes dealt with the same ailment, such as the pair of recipes labeled as "the chilldrens drink by Mr Balles." In other cases they are united by similar production methods, as in the pair of instructions to make black cherry and lemon water. In another instance, Boscawen or Fortescue pinned together different versions of the same recipe as they tested and refined the recipe. The pair of paper slips has on the top a slip labeled "my one drink from Mr Baller when with child." This version lists sarsaparilla, china root, hartshorn, raisons, alehoof, and sassafras as ingredients (fig. 2.2). On the second slip, two very similar recipes are copied out by different hands, both requiring only sarsaparilla, china root, hartshorn, and sassafras (fig. 2.3). The second of these recipes, though, required three rather than two ounces of china root and hartshorn. It was not unusual for householders to work through and modify recipes, and the three versions here clearly represent the Boscawen/Fortescue family's efforts as they adjusted the recipe to suit their own changing needs.

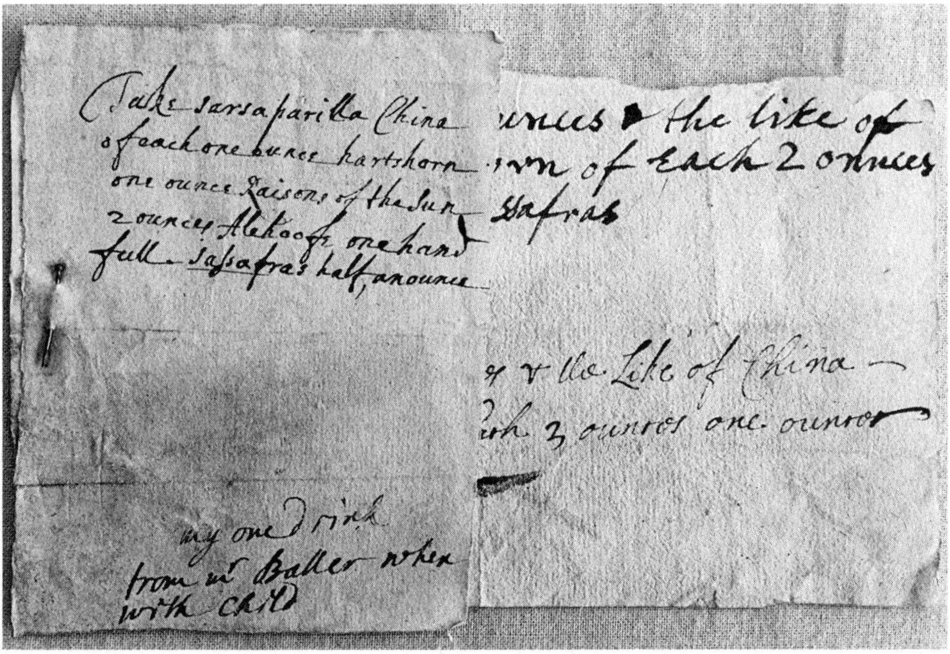
Take sarsaparilla China
of each one ounce hartshorn
one ounce Raisons of the Sun
2 ounces Alekoofe one hand
full Sassafras half an ounce

my one Drink
from mr Baller when
with child

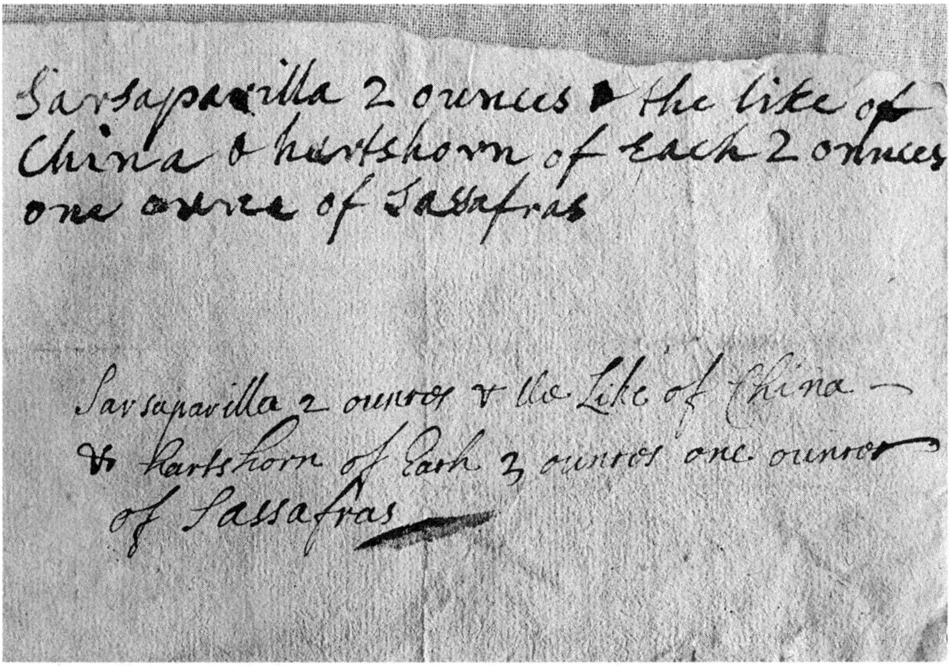
Sarsaparilla 2 ounces & the like of
China & hartshorn of Each 2 ounces
one ounce of Sassafras

Sarsaparilla 2 ounces & the Like of China
& hartshorn of Each 3 ounces one ounce
of Sassafras

Figures 2.2 and 2.3. Pinned loose paper slips showing how the Boscawen women used paper technologies to record their recipe trials. 1262M/0/FC8, Devon Heritage Centre.

As noted above, systems of loose paper slips were common information management strategies in the scholarly world.[67] The desire to continually sort knowledge into different systems of order was common across diverse communities. Loose leaves can be easily rearranged to recategorize information and recodify knowledge. They can also be easily added to existing material in bound notebooks either by copying the information directly onto the page or, as in the case of Aldrovandi, glued into the books themselves. Boscawen and Fortescue also adopted similar practices. For example, on the back of one slip with a recipe for a rickets remedy for children, someone, likely Fortescue, noted that the information could also be found in "my mothers Large boke."[68] More reversible solutions were also used—in one instance, they simply pinned the additional recipe to the relevant page.[69]

From alphabetization to the use of multiple notebooks and paper slips, householders adopted various paper technologies, usually connected with learned male scholars, to organize their recipe collections. In doing so, they constructed boundaries between different kinds of knowledge and expressed ideas about the human body. Recovering these practices through paper encourages us to recognize that while householders might be writing medicine outside academies and institutions, they were consciously utilizing scholarly paper technologies to repeatedly and continuously recodify medical knowledge. Domestic medicine and with it, women's medical activities, has often been presented as old fashioned and static; in this case, it seems that it was definitely far from being so.

The early modern household was filled with paper. In kitchens, stillrooms, and bedrooms, paper was central to a range of everyday technologies. It lined cake tins and glass funnels, delivered ointments and salves to the body, and preserved precious fruits, cakes, and materia medica. In the library, the study, and the closet, not only did paper serve as the carrier for inscription practices but paper technologies such as notebooks and loose slips enabled householders to sort, categorize, and express their ideas about the human body and delineate boundaries between epistemic spaces.

Our paper trail brings together a series of connected epistemic and hands-on practices used by householders to shift and filter, contain, and shape both knowledge and things. It emphasizes the continuities between practices of making and doing and practices of reading and writing, offering a new lens through which to view the intersections between the study of production of material goods and the study of knowledge codification. The connection between these two threads is significant for several reasons. First, it reminds us that housewifery consisted not only of practical manual tasks but also of knowledge gathering, organizing, and codifying, tasks usually associated with the work of learned men. Secondly, the strong similarities in paper use and paper-based practices across the activities of housewives and learned scholars, surgeons, apothecaries, and physicians, prompt us to envision knowledge practices in and out of the household as a continuum. In doing so, we recover not only a different set of historical actors but also further probe our historical actors' and our own gender- and class-based constructions of authority and work. Continued investigations into the sociomateriality of paper and gendered labor offer opportunities to further tease out the complex knowledge practices of the early modern household.

THREE

The Sociomateriality of Waste and Scrap Paper in Eighteenth-Century England

Simon Werrett

To how many uses does paper now serve
for which nothing else would be so convenient?
Joseph Priestley

An intense focus in recent years on the history of the book, publishing, and journals has done much to reveal the role of paper as a technology of writing and text.[1] Undoubtedly this was one of the principal uses of paper in the early modern period. But as this chapter will argue, paper served many other uses and, at least in the English case, it was exactly the adaptability and serviceability of paper, the fact that it was *not* just a carrier of text, that may have given it such value for early moderns. Paper was exemplary of the kind of materials valued by thrifty early moderns, who sought to make the most of their possessions and gave credit to materials that offered a variety of services in the home and elsewhere. Householders approached paper as one among many open-ended, "incomplete" objects in their homes, the uses of which were not fully determined. They made use of paper not only as a medium for text but also as a material for cookery, packaging, hairdressing, building, and many other functions. Paper was a kind of early modern plastic, adaptable to a wide variety of shapes, forms, textures, strengths, and uses.

Different paper practices were associated with different genders in an ongoing interaction between identity and material, which I refer to as *sociomateriality*.

The term, originating with the organization theorist Wanda Orlikowski, indicates that the social and material are inseparable: there is a mutual interaction between people and things in which the identities of both are transformed.[2] Gender, then, is always linked to a set of material conditions. The places, technologies, and bodily practices through which people operate have a fundamental effect on their identities, just as people in turn shape and give meaning to their material environments.[3] For early moderns, the diverse uses of paper, often linked to particular sites or occasions, gestured toward particular negotiations of gender. The examples explored below consider practices associated with "waste" or "scrap" paper, and take examples from hairdressing and cookery to reveal the nature of these practices' sociomateriality. In these examples, the gendering of paper practices was used to claim divisions between the sexes. Male scholars represented nontextual uses of paper by women, artisans, and children as evidence of a lack of erudition, intellect, or ingenuity, contrasting with their own efforts to save texts from oblivion. The notion that paper should be understood primarily as a bearer of text thus served a masculine agenda that deserves questioning.

DEFINING WASTE AND SCRAP PAPER

As Heather Wolfe notes in this volume, early modern paper was in plentiful supply. There were many types and qualities, but the most common type cost, between 1570 and 1630, fourpence for a quire, or a penny for six sheets. A laborer could afford seventy-five sheets on a single day's wages. But if paper was cheap, people still made good use of the paper they had. Early modern letter writers folded a piece of paper in two, then wrote on one half, leaving the other side blank. Although one side was blank (since this would provide the outside of a paper packet once folded further), writers avoided wasting space in the remainder. Sometimes they continued a letter in the side margin rather than move to the blank part of the page, wrote vertically and diagonally across the page, or just stopped writing when the page was full.[4] There were occasions for leaving paper blank, but people also filled up sheets to avoid wasting them.

Such practices might be seen as examples of early modern thrift. Today "thrift" implies saving money but in early modern England it had different meanings. Thrift was a positive virtue, related to *thriving*. It might mean temperance and moderation. The author of a book of advice to young gentlemen insisted they should "expend[] profitably, avoid[] unnecessary Expences."[5] It also entailed a balance, between the purchase of new things with making the best use of what one already had. Thrift was part of *oeconomy*, or household management. An essay examining "Oeconomy and extravagance" in the *Gentleman's Magazine* of 1731 explained that oeconomy was "Wisdom applied to the Practice of private Life; it is situated betwixt Profuseness and Avarice, and consists in a just Medium of Concern, as to exterior Goods, between being over Careful and having no Care at all."[6] Thrift then entailed buying, for example, new clothes, but also making use of what one already owned. A lady might purchase fashionable items and simultaneously patch and rework old garments. It was fine to leave *some* paper blank, but ideally it should be put to good use.

Since old things were often used and reused, there was not much sense of things becoming "waste" among early moderns, so the term was not common, at least in the sense we use it. The term appeared more often in medicine, in talk of muscles, bones, or spirits wasting or being consumed. Waste paper, like wasteland, was consequently not a useless thing but rather paper that had *not yet* been used. "Waste" paper was not, in fact, wasted. It is better to think of it as a "genre" of paper in the early modern period, a certain kind of paper. Printers might remove as "waste" the top and bottom sheets of a block of paper used in printing, if they became dirty or creased; old documents or unwanted writing paper could also be designated as waste. A great deal of waste paper was purchased from printers, to judge by newspaper advertisements. An auction sale in 1787 of goods belonging to a bankrupt tea dealer included "about two tons of Waste Paper" together with "a large quantity of old Iron and Brass."[7]

While waste paper was not thrown away but used for a variety of practices, for early moderns, scrap paper was not the same as waste paper, as it is today. The idea of scraps belonged to a medley of words that signified the potential uses of some leftover substance. The debris of knocked-down buildings was "rubbish," perhaps related to the word *rubble*, while *trash* meant bits of wood and straw commonly used to light a fire. The term *scrap* was similar. Coming from the Old Norse, it referred to a remnant, originally one of food or linen. This did not mean it had no use but simply that it was a piece of something that was formerly more substantial. Scraps of food, then, had many uses. The same was true for scraps of paper.

INCOMPLETE OBJECTS

There is no need to be too particular about the definitions of waste and scrap paper. Following the work of sociologist Karin Knorr-Cetina, I have elsewhere argued that early moderns entertained a view of material goods, including paper, as what I call "incomplete objects."[8] Knorr-Cetina uses this term to describe laboratory artifacts that do not yet have a defined behavior or purpose, and are open-ended in their meanings and functions.[9] The notion of an incomplete object applies well to paper and captures a widespread, thrifty, early modern attitude to things in general. In this period, production and consumption were not as distinguished or separated as they would later be in the Industrial Revolution. Instead, householders created, reworked, and reused their possessions on a regular basis (which is why early moderns had no term for "recycling"—it was going on all the time). The home was a place where a division between production and consumption made no sense. This gave possessions a sense of being incomplete, since it was not obvious how a thing might be used in the future, or that the various uses of a thing would come to an end. A shirt, for example, might be altered into a different shape, patched and repaired, or turned into a completely different garment. When it became ragged, it would then be used for cleaning in the home, or sent out to collectors who used the rags to make paper.

Such material circulations were also moral, since the status of paper and the rags used to make it was inevitably tied to people and judgments about class, race, age, and gender. Consider, for example, this account of papermaking addressed to "juveniles" by John Aikin, editor of the *Monthly Magazine*, in 1795: "This delicate and beautiful substance [paper] is made from the meanest and most disgusting materials, from old rags, which have passed from one poor person to another, and at length have perhaps dropped in tatters from the child of the beggar."[10] Different kinds of paper carried certain associations of identity, which Aikin expected his readership to be familiar with. Both people and paper could be "quality" or "mean" and "disgusting."

Domestic work with textiles was often associated with women, who were expected to make, clean, and rework garments. Moral and material "clothing" were both the housewife's responsibility. Gervase Markham advises in *The English Hus-Wife* (1615), "Our English housewife, after her knowledge [of] her family, must learn also how, out of her owne to clothe them outwardly and inwardly."[11] Other labor was gendered male, including, as the name implied, husbandry, or the work associated with keeping an estate in good order and keeping livestock. Treatises on husbandry proposed diverse uses of paper: to cover grafts on trees, to make matches by soaking brown paper in the water used to boil potatoes, or to cover holes in a wooden box for distributing seeds.[12] Boundaries were not clear-cut, however, as the historian of domestic oeconomy Karen Harvey has argued.[13] While books of advice distinguished men's and women's labors, in practice these might cross over; for example, cookery (identified as women's work) often entailed gardening (men's) to grow herbs or vegetables. It was a similar case with paper.

The open-endedness of domestic practices was matched by the material open-endedness of paper: it was a material that was flexible, adaptable, and valued for being so. Very soon after its introduction into Europe, paper was given many uses, in writing letters, registers, books, and notes, for making book covers and for packaging, for printing maps, charts and diagrams, banknotes, playing cards, engravings, and so on. Early moderns celebrated the transformative capacities associated with paper and the materials it was made from. An early modern German verse proclaimed,

And if the canvas becomes worthless
Or the flag is lost
Then its true value is realized
Paper rustles in the ears
One prints the Word of God on it
And writes on it with ink.[14]

To complicate matters, early moderns did not speak of "use" so much as "service." An eighteenth-century book on accounting noted that "pieces of paper may serve instead of counters, and brown paper may serve for pence, and white, for shillings."[15] Writing in 1606, Alexander Craig described lovers exchanging letters. If a lady did not like her suitor's letter she might make the paper serve her closet,

Take then my faultles Sheet,
bedewd with mourning Inke,
And if thou wilt not view my Verse,
to know the thing I thinke;
Yet shall the Paper serue
(O faire and matchles Dame)
To be a Bottom to thy Silke,
or safftie [safety] to thy Seame.[16]

In other words, an unwanted love letter could be put into service lining a drawer or as support in clothing. Service was of course also a social category in the early modern period, so again in the use of this term there was no strong division between people and things, between the morals and the material. Both people and possessions were described as "bodies" and attributed similar kinds of agency and behavior. For instance, instead of saying that a thing was damaged early moderns said it was "hurt" or had "suffered an injury" that needed to be "remedied." Injury was as much a moral as a material matter, so that paper, like some destitute unfortunate, could be quite "ruined" or suffer from "defects."[17] Paper was thus an open-ended, incomplete object, and a body capable of all sorts of unanticipated uses, behaviors, and agencies.

Paper thus moved through different uses over time. As the quotations above suggest, rags became paper, which could be used for letters; then letters could be put into service lining drawers or filling out dresses; and when these became rags they would serve to make paper again. In this course, the social, moral, and material were constantly interacting, and indeed, were scarcely divided. The process might be described as being *sociomaterial,* in that it involved this continuous negotiation between people and things, the outcome of which could shape the identity of all the bodies involved. All sorts of domestic labors were sociomaterial. In performing them, both material things and the people engaging with them were changed. People were incomplete, just like objects.

Paper was not entirely incomplete, however. As several chapters in this book attest, specific types of paper were made to serve some ends better than others. In the seventeenth and eighteenth centuries, one could purchase, for example, demy, royal, Rochelle, Venice, black, white, or brown paper, depending on what size and fineness were required for a particular task such as writing a letter. Different properties in these various kinds of paper could be harnessed to new ends by a thrifty householder. Black paper or card was used in the seventeenth and eighteenth centuries for letters of mourning, drawing in chalk, or making silhouettes.[18] Edward Fountaine used it for writing with silver ink.[19] The blackness was ideal for Isaac Newton, who made use of it in his optical experiments to investigate refrangibility. To create distinct colors he took a "black oblong stiff paper," painted one side red and the other blue, then viewed them through a prism.[20]

PAPER AND HAIRDRESSING

One area that indicates how early moderns made use of paper in a variety of gendered practices was the decoration and ornamentation of the body. Paper served to provide support for various forms of dress, while print media communicated news of fashions and fads. Hairdressing, an activity that scholars have long recognized as important in constructions of gender, also involved making use of paper, and will here provide a focus for considering the sociomateriality of paper.[21]

Hairdressing made use of paper first in a series of printed books that appeared in the eighteenth century in Britain and France. Although hairdressing was practiced by both men and women, it was principally male authors who articulated its craft secrets in print. In 1768 the French court hairdresser Legros de Rumigny published *L'Art de la coëffure des dames françoises, avec des estampes,* richly illustrated with engraved designs for women's hair.[22] In Britain, the Bath hairdresser and perfumer William Moore offered *The Art of Hairdressing* in 1780, intended, as he explained, to provide a "competent Idea" of the art to the public.[23] Another male hairdresser, Alexander Stewart, published two short pamphlets in 1788 on the art of dressing gentlemen's and ladies' hair.[24] The brevity of these works was perhaps a response to the gargantuan 450-page *Plocacosmos* of James Stewart, an extensive essay on hairdressing published in 1782 that also included a scientific account of hair, a history of hairstyles, a study of hair in theatrical performances, and, as the subtitle had it, "complete Rules for the Management of Children and Education of Youth; and excellent Rules for the Preservation of the Health and Happiness of Age."[25] Once again the material and moral were intimately connected. Paper then secured hair and the status of hairdressing. Styled a "guide to the seven ages of Man," *Plocacosmos* was likely intended to raise the status of the art by presenting it in erudite form at a time when grand hairstyles were inviting accusations of vulgarity and excess. Addressing both male and female audiences, Stewart insisted that there was nothing about hairdressing that would offend any sort of lady, not "the blooming virgin, pious wife, or happy mother."[26]

Print thus helped to reveal and gain credit for the methods and styles of hairdressing in the eighteenth century. But paper also played a significant role in its practice, put into service by hairdressers in wig making and in the curling of women's hair. Wigs, worn primarily by men, were made with batches of human hair by male or female peruke makers.[27] Parcels of hair were rolled up on clay pipes, then boiled in water in a pot over a fire, which set the curl, following which they were dried out on sheets of brown paper.[28] The curls were assembled into a wig, and when a man put his wig on, he covered his face with a paper bag as powder was puffed onto the wig to whiten it.[29] Wigs, then, served to mark masculine identity. As Margaret K. Powell and Joseph Roach have noted, hair in the eighteenth century was "one of the most visible ways of marking different social roles, occupations, aspirations, and conditions."[30] Hair was "a performance, one that happens at the boundary of self-expression and social identity, of creativity and conformity, and of production and consumption."[31] Powell and Roach cite Stewart, author of *Plocacosmos,* who explained that between the reigns of Charles II and George II, "as the perukes became more common, their shape and forms altered. . . . The merchant, the man of business and of letters, were distinguished by the grave full bottom, or more moderate tie, neatly curled; the tradesman by the snug bob, or the natty scratch. . . . All conditions of men were distinguished by the cut of the wig."[32] Powell and Roach also examine the phenomenon of "Big Hair" in the eighteenth century: the tall, fashionable hairstyles worn by women, particularly during the 1770s. Women wore their own hair rather than wigs, and had it dressed by a hairdresser, who might be male or female.[33] Women's hair became increasingly large and elaborate, culminating in the 1770s, when it became so big that satires and scoffing proliferated. Only in the 1790s, when a tax on hair powder was introduced by William Pitt, did the fashion come to an end. Subsequently men wore their hair short, while women's hair switched to hanging curls.

Paper played a role in shaping these hairstyles. Tall hairstyles and later natural hair made use of "French curls," which were, as the name suggests, a fashion originating in France. To make them, hairdressers wound a length of hair into curls and held these together with twisted pieces of paper. In French the curls were known as "papillotes," or "parchments," making reference to the use of paper to create them. In *Plocacosmos,* Stewart explains the elaborate process of curling, which he reckoned needed an appreciation of the qualities of paper to succeed. First the hair was separated into strands and each strand was coiled up. A sheet of paper was placed behind the coil, then folded over it several times until it was tightly enclosed in the paper wrap. At this point, iron tongs heated in the fire were pressed onto the paper and held steady for a few moments before being released. Singed paper was a sign that the tongs were being held on too long. After the heated wrap was allowed to cool, it was untwisted, leaving a coil of hair that fell down in a curl. Repeating the process across the whole head left the hair fully curled.

The qualities of some types of paper evidently served curling better than others, and Stewart expected readers to be sensitive to the best qualities of paper for achieving this style, stating, "You are therefore, it is supposed, furnished with a proper quantity of French curling paper, which is certainly the best, being tougher, bearing the iron better, and, at the same time, thinner, so as the heat gets sooner through than any other paper, except India paper; but that is very insipid and limber, and has not enough strength to twist."[34] French curling paper may have been "French paper," a fine paper used in printing, considered to be better than "English paper" but inferior to "India paper." The latter was a soft and absorbent paper used for printing engravings, made with fine bamboo fiber and imported by the East India Company from China.[35] To shape a woman's hair necessitated a nuanced appreciation of these different papers' strength, thickness, and their capacity to bear heat and be twisted.

While Stewart allied print and hairdressing to enhance his reputation, other authors used the same connection to denigrate women's capacity for learning. A number of eighteenth-century writers exploited the image of paper being used to curl women's hair to make points about men's and women's capacities for erudition. For example, an anonymous poem titled "Woman," appearing in the *European Magazine* for 1824, records the activities of various ladies:

> Whilst Ellen, quite a blue—, with Lady Di,
> Exalts some flashy author to the sky.
> The confab ended, lo! his pages fair
> Lights Ellen's lamp, or curls my lady's hair.[36]

The verses capture the sense that paper could be turned to different uses (reading, lighting, curling), and mock the idea that two female bluestockings' discussion of a book could be anything more than whimsical, as the book is reduced to a hair-curling accessory soon after the enlightened conversation.

The essayist Philip Dormer Stanhope, Earl of Chesterfield, also linked print, learning, and hair curling in a 1755 issue of his periodical the *World*. Stanhope reflected on the fate of his publication, as it passed through the hands of several persons belonging to a well-to-do family. He marked differences in class and gender through different responses to the journal. A porter to a family of condition, he wrote, would cast aside the *World* on finding neither politics nor the "casualties of the week" inside it.[37] Similarly, "Mrs Betty" (a generic name for a maid) would "find it deficient in point of news" but prefer a *Daily Advertiser* to read about the invitation of "single gentlemen of undoubted characters, to agreeable young women of unblemished reputations, to become either their wives or their companions." For this Mrs Betty would exchange "a fine wholesome breast of milk."[38] In contrast to the servants, in "my lady's dressing-room," a more refined woman might pronounce the contents of the journal "charming" or "poor stuff" if read to her by a young companion. Stanhope was proud that the circulation of the *World* amounted to some two thousand copies, but to teach himself some humility he considered the source of such good sales: it was a cheap source of paper for curling hair. "A head that has taken a right French turn, requires, as I am assured, fourscore curls in distinct papers. . . . Four of my papers are sufficient for that purpose, and amount only to eight pence, which is very little more than what the same quantity of plain paper would cost. Taking it altogether, it seems not inconsistent with good oeconomy to purchase it as so small a price."[39] It was comforting, Stanhope concluded, to know that "of the prodigious number of daily and weekly papers that are now published, mine is perhaps the only one that is ultimately applied to the head."[40]

Another writer who linked paper, hair, and gender was Alexander Pope in the *Rape of the Lock* (1714). The poem dramatizes a scandal over a suitor cutting off a lock of hair from a woman named Belinda without her consent. Pope highlights the injustice of the "rape" through the labor needed to make Belinda's lock:

Was it for this you took such constant Care
The *Bodkin, Comb,* and *Essence* to prepare?
For this your Locks in Paper-Durance bound,
For this with tort'ring Irons wreath'd around?
For this with Fillets strain'd your tender Head,
And bravely bore the double Loads of Lead?[41]

Pope presents curling with paper as a sign of the effort to which Belinda would go to make herself beautiful, only for her to experience an injustice.

Paper practices connected to hair and hairdressing reinforced gender stereotypes of women as overly focused on beauty and not on education. Women too easily turned the products of enlightenment to superficial uses such as curling hair. But paper practices could also challenge accepted gender roles. Hairdressing was often represented in printed texts and engravings as a profession of gender ambiguity, since the men who practiced it were considered effeminate for their interest in women's hair. Male "macaronis" were also considered effeminate for their own elaborate hairstyles. Printmakers used paper to assert what they considered correct and incorrect versions of masculinity. Compare the very different masculinities on display in the London printmaker Matthew Darly's "Ridiculous Taste or the Ladies Absurdity" of 1771 (see plate 6). A macaroni-like hairdresser wearing an elaborate wig is holding curling tongs to curl the enormous hair of a lady, having to stand on ladders to do so. Her husband is represented as a more robust English gentleman with a sextant, needed, so the image suggests, to see so high up.[42] The image recognizes that hairdressing involved ambiguously gendered individuals, and sought to reinforce traditional gender boundaries. Between the two men, the lady is again an object of ridicule for her excessive submission to fashion.

WRAPPING AND GENDER

Hairdressing thus offered opportunities to make strong connections between paper and people because hair was such a potent sign of gender in the eighteenth century. One of the particular techniques of paper use involved in curling—that of wrapping—also had complicated relations to gender. Wrapping—enfolding things inside packages or envelopes—was a widespread practice using paper that was performed by men and women in all manner of situations. Its commonality perhaps explains why the practice has not been studied historically.[43] Yet wrapping was a complicated sociomaterial practice, because acts of wrapping and enfolding things were allied to a variety of gender performances and moral associations. Wrapping transformed both objects and people, folding them together. In Tobias Smollett's novel *The Life and Adventures of Sir Launcelot Greaves*, for example, the title character suffers rejection from his beloved Aurelia in a letter. He resolves to read it again so that "he might once more banquet his spleen by reading the original billet, which . . . he kept in a small box. . . . This being instantly unlocked, he unfolded the paper and recited the contents. . . . [Then] he hastily replaced the cruel scroll, and . . . threw himself into his bed in a transport of despair."[44]

Acts of wrapping could be highly moralized. Eighteenth-century oeconomists invoked Jesus as an exemplar of good oeconomy for folding up the cloth that covered his face in the grave. But in the parable of the talents, a servant who buried his money wrapped in a napkin and put it to no good use was condemned.[45] Numerous books noted the ways different wrappings of cloth around the body marked gender among non-European peoples. "Of all the . . . Asiatics, the Chinese have . . . the best title to modesty. Even the men wrap themselves closely up in their garments, and reckon it indecent to discover any more of their arms and legs than is necessary. The women, still more closely wrapped up, never discover a naked hand even to their nearest relations."[46]

Gender roles might also be consolidated through wrapping performances with paper. The Norwich botanist Charles Bryant described how the chemist Andreas Marggraf purified a sugar substitute made from skirret roots. Marggraf wrapped the sugar made from the roots "up in a piece of blotting paper, and then gently pressed it with his hand; the effect was that the paper sucked up much of the viscid moisture." This left the sugar "neat" and divested of impurities. But it also transformed Marggraf, who Bryant now claimed was "ingenious" for discovering this useful substitute, a term commonly used in the eighteenth century to praise men but more rarely used to describe women.[47] The chemical identity of sugar and the gender identity of Marggraf were constructed simultaneously.

In other cases, discussions of wrapping allied paper with class as much as gender. A famous poem by Benjamin Franklin plays on the associations of different kinds of paper with different identities.

> The wretch whom av'rice bids to pinch and spare,
> Starve, cheat and pilfer, to enrich an heir,
> Is coarse *brown-paper,* such as pedlars choose
> To wrap up wares, which better men will use.
> [. . .]
> Observe the maiden, innocently sweet,
> She's fair white-paper, an unsullied sheet;
> On which the happy man whom fate ordains,
> May write his name, and take her for his pains.[48]

Franklin makes use of paper here as a metaphor for purity and, one assumes, virginity, the signature on paper standing for male ownership of a woman presented again as a superficial object. Once again we see here an intimacy between the material and the moral. Franklin allies the coarse texture of wrapping paper with the imperfections of the "wretch."

Different objects were wrapped by people of different genders in different ways. Wolfe provides evidence of this in relation to the folding of letters in chapter 1 of this volume. In the eighteenth century, in the masculine space of James Watt's workshop (now reconstructed in London's Science Museum), there are numerous wrapped-up packages of screws, parts of tools, pieces of iron wire, washers, pieces of brass, felt, pasteboard, steel springs, molds, and mineral and chemical powders of various kinds.[49] Watt's techniques of wrapping, and their gender specificity, require further investigation. Most are wrapped in coarse paper, of the kind Franklin associated with wretches. Perhaps Watt wrapped these up himself, but he may have relied on servants or family members to do it for him.

Certainly many wrapping practices must have been pursued by both men and women, being as open-ended, in terms of gender, as were the uses of paper itself.

In *A Tale of a Tub,* Jonathan Swift considers further uses of waste paper: "What is then become of those immense Bales of Paper, which must needs have been employ'd in such Numbers of Books? Can these also be wholly annihilate . . . ?" Swift does not attribute any gendering to the different uses to which books were applied, dividing them instead in terms of sites or functions, being applied in "a jakes, or an oven; to the windows of a bawdy-house, or to a sordid lanthorn. Books, like men their authors, have no more than one way of coming into the world, but there are ten thousand to go out of it, and return no more."[50] Books might be written by men, but everyone might use them to light fires and to make lanterns. They were oiled to cover windows as a cheaper alternative to glass, and used as toilet paper, applied in "a jakes."

Both men and women made use of paper in cookery, which could be the province of housewives, cooks, bakers, butchers, grocers, and many others. White and brown paper were put into service in a variety of recipes to cover pies, tarts, and meat roasts to prevent them burning (akin to the use of tinfoil and baking paper today).[51] "Brandy paper" was paper dipped in brandy and sealed over a jar with a bladder to act as a preservative—presumably the alcohol kept out bacteria.[52] Books on cookery were authored by both men and women, and they were often silent on the gender of readers and users. Richard Briggs, whose *English Art of Cookery* was published in the 1790s, said that his directions were based on his own experience as a cook, but he only connected recipes explicitly with gender in one chapter, on foods prepared by "seafaring men," while the others were written in the form of instructions that could be followed by anyone. In cooking veal cutlets, for example, Briggs advises, "Take as many half Sheets of Writing Paper as you have Cutlets, and butter them with melted Butter on one Side; likewise dip your Stake in melted Butter, and . . . [put] a Stake on each Half Sheet of Paper cross the Middle of it, leaving the Bone about an Inch out; then close the Two Ends of your Paper, closing it on the Sides as you do a Turn-over Tart. Cut off your waste Paper; So broil them half an Hour."[53]

One way that such culinary practices were allied to gender and class was in the accounts of scholars who claimed to have saved old manuscripts from being used for cookery and food preparation. Making use of paper was fine until it threatened scholarship. Male scholars recounted their heroic attempts to save letters and documents from being used to line pies and pastries by cooks. In the 1790s the *Encyclopaedia Britannica* recorded, "It is very remarkable that the manuscripts of [renaissance scholar Sir Julius Caesar] were offered (by the executors of some of his descendants) to a cheesemonger for waste-paper" though they were saved and later sold at auction for £500.[54] Scholars imputed that an intellectual use of old papers was "better usage" than employing them in more artisanal trades. Retaining papers in the male domain of scholarship was to "rescue" them.

John Aubrey said that he sifted through the papers of the deceased mathematician Seth Ward and "rescued [them] from being used by the cook since his death . . . to be put under pies."[55]

So while women were represented as using scholarly papers for unscholarly labors such as hairdressing, men were represented as saving such papers from unscholarly uses, returning them to the world of learning. Artisans, in this regard, were looked down upon as women were, or indeed anyone who made use of paper in unscholarly ways. The seventeenth-century writer on husbandry Gabriel Plattes lamented of printed newspapers that they were "torn in peeces to light Tobacco" instead of being read by more people.[56]

Children also figured in these sociomaterial judgments. Of the physician and mathematician John Arbuthnot it was said, "He was so neglectful of his writings, that his children tore his manuscripts and made paper-kites of them."[57] Alternatively, children's use of paper for writing and drawing could be taken as a sign of nascent ingenuity. The biographer of Sébastien Leclerc identified early signs of greatness in the artist's childhood, when he busied himself "in making little portraits with a pen on such scraps of paper as he found about the kitchen."[58] Such stories were told of boys and girls. The Utrecht theologian Anna Maria Schurmann was similarly celebrated, "for at the age of six years she cut with her scissars [*sic*], out of paper, all sorts of figures."[59] Playing with paper might even be singled out as an ideal form of education in "making use" of things. In *Some Thoughts Concerning Education* (1693), John Locke noted, "A smooth Pebble, a Piece of Paper, the Mother's Bunch of Keys, or any Thing they cannot hurt themselves with, serves as much to divert little Children as those more chargeable and curious Toys from the Shops, which are presently out of order and broken."[60] Not only were purpose-built toys liable to break but they failed to teach a child ingenuity. Inventing new uses for paper, on the other hand, was to be encouraged, "when they once begin to set themselves to work about any of their Inventions, they should be taught and assisted."[61] For Locke, like other contemporaries, playing with paper was material and moral: "Whatever they do leaves some Impression on that tender Age, and from thence they receive a Tendency to Good or Evil."[62]

This chapter has focused on just a few of the nontextual uses of paper by early moderns, and their role in negotiating and consolidating gender identities. Thanks to the values of thrift and good oeconomy, the uses of paper were manifold in the period, and different practices with paper were allied with different genders and gender representations. Here hairdressing, wrapping, cookery, and child's play have offered examples. Materials, morals, and identities were folded together in these practices, so that paper was sociomaterial. If some generalizations emerge from these examples, they point to the way male scholars criticized others—women, artisans, the vulgar, and children—over the inappropriate use of paper. To use paper for scholarly writing was presented as valuable and masculine, while other uses of paper, in curling or cookery or play, might be dismissed as trivial and distracting from learning. If making use of paper was "ingenious" as in experimental uses, then this was to be applauded, but other uses were denigrated as wasting a valuable opportunity for study. Male scholars "rescued" paper from inappropriate uses, such as wrapping up pies and pastries, to restore it to learned use. The implication is that a focus on paper as a carrier of text was itself a gendered practice, allied with male scholarship and asserted to diminish the value of the diverse uses of paper by others. This should be recognized in the present historical focus on the history of the book and textual strategies in science, since it occludes a variety of paper practices carried out by persons of varied classes, ages, and genders.

No doubt many other nontextual uses of paper might be considered and their links to gender explored. Paper was used to stiffen and support clothing and bedclothes, to make cases for fireworks and, in the form of papier-mâché, to produce all manner of commodities. Priestley's remark at the beginning of this chapter indicates how paper served a role for early moderns similar to plastic today. If paper was a carrier of text this was not its only significance, and it may be its plasticity that accounts for why paper was so successful as a medium beyond its place in the world of publishing. Furthermore, print culture was never disconnected from these more plastic uses. When women, children, and artisans invented new uses for paper, to curl their hair, make paper kites, or bake pies, scholars looked down on them. But when men sought to do the same, they celebrated it as a great achievement. Many people knew that paper was fit to be used in ways other than in writing and printing, and the plasticity of paper was widely valued. To obscure this fact, however, served an elite, masculine vision of paper as a primarily textual medium.

FOUR

Paper Trials, Multiple Masculinities, and the *Oeconomy* of Honor

Gabriella Szalay

In 1765 the Protestant pastor Jacob Christian Schäffer (1718–1790) published the first in a series of six books recounting his paper trials.[1] *Versuche und Muster ohne alle Lumpen . . . Papier zu Machen* opens by describing his reasons for undertaking such trials, which sought a material substitute for the linen rags that were used to make paper. If we are to believe Schäffer, his chief motivation was to resolve the paper crisis at the time.[2] He was particularly concerned about how difficult it had become to procure the kind of fine white writing paper that was used to publish learned treatises like his own.[3] Sensing a threat to his intellectual ambitions, Schäffer cast the paper crisis as an event that promised to have dire consequences for knowledge production, if allowed to continue unabated.

Over the past three decades, historians have questioned such claims to resource shortages on the grounds that they were often rooted in ideological considerations that had little to do with fact.[4] Schäffer's efforts to combat the paper crisis in eighteenth-century Germany are no exception, as can be seen by his decision to turn not to the men and women who made paper for a solution but to his fellow men of learning. Or rather, he turned to their natural historical writings, which offered numerous recommendations for making paper out of materials other than linen rags. The prominent French naturalist René Antoine Ferchault de Réaumur, for example, had recently published a treatise on wasps that built nests out of the fibrous components of plants and trees. Arguing that these nests have a color and consistency similar to fine white writing paper, he observed that papermakers should try their hand at such materials.[5]

To suggest that there was an inherent divide between Schäffer and papermakers is to enter a long-standing debate in the history of science over the relationship between scholars and artisans. This debate first began in the 1940s when Edgar Zilsel argued that the "real pioneers of empirical observations, experimentation and causal research" were early modern artisans and not scholars. Due to their lowly social status, he continued, these men had been unable to properly lay claim to their ideas, which were instead ascribed to the "great men" of science.[6] While Zilsel's thesis has recently been the target of some critique, as his preference for individual case studies reinforces rather than upends the role of "great men," the gulf between scholars and artisans throughout much of the early modern period continues to be discussed by historians as a problem of knowledge production.[7] However, as I will argue, it was also a problem of gender, particularly of competing—or multiple—masculinities.[8]

Historians of gender such as Ute Frevert, Karen Harvey, and Alexandra Shepard have persuasively argued for "multiple masculinities," contending that the two-sex model that casts gender as a category of biological difference does not sufficiently account for the varied expressions of masculinity that we find in this period.[9] The household patriarch, the polite gentleman, and the fop were just some of the many kinds of masculinity that existed alongside one another. In fact, a single individual could perform multiple masculinities, depending on the context(s) in which he (or she) operated.[10] Thus, a more fruitful avenue of inquiry is to ask when specific conceptions of masculinity emerged, how they took root within a particular milieu, and in what ways they informed and played off one another.

Paper trials provide a case in point for this dynamic, as they illuminate what it meant to be a man in the late eighteenth century. Scholars like Schäffer shared in a conception of masculinity that rested on a combination of book learning, knowledge gathered during excursions into the "field," and an overarching mission to contribute to the common good.[11] Artisans like papermakers ascribed to a model of masculinity that prized physical strength, "gestural knowledge," and commercial savvy.[12] What makes the case of paper particularly interesting is that both learned and artisanal concepts of masculinity placed considerable weight on honor.[13] Each group saw their ability to belong to a community of their peers, which they described as a republic, as contingent upon honorable conduct. They did not, however, agree on what constituted such conduct. Scholars sought honor in intellectual accomplishment and valued behavior like reporting one's results accurately. Papermakers sought honor in labor and emphasized using one's skills to make the finest product possible.

These differing conceptions of honor are evident in Schäffer's paper trials, which occurred at a moment in which papermaking was undergoing a paradigm shift. Once a practice that lay largely in the hands of artisans, papermaking was becoming wrapped up in the quest for improvement that characterized many European states in the eighteenth century. As such, it was placed under the watchful eye of the men of learning who occupied bureaucratic positions within these states.[14] The methods of papermaking, which had been more or less fixed for centuries, were thus suddenly opened up for debate and discussion.

Especially heated debates took place in the German-speaking territories, where universities had begun to offer instruction in the police and cameral sciences.[15] These sciences were designed to offer practical training to bureaucrats in areas like agriculture, mining, forestry—as well as manufactories—so that they could better supervise both laborers and production. Although the police and cameral sciences were hotbeds of multiple masculinities, as they taught scholars how to assert their newfound authority and drove artisans to fight for their independence, they have yet to receive much attention from historians interested in the study of gender.[16] This essay will address this lacuna, by carving out how competing concepts of masculinity were forged, contested, and mediated with and through paper.

SCHOLARS, ARTISANS, AND MULTIPLE MASCULINITIES

A great admirer and longtime correspondent of Réaumur, Schäffer tested the elder scholar's theory by having paper made not only from plants and trees but also from specimens culled from the animal and mineral kingdoms. At first he limited his investigation to materials that could be found in and around his home in Regensburg; his earliest experiments explored the possibilities of making paper from indigenous materials like the seed tufts (wool) of the poplar trees that grew in nearby forests and the cotton grass that covered neighboring meadows.[17] But then as his paper trials started to gain notoriety, Schäffer began to explore the possibilities of making paper from more exotic materials, like the leaves of aloe vera plants and the bark of mulberry trees, which he procured from his extensive network of foreign correspondents and local informants.

Schäffer's experiments were as much about participating in a given intellectual tradition as they were about making paper. By the time he published his *Versuche und Muster*, the members of scientific societies all over Europe had been regularly reporting on their own paper trials for almost a century.[18] That such societies were highly gendered organizations that not only excluded women from their ranks but also admitted only men of a certain social standing is well known and does not need to be rehearsed here.[19] What does need to be underscored is that admission into these societies constituted a rite of passage for the men who could aspire to such a privilege, as it signaled their acceptance into learned circles. We see this in Schäffer's nomination to some of the most important scientific societies of his day, which recognized his extensive contributions to fields like botany, entomology, mycology, and zoology, and helped to elevate him from a humble country pastor to a scholar respected across Europe and into the Americas.[20]

His close ties to the Bavarian Academy of Sciences are particularly important within the present context, as it helped to finance Schäffer's experiments. A founding member of the academy, Schäffer had convinced his brethren of his enterprise by showing them the result of his first trial with poplar wool, which yielded paper that had a density similar to paper made from linen rags (fig. 4.1).[21] He sent further samples of paper made from algae and hops plants—even pine cones (see plate 7)—to other scientific, economic, and artistic societies in a bid to "lend more credibility" to his experiments.[22] He also included samples in the books with which I began, appended to texts that describe the circumstances of their production so accurately that it would be easy to repeat Schäffer's experiments.[23] It was in fact Schäffer's hope that his books would inspire emulation, especially on the part of papermakers: "As soon as I see or hear that one or another papermaker has begun to work after my example, I will abandon the rest of my trials and lay aside my pen, since I will have then achieved my goal."[24]

Papermakers, however, were less than enthusiastic about Schäffer's project. They were not convinced that there was a paper crisis and saw no need to alter their working methods. They were also skeptical of paper made from materials like wasps' nests, which in their eyes was little more than a curiosity with no real commercial application.[25] One might dismiss their reaction as an example of "craft conservatism" had recent scholarship not done so much to dispel the notion that early modern artisans were resistant to change. Indeed, many artisans were happy to test out new tools and techniques if they thought that it could help them to improve the quality of their wares.[26] For this reason, the continued preference of papermakers for linen rags, even after seeing and handling the samples sent to them by Schäffer, deserves more detailed consideration.[27]

MAKING PAPER

Most early modern papermakers worked in commercial paper mills, which, while varying slightly from one another in terms of their size, layout, and management, all operated according to the same basic principles.[28] First, used articles of linen were delivered to the mill by rag pickers and/or dealers. These were then sorted and washed by women, our first indication that paper mills were multigendered

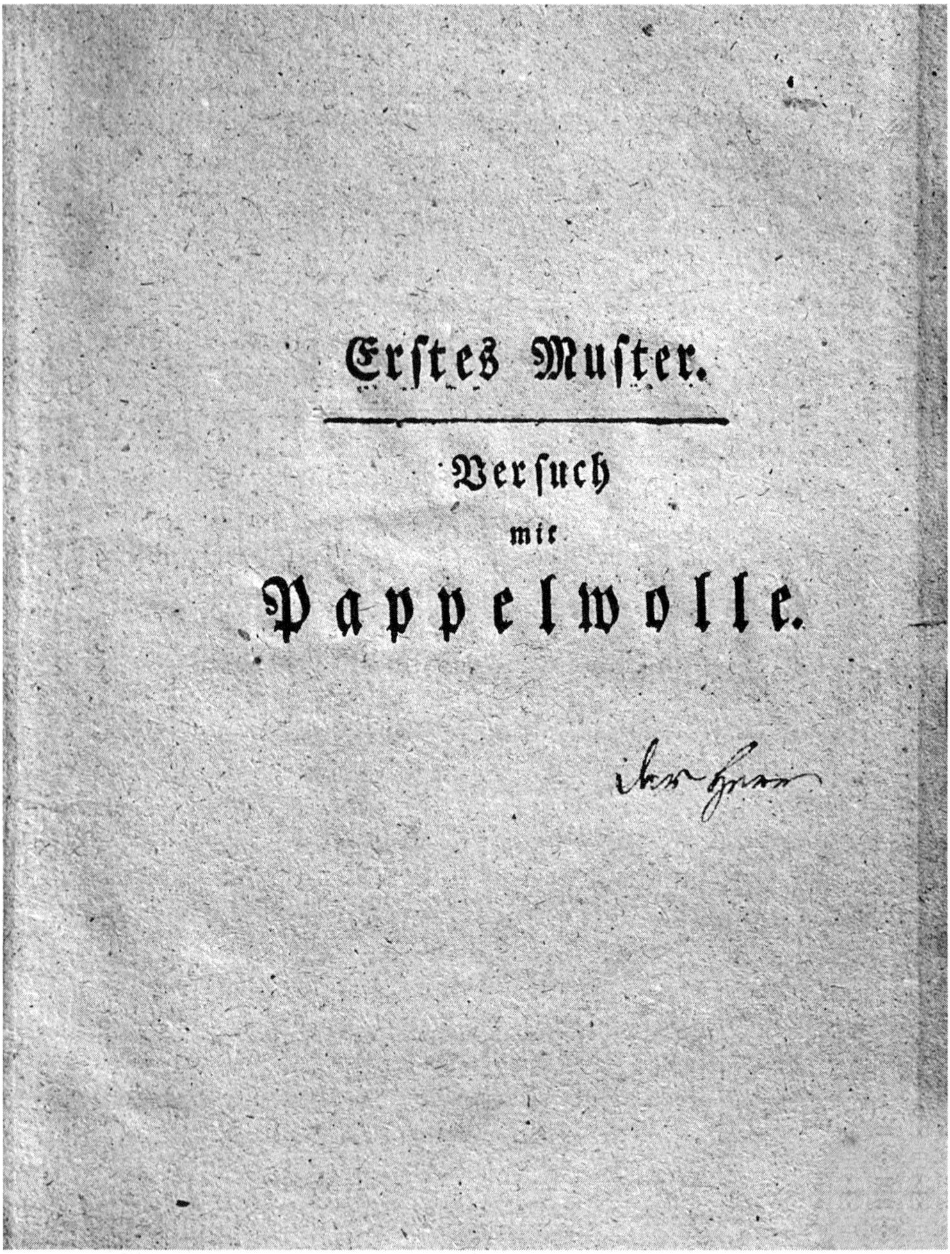

Figure 4.1. Each paper sample that resulted from Schäffer's paper trials was printed with a text, which indicated its place in the larger sequence of his experiments, as well as the material that had been used in its manufacture. It also assumed the function of tangible evidence that his experiments resulted in paper, which could immediately be put to use. Here the text in question reads, "First Sample. Trial with Poplar Wool" and comes from a sample that appears in *Versuche und Muster ohne alle Lumpen . . . Papier zu Machen* (1765). 037/4 Lw 602-2, S. 3, urn:nbn:de:bvb:12-bsb11217981–4, courtesy of the Bayerische Staatsbibliothek München.

spaces by definition.[29] Once sorted and washed, the rags were deposited in pits or storehouses and left to ferment for days, weeks, even months at a time. This process helped to break down their fibers, which were typically composed of spun flax plants. The rags were then cut up and transferred to a machine that was outfitted either with large wooden hammers (a stamper) or with rotating metal blades (a Hollander beater). Here they were mixed with water, which necessitated that the mill be close to a reliable water supply, and hammered or beaten into a pulp.

The pulp was then divided into vats, which were overseen by a vatman, whose primary task was to dip a large, rectangular mold into the vats in such a way that a sheet of pulp formed on its surface. With an even layer of pulp on the mold, the vatman handed it off to the coucher, who laid it down on a piece of felt and pulled, or couched, off the wet sheet of paper. The combination of strength and stamina required to carry out this task from anywhere between five and ten thousand times over a fifteen-hour workday makes it easy to understand why physical prowess became an important marker of masculinity among papermakers.[30]

When enough alternating sheets of felt and paper had been gathered, the entire stack (called a post) was sent through a press in order to rid it of any water. The individual sheets were then dried and sized, which was done by either brushing or dipping them into a mixture of water, alum, and animal bones that prevented ink from bleeding into their fibers. After this they were dried again and smoothed out with the aid of a metal press or polished with special stones. Finally, each sheet was scrutinized for impurities and if it met the grade was gathered into a ream (a bundle of five hundred sheets) and sold.

From this brief description, it should be clear that many different hands and skills could be found in early modern paper mills. What is perhaps less evident is that these skills were arranged into a distinct hierarchy that reflected their difficulty and the time required to master them. It also determined one's rate of pay and ability to command authority within the mill.[31] At the top was the master papermaker: a man who had himself apprenticed in a mill, which often belonged to his father or a close relative, as papermaking, like other artisanal practices, was a family affair.[32] This experience gave him insight into the various roles in a mill and prepared him for supervising other papermakers. Before he was ready to assume this responsibility, however, he was expected to acquire the rudiments of bookkeeping, so that he could keep the mill's finances in order, and a basic knowledge of chemistry, so that he could control the sizing.[33]

Still, the master papermaker's authority was not absolute: journeymen were prone to strike when dissatisfied with their working conditions.[34] Two of the most volatile figures in this regard were the vatman and the coucher, who besides exhibiting the kind of skills outlined above were also expected to possess considerable gestural knowledge.[35] Gestural knowledge, described by historians of science as "the complex of skills and forms of mastery developed in . . . real-time performances," is something that cannot be learned from books but is the result of lived experience.[36] It can, however, be measured and judged by others, as in the case of the vatman and the coucher, who could cost a mill owner considerable expense if their gestural knowledge proved to be insufficient.

A vatman, for example, could make "a garbled sheet of paper, when he does not spread his material out enough. . . . when his arm is too stiff, when he pushes the wrong way."[37] A coucher meanwhile needed to take care that he did not lift the sheet "so forcefully, that the water from the felt pushes into the sheets of paper already lying there, which causes the paper to be water damaged."[38] In order to be respected as artisans—and as men—papermakers needed to demonstrate their gestural knowledge on a daily if not hourly basis, which meant gaining command not only over their materials but also over their bodies.[39]

GOVERNING LABOR

Paper mills were spaces that invited the performance of multiple masculinities. The kind of masculinity that was performed by the master papermaker—defined by the ability to control both material and human resources—was different from the one performed by the vatman or the coucher, which rested on strength and stamina, to say nothing of a deft hand. Scholars have contended that early modern workshops were spaces of not only labor and commerce but also identity formation.[40] In the context of the paper mill, we see such identity formation in the "trials" that an apprentice had to pass before becoming a journeyman.[41] First, he had to prove that he came from a family that was not involved in any dubious deeds or trades, as their shame would transfer to him and his work.[42] Second, he had to learn how to read and write, typically at the mill owner's expense, with special emphasis on learning the catechism. Third, he was to spend four years working under the direct supervision of a journeyman, during which time he performed any number of menial tasks in and around the mill and received little recompense for his efforts. This made it difficult to pass the fifth and final trial, which involved hosting a feast at the end of his training for all of the journeymen that could go on for six days and cost as much as two hundred Reichstaler. In order to pay off the debt that he accrued in the process, he had to work for several more years without pay.[43]

Such excesses caused considerable consternation among the bureaucrats charged with overseeing paper mills. These men had come to play an increasingly important role in paper production over the course of the eighteenth century, as European monarchs took a more active interest in papermaking. One reason for this interest was that importing paper (as well as linen rags) was an expensive proposition. If local resources could be used to establish or improve upon existing manufactories, then the money being spent on such products could be invested elsewhere. Such an approach toward fiscal management was informed by *oeconomy*, a term that finds its roots in the Greek words for home (*oikos*) and law (*nomos*) and was used to describe the practice of maintaining order in one's household through the prudent management of resources.[44] We need only turn to Simon Werrett's discussion in this volume of how printed paper was recycled and reused as filler for dresses and a tool for hairdressers to realize the importance of oeconomical thinking in most eighteenth-century households.

Practicing oeconomy was not always easy, though, as the various members of a household—at the individual or the state level—could become unruly. In such cases, the head of the household, or *Hausvater*, had to step in and lay down the law.[45] Thus, we find local and regional governments issuing a host of rules and regulations in an effort to rein in papermakers. The most famous of these was the edict passed by Holy Roman Emperor Charles VI in 1731 in a bid to correct the "abuses" of the empire's guilds.[46] Although papermakers did not belong to a guild—which was listed as one of their abuses—they were ordered to do away with the feast at the end of an apprentice's training, as it brought financial ruin to the individual in question and caused work to be suspended for days on end.

Papermakers, however, fought this edict tooth and nail and the issue remained active for decades. Discussions in publications like the Cameralist organ *Journal von und für Deutschland* indicate that bureaucrats blamed the celebratory feast of papermakers for the lack of "good paper" in the region.[47] This lack, however, had just as much to do with the invention of the Hollander beater in the 1670s, which increased the production rate and quality of Dutch papermaking.[48] Government reforms and stricter regulations in France had also made locally produced writing paper coveted all over Europe.[49] German rulers wishing to correct the situation in their own lands appointed Cameralists to oversee paper production, adding them to the kinds of men jockeying for power in paper mills.

CIRCUMSCRIBING KNOWLEDGE

Educated at the University of Halle an der Saale, which was among the first to offer classes in the police and cameral sciences, Schäffer was well aware of the changes taking place in paper mills. He did not, however, believe that appointing an inspector was enough to improve the state of papermaking. Instead, he argued that papermakers had to stop using linen rags as a source of raw material, since "not only in the immediately surrounding area, but also in most of Germany, the general shortage of so-called linen . . . has for some time, and now more than ever before, been the cause of considerable grievances."[50] Read before the Bavarian

Academy of Sciences in 1764, this passage constitutes the first of many public statements by Schäffer on the paper crisis of the time. Schäffer was convinced that this crisis was an outcome of the "the last war," by which he meant the Seven Years' War (1756–1763), a conflict that is well known to have wreaked havoc in the region and even led to famine.[51]

What is less well known is that the same war destroyed countless manufactories, including paper mills, and necessitated that large quantities of linen rags be repurposed for casings for gunpowder and burial shrouds for the dead.[52] Thus, Schäffer's claims as to the difficulties of finding locally produced writing paper were not entirely unfounded. Many landgraviates and principalities responded by enacting strict laws in order to control the production and consumption of linen, and even levied hefty fines against linen smugglers, who collected, transported, and sold linen rags to customers abroad.[53]

Similar measures, however, had already existed in the seventeenth century and paper prices remained relatively stable during the period under consideration.[54] Papermakers like Georg Christoph Keferstein, a mill owner outside of Halle, also denied such difficulties in acquiring linen rags, arguing that "as long as there are human beings on German soil, they will be in need of clothing; and when one assumes that each person tears two shirts per year, and that these are then directly brought to a paper mill, I have no need to cause myself concern over it."[55] Published as part of a rather damning review of Schäffer's experiments, Keferstein's claims, while exaggerated, point to the resistance of papermakers to paper trials. They also suggest that Schäffer's experiments may have had a different motivation. By arguing that there was an urgent need for a material substitute for making paper, Schäffer was realigning the hierarchy of skills that had come to define the papermaking process. "Good" paper was no longer a question of knowing how long to allow used linen to ferment or the best way to hold a mold as you pulled it out of the vat. It was now a matter of knowing which trees, bushes, and plants had the right kind of fibrous structure.

Such knowledge had a long history among learned circles in Europe. Regensburg itself had been a major center for the production of botanical knowledge since the thirteenth century and in 1790 would become home to the first scientific society dedicated exclusively to the study of botany.[56] This is not to say that an interest in botany was limited to men of learning. Artisans likewise saw the plant kingdom as an invaluable source of food and medicine and sought to learn more about local plant life. Some even approached men of learning with questions about particular species.[57] These men, however, faced distinct challenges and limitations to the expansion of their knowledge, such as their poor or nonexistent Latin, which continued to be used in taxonomies and herbals well into the nineteenth century, or their lowly social status, which made it difficult for them to access the libraries and natural history cabinets that were crucial to the production of botanical knowledge, as it was here that one could consult classifications and compare specimens gathered over generations.

We find a similar chasm in Schäffer's description of herbal merchants (*Kräutermänner*) in a text published just two years before his first book on paper trials. Kräutermänner, Schäffer argued, were excellent guides for those who were unfamiliar with a given territory and could be given explicit instructions to procure particular specimens. Under no circumstances, however, should they be asked to identify or distinguish between species, as only men of learning could be "trusted" to make an informed judgment in such matters. Or to put it more bluntly, "The unlearned man must find, carry, and deliver into the hands of the learned man" who would then "observe, inspect, test, describe, concur, detract and bring to order."[58]

CONTROLLING RESULTS

How such a division of labor influenced Schäffer's paper trials can be seen in his description of his first experiment, dedicated to making paper from poplar wool. Like the men of learning who had inspired his experiments, Schäffer began by listing the botanical properties of his specimen, which he identified as the black poplar, or *Populus nigra,* that had recently been "discovered" by his friend and correspondent the Swedish naturalist Carl Linnaeus. He also described its preference for damp, swamplike growing conditions and tendency to grow high and wide with thick, long branches.[59] But as soon as it came time to explain how he had turned poplar wool into paper, Schäffer became much more measured in his language: "I allowed both kinds of tufts to be gathered, carried home, and discussed it with the local papermaker Meckenhäuser."[60]

In forgoing mentioning his other assistants by name, Schäffer behaved much like the English natural philosopher Robert Boyle, who seldom identified the men who helped him to carry out his experiments.[61] Indeed, the impression that Schäffer considered such men to be little more than "invisible assistants" is underscored by an engraving that serves as the frontispiece for his book and shows them in the guise of putti (fig. 4.2)—a device that not only obfuscates their labor but also obscures their identity. The same can be said of his description of Meckenhäuser. All we know about this man is that as soon as he was approached by Schäffer, he warned Schäffer that it would take at least twenty-five pounds of poplar wool to make paper at the commercial level.[62] But Schäffer insisted that they use the far more modest amount that his assistants had gathered, and when the resulting sample proved to lack "the right kind of consistency and stiffness and was additionally filled with brown knots, which were the remains of the ground seed tufts" he blamed not himself, but Meckenhäuser.[63]

Figure 4.2. Frontispiece, *Versuche und Muster ohne alle Lumpen . . . Papier zu Machen* (1765). 8 TECHN II, 2175 RARA, courtesy of the Niedersächsische Staats- und Universitätsbibliothek Göttingen.

Following his dispute with Meckenhäuser, Schäffer decided to conduct all future paper trials "from beginning to end, in my own house, and under my constant watch and supervision."[64] Here he had ready access to the books that had inspired his investigation, to the specimens his assistants had gathered, and to extra sets of hands, should they be required. Indeed, we know from his many publications that Schäffer's household was a frequent site of experimentation, whether the object of investigation was the nature and uses of electricity or how to prepare foodstuffs new to the European diet, like the potato.[65]

In each case, his ability to conduct his experiments was contingent upon having the right tools and instruments, as well as sufficient manpower to operate them. The first requirement was easily met in that Schäffer was a capable instrument maker, who was celebrated for improving wood-burning ovens and building the first washing machine.[66] But the second could be a challenge, depending on the skills involved. We see this in the miniature stamper that Schäffer installed in his house (fig. 4.3).[67] Much like the larger commercial models after which it was patterned, it sports a series of hammers, seen here to the left, which break down any raw materials placed in the trough, situated to the right. These hammers are activated by a serrated drum, which is operated by the hand crank visible in the foreground that also helps to feed water from the tank in the background into the trough. An accompanying legend clarifies the individual parts for the reader.[68] What it does not explain, however, is who was to operate it upon arrival.

Such an omission is curious when we consider that multiple hands were clearly needed to add the specimen(s) to the trough, while activating its hammers and pump. Anticipating this need, Schäffer hired a young journeyman "to show my servants all of the techniques and gestures of papermaking."[69] Here we return to the motif of gestural knowledge, although in this instance such knowledge is not cast as remarkable but rather as something that could easily be taught to others.[70] Once again Schäffer is silent about the journeyman. All we know is that Schäffer conceived of him as a "one-time expense," whose sole purpose was to train the men (and women) in his employ.[71]

THE OECONOMY OF HONOR

The journeyman whom Schäffer hired did not take kindly to being treated in this fashion and spread rumors that the resulting paper samples were not free of linen rags.[72] This was a serious accusation to make in an age in which honorable conduct held the key to one's admission to scientific societies, as well as to the Republic of Letters.[73] Any indication to the contrary—like falsifying one's results—could ruin a man's reputation and cost him his fortune.[74] To make matters worse, the accusations against Schäffer found their way into the various *Höfer Intelligenz-Blätter* as well as other popular journals.[75]

Schäffer issued a public statement on the matter in the second volume of his books on paper trials, also published in 1765.[76] Here he admitted that at times it had been necessary to add a small amount of linen rags to the pulp, in order to provide paper samples with the right kind of stiffness and consistency. But he also maintained that the charges against him were grossly exaggerated and were

Figure 4.3. Stamper, ca. 1763. Previously owned by Jacob Christian Schäffer and used in his at-home paper trials. Photo: Museen der Stadt Regensburg, Germany.

the handiwork of jealous rivals, who wanted to detract from the attention that his experiments were receiving.[77]

In the ensuing debate, mill owner Kerferstein remained one of Schäffer's most strident opponents. Keferstein saw Schäffer as meddling in a matter that lay outside of the boundaries of his knowledge, a point that he underscored by mockingly referring to Schäffer as a theologian.[78] That such boundaries were of concern to artisans as much as they were to men of learning shows that papermakers were also part of a community; one that carefully controlled who could and could not join its ranks. The reasons they gave for drawing such distinctions, however, differed substantially from those of men of learning. Their concern lay not in demonstrating that they were operating within the parameters of a given intellectual tradition or adhering to certain standards of truth telling. Rather, they wanted to guarantee the quality of their wares.

As one of the first papermakers in the German-speaking territories to employ a Hollander beater, Keferstein was clearly interested in producing the best possible product that could compete with high-quality foreign wares.[79] He was not, however, prepared to risk making "bad" paper by having his journeymen work with materials whose properties were unknown to them. Having nothing to gain and everything to lose, Keferstein chose to ignore Schäffer's recommendations and guard his own carefully cultivated reputation. The full import of his decision comes into focus when we realize that his review of Schäffer's paper trials appears in an instructional manual addressed to his sons. Here he advises the next generation to "make the last sheet in a post with the same diligence as the first" and to exercise care in every step of the papermaking process, as even the smallest detail mattered.[80] Merchants, for example, noticed things like when a ream of paper was bound together too hastily, leaving the string that held it together to lie crookedly on its surface.[81]

Keferstein's words are indicative of an individual who possessed gestural knowledge as well as commercial savvy, both key criteria used by papermakers to define their masculinity. Such an image is far removed from the one propagated by men of learning in the guise of Cameralists, who depicted papermakers as merrymakers in search of the quickest way to turn a profit.[82] While, undoubtedly, some papermakers were more intent on offering a high-quality product than others, Keferstein's remarks show that artisans cared about their reputation and knew that their welfare depended not only on the prudent management of resources but also on the daily performance of pride in one's workmanship and honest conduct in matters of trade. Thus, they had an understanding of oeconomy that took into consideration both material resources, like linen rags, and intangible values, like honor.

The paper trials of eighteenth-century men of learning like Schäffer were not first and foremost an attempt to resolve the paper "crisis" that emerged at the end of the Seven Years' War. Rather, they were geared toward gaining control over the papermaking process in an age in which manufactories were becoming of increased interest to monarchs. While Schäffer never occupied the kind of bureaucratic positions that grew out of this interest, he was in communication with local papermakers and approached scientific and economic as well as artistic societies with proposals for making paper out of a range of materials.

His decision to pursue the possibilities of making paper from botanical specimens in particular was not unusual when we consider that European papermakers had long been using flax plants—as linen rags—to make their wares. Moreover, other men of learning, such as Réaumur, had already published on the suitability of various plants and trees for making paper. Schäffer's paper trials, however, are unique in being so well documented, which allows us to see how the question of which material(s) was most suited to making paper was really a question of who had the right kind of knowledge for making such a determination: scholars or artisans?

The tensions between scholars and artisans in the early modern period long familiar to historians of science have here been presented as equally informed by competing—or multiple—conceptions of masculinity. We see this in how Schäffer refused to listen to Meckenhäuser's advice that the poplar wool his assistants had gathered would result in "bad" paper, as this would mean conceding one of the cornerstones of his masculinity, his book learning. We also see it in the refusal of papermakers to follow the rules and regulations that Cameralists were trying to impose on them in the interest of improving production. As Keferstein noted in the manual he dedicated to his sons, papermakers formed their own "little republic" and had to protect its borders from intruders.[83] That these borders were drawn not only on but also through paper demonstrates its unique ability to serve as a site of inscription and experimentation.

PART II

Transcending Boundaries

Tools and Technologies

FIVE

Bookkeeping for Caring

Notebooks, Parchment Slips, and Enlightened Medical Arithmetic in Madrid's Foundling House

Elena Serrano

On March 29, 1801, Juliana de Baza bundled up her five-month-old niece, María Salomé, and left her home in the center of Madrid. She walked down the bustling La Montera Street and headed for the nearby foundling house, where she left her brother's baby.[1] Just as soon as the nun at the door received the baby, the paper machinery of the house swung into action.[2] In the chilly office that never closed, by the light of an oil lamp, the clerk perused the letter that had accompanied María Salomé, searching for confirmation and evidence that the child had been baptized.[3] Checking for proof of baptism was certainly an urgent issue, as infant mortality rates in foundling houses oscillated between 70 and 90 percent and one of their main missions was to ensure that all children were baptized before dying, thereby safeguarding their entrance to heaven.[4]

María Salomé's letter was signed by a clergyman, who certified that she was the legitimate daughter of the Baza family.[5] Satisfied, the clerk reached for the bulky Book of Entries number 169. He unbuttoned the wooden closure sewn on its yellow parchment, opened the flap that protected its folio pages, and flicked through until he found the first blank page: page 326. Although it was only spring, he had already registered 325 children and still another 825 more would follow before the year ended. In total, abandoned children amounted to a quarter of all the annual births in Madrid.[6] As always, the clerk made a margin of a quarter-page width, leaving the page number perfectly visible, and wrote a beautiful capital *M*. He then underlined the name and began a new line. Indeed, this came as habit. For centuries, all entries had begun in the same way: "On the [day] of [month], it was received in this Holy House, a [boy or a girl] that . . ." The clerk then wrote down the scarce information known about the child: who had brought her and where this person lived, the name of mother and father, date, time, and place of birth, and finally, the parish in which he or she was baptized. The Book of Entries thus functioned as all church parish books did across villages in eighteenth-century Spain. It solemnly and publicly registered the entry of a new baby into the Christian community. Once registered, María Salomé was officially under the care of the foundling house. From then on, the information entered in the book would be linked to the actual body of the child: she would be labeled with a little piece of parchment tied tightly to her waist.

Such was the protocol for admitting a new arrival to the paper panopticon of the foundling house. It entailed the use of different kinds of paper tools such as hastily scrawled notes of abandonment, formal letters by clergymen, heavy notebooks kept in the clerk's office, and light ones to be carried around to the wards. But it also involved both human and nonpaper instruments: durable parchments, numbered cradles, and the foundling house employees themselves. The procedure described above illustrates the bureaucratic practices and the sociotechnical system of surveillance of children that were in place during the administration of the Junta de Damas de Honor y Mérito (Board of Ladies of Honor and Merit), which, in 1799, officially took over the foundling house, replacing the former administration, which had been entirely male.[7]

When the Junta de Damas took charge of the house, it set an optimistic five-point agenda for lowering mortality rates. It aimed to improve the conditions under which babies arrived, improve hygienic practices while they were in care, fight "hereditary or epidemic illness," avoid bad wet nurses, and, finally, reduce "mistakes in weaning."[8] To implement these ambitious changes, the Junta de Damas doubled its staff, carefully planned its everyday workflow, and moved the house to a more salubrious location.[9] However, the deepest changes did not occur in the wards of the foundling house but rather in its office.

Focusing on the paper tools for tracking children, this essay makes two claims. First, it argues that the books for registering the children were used as material proof that a learned female aristocracy could play an important role in Spanish politics. The tidy, careful manner in which books were written attested to the fact that an all-female association could manage an institution considered by contemporary political philosophers and economic writers to be central for maintaining the social order. Foundling houses played a decisive role in the economies of the poor, as well as in protecting the reputation of families.[10] At the same time, the houses were targets of policies to increase the working population, halt social revolt, and achieve political control.[11] However, the alarming mortality rates at these establishments posed a major threat to the community. The Junta needed on the one hand to decrease infant mortality rates, and on the other to publicly demonstrate that its calculus was accurate. As Andrea Rusnock put it, the uses of medical arithmetic underlined "a widespread belief in the agency of humans to bring about change."[12] The carefully maintained books, with their clean, ordered pages, were key in assuring the credibility of its figures.

During the first two years of its administration the Junta coped with the old system of registering children that had been in vogue since the seventeenth century. But in 1802 they made a crucial modification to that system. Instead of using one book for registering the entrance of children (as we have seen in the case of María Salomé, the Book of Entries) and a second book for recording the placement of children outside the house with external wet nurses (the Book of Exits), the Junta decided to create one big record book that combined all the information that the institution had about each child. My second claim, then, is that this new method of bookkeeping reflected contemporary gynotechnics, in the sense understood by Francesca Bray: a "set of technological attributes and practices that combine to shape and define female roles and hierarchies."[13] The Junta subverted the former accounting regime that had been modeled on commercial accounting, and set up a new regimen based on caring, on the consideration of each child as an individual for whom the Junta was responsible.

The design of the new books was undoubtedly driven by a question of accuracy in the calculus of mortality rates. However, the revised layout also reflected the urgency of knowing when and where the system of tracking children failed (what Christine von Oertzen calls the "frictions" of the system in her chapter in this volume). But, above all, the new books conveyed not only a dramatically new way of managing the Madrid foundling house but a novel way of conceptualizing the children themselves. These innovations were based on skills and practices considered to be "feminine," and they served to redefine upper-class women's role in the welfare of the community. The paper technologies that the Junta introduced were based on traditional domestic gynotechnics put into the service of the nation. Humble paperwork is replete with political and moral values, as scholars have shown.[14] But in particular, the way foundling houses registered the abandoned children reflected societal moral values.[15] Under the rule of the Junta, the sociotechnological system of surveillance reflected the power of gynotechnics.

As a consequence of these material changes, the new record books afforded novel ways of producing knowledge. They allowed for the tracing of a number of new categories, which in turn enabled the women to classify the overall condition of the children and their illnesses. The way the information was laid out on the pages—the blanks left, the lines traced, the columns that divided the pages—displayed the information in a different way than had been demanded by the form of previous books. Hence the new books afforded more sophisticated ways of producing statistics and interpreting mortality rates.[16]

TAKING OVER THE MADRID FOUNDLING HOUSE

The Junta de Damas was set up in 1787 as the female branch of the Madrid Economic Society. By 1799 it had sixty-four members, mostly from the high aristocracy, who met regularly each Friday afternoon in the Madrid Council Hall.[17] It would appear that the Junta was a peculiarly Spanish phenomenon.[18] Among other projects, the Junta ran the Monte Pio de Hilazas, an organization that loaned raw textiles to poor women; they were also deeply involved in the education of girls and set up schools to teach spinning, embroidery, and the making of textile flowers.[19] They also experimented with dyes, tested textiles, and supported the production of "useful" translations, such as agricultural treatises and pedagogical books.[20]

The administrative methods of the Madrid Foundling House were shaped by the special place that the Junta occupied in the scientific arena and the circumstances surrounding its establishment. The process by which the Junta became the female branch of the Madrid Economic Society was a long one that inspired passionate public debates. As Spain lacked a central academy of sciences similar to those in France or Britain, the Madrid Economic Society (founded in 1775) played some of these roles, such as advising the Crown on technical issues, circulating novelties, translating foreign texts, fostering "useful" inventions, and organizing public prizes.[21] Discussions about the admittance of women began soon after its creation and lasted a decade. As feminist scholars have thoroughly argued, what was at stake was the role of upper-class women in the modernization of the country.[22] Some of the more than two hundred fellows of the Economic Society defended the admission of women, although they widely disagreed on how women should be involved in the society's activities.[23] Meanwhile, however, women were in favor of joining the male society on equal footing and made a determined defense of their position. For instance, Josefa Amar y Borbón (1749–1733) argued that women would contribute to the advancement of knowledge by claiming that this was well within the capacity of ordinary women: "Compare the foreign works with ours to see what can be improved . . . ; make new inventions in the arts, improve the known ones, . . . although these are things that demand careful consideration and awareness of the new innovations, they are not such abstract matters that a woman of normal talent cannot understand."[24] However, it was finally decided that women would join a separate branch of the society and be allowed only to engage in "feminine issues" (*de acuerdo con su sexo*).[25]

Forced to renounce their equality with men, the Junta developed a strongly gendered identity. Women stressed their maternal, domestic, and practical skills in order to secure a position in the male scientific and political arena. According to the Junta, these "feminine" qualities made their organization more suitable than the men's societies for handling certain issues such as the education of girls. At the same time, the Junta strongly stressed its nature as an enlightened society, which set it apart from other women's associations like religious or charitable ones.[26]

This "feminine" strategy was put to use by the Junta in their struggle to take the Madrid Foundling House from the hands of the all-male committee. The reports that the Countess of Montijo (1754–1808), then secretary of the Junta, prepared in 1789 to be sent to the Crown asking that the foundling house be put under the Junta's authority highlighted the purported maternal qualities of women, such as patience and natural affection for children.[27] According to the countess, these feminine characteristics made the Junta more suitable than an all-male committee: "Men governed the republics with their wisdom and intelligence. The way women can contribute to the public happiness is by taking care of these innocent creatures."[28] But the countess also used arguments based on medical arithmetic, the very arguments used by doctors, politicians, and philosophers in their urgings for action.[29] She pointed out that according to the available tables of general mortality (*tablas generales de mortalidad*), which tabulated the mortality rates of the "general" population, half of all children born would reach the age of ten. Yet in the Spanish foundling houses, according to the countess, a mere 10 percent of children reached ten years.[30] She then quantified the number of lives lost for the Spanish state and concluded: "Such exorbitant and distressing loss [of lives] needs the most serious attention of a patriotic society, to find out and bring to an end the causes that produce it."[31] Thus, the countess cunningly placed the Junta in a sphere separate from that of men (men governed the republic), while at the same time stressing that only an enlightened society was capable of finding rational solutions to the enormous problem of infant mortality.

From the very beginning infant mortality rates played a key role in the way that the Junta managed the house, an approach that was entirely new.[32] In 1798, while the house was still in the hands of the all-male committee, the Crown authorized the Junta to visit the foundling house and review its archives.[33] Besides taking note of the awful hygienic conditions, the Junta stressed that 313 children appeared to be missing from the balance sheets.[34] Because of the "contradictions" it found in the children's record books, the Junta reported that it could not accurately assess "the issue that most interests us, the mortality rates of these innocents." However, looking at the general figures in the record books, they estimated the mortality rates to be on the order of 96 percent.[35] Finally, in October 1799 the Crown granted the Junta de Damas governance of the foundling house.[36]

THE JUNTA DE DAMAS'S REFORMATIVE PROJECT

Two months after taking over the administration of the foundling house, the Junta staged a performance in honor of the Spanish queen. In the discourse read by one Josefa Díez de la Cortina, the old foundling house was called an "insatiable sepulchre of children" and was contrasted with the foundling house as it was now run by the Junta: "Economy has followed disorder, intelligence [has followed] ignorance, and an active and protective care [has followed] mercenary negligence. . . . Humanity congratulates itself for such an interesting project: the state will have new individuals, and the arts will have new arms."[37] The Junta construed its successful management of the foundling house as a result of a feminine difference. Díez de la Cortina opposed the former board's financial motivation with the women's caring attitude, and she summarized using the concept of "economy."

In their role as wives and mothers, women were supposed to keep an "oeconomic" household, understanding the term with the rich moral and political connotations that it had in the eighteenth century.[38] As Amar y Borbón put it, the obligations of wives were "the prudent arrangement of rents with necessities" and "the continuous vigilance over personal property and its interests, and the good governance of the family."[39] According to a well-known treatise on the education of girls, "A prudent woman . . . is the soul of even the biggest of houses: she puts order in the economy, mends the souls, and fortifies the health of the family."[40] Far from being a private issue, the "prudent" governance of households was thought to reflect on the overall improvement of the country, as Simon Werrett shows in this volume.[41] As a patriotic society, the Junta was seriously committed to promoting the ideals of oeconomy, and in particular accounting skills, to the female population.[42] It supported the translation and printing of pedagogical texts such as Madame de Genlis's famous book on education, *Adèle et Théodore*, which shows how to effectively teach accounting to young girls.[43] In 1795 the then director of the Junta de Damas, the Countess of Trullas, suggested publishing a collection of writings on the subject of women's education; one of the essays, submitted by the Marquise of Ariza, presented a practical, original way to keep household accounts.[44] Additionally, in the Junta de Damas spinning schools, where the pupils were exclusively poor girls, annual prizes were given to the best pupils in arithmetic.[45]

It does not come as a surprise, then, that the Junta de Damas organized the foundling house according to the norms of a well-managed oeconomic household, emphasizing the skills that were the usual domain of mistresses and housewives. In particular, these skills were evident in the keeping of accounts, the management of servants, and in taking care of the emotional, moral, and physical welfare of children and employees of the house. One of the major duties of a good housekeeper was to precisely define the tasks of employees and servants. The Junta's written contracts with doctors, surgeons, and nuns are exemplary in their meticulous, careful, and comprehensive description of tasks and duties. Among these they specified how beds were to be made, how wards should be daily ventilated and cleaned, how babies should be dressed and fed, and how medical visits were to be carried out. They were especially careful in stressing their authority over all the personnel in the house, including religious and medical authorities (clergymen, nuns, surgeons, and doctors).[46]

During the first months of the Junta's tenure, a member made daily visits to the foundling house and deemed it necessary to double the personnel employed there. The Junta also engaged a second clerk for the office, a second doctor, and twelve nuns from the French order of the Sisters of Charity. They improved the stock of basic household items such as dishes, wool mattresses (considered easier to clean), and new bedclothes.[47] The Junta reallocated spaces according to function and efficiency: they established a room for dirty clothes, a room for sick children and rooms for healthy ones, a room for children still not baptized, and a room for the bodies of dead children. All the cooking was centralized in the kitchen and wet nurses were forbidden from cooking their own meals in the house.[48] Since laundry was thought to be bad for children's health, the Junta outsourced it. A moral code of behavior was imposed upon all personnel.[49] For example, wet nurses were to dress the children according to a set schedule, handle them hygienically but with "tenderness," and all under the oversight of managing nuns: "The ward sister will have as her major concern that the wet nurses dress the babies in the stipulated hours, with the correct order, tenderness, and silence; correcting them if something was done carelessly or in a hurry, or in case a child was too tightly dressed or badly wrapped, or not properly clean."[50]

Accounting books were essential paper tools for successfully managing the economy of households. As Amar y Borbón put it, without them "one can hardly keep an exact reckoning of incomes and outcomes to organize expenses."[51] Moreover, keeping the accounting books in order was a symbol of good housewifery. In Fénelon's words, "it is well-known that exactness in keeping accounts often produces good order in households."[52] Every Saturday, the two ladies who directly ran the foundling house sat with the highest-ranking nun to revise the accounting books.[53] Clothes, medicines, kitchen expenses, wages paid to employees, and so on—all was carefully written down in accounting books and regularly reviewed.

Still, the voluminous children's record books stood out among all the accounting notebooks kept by the foundling house. In the prudent management of incomes and expenditures, the most precious goods to preserve were children. In the next section we will explore in detail how the system of tracking babies functioned and what changes the Junta introduced to bookkeeping by closely following the baby María Salomé de Baza (whom we met in the introduction) on her journey through the foundling house.

SUBVERTING THE REGIME

Fears of losing babies in the chaotic traffic of the foundling house were warranted. Often, three babies per day were abandoned, some days seven. In the month that the baby María Salomé arrived (March 1801), 115 new babies were left in the care of the house.[54] To avoid misplacing babies, foundling houses used the same method that naturalists or librarians used to keep track of their specimens and books; that is, labels.

Once María Salomé was registered in the Book of Entries, the clerk prepared her label, the *pergamino,* which was a small rectangular piece of parchment of around four centimeters by ten centimeters similar to those shown in figure 5.1.

The clerk wrote María Salomé's name in the center of the parchment, the book number (169) in the upper left corner, and the page number (326) in the lower right one. As long as she was under the care of the institution, these two numbers would always identify her. Once the parchment was tied to the baby's waist, María Salomé was ready to circulate through the foundling house system.[55] First she was sent to the in-house surgeon, and after he had checked her over and determined that she was not ill, she continued to the wards for healthy children. Here, the ward sister noted María Salomé's cradle number in her notebook and assigned her to the in-house wet nurse Teresa Facho.[56] Luckily, as infant mortality rates were six times higher in the house than outside of it, after only three days in the crowded ward María Salomé went to an external wet nurse.

Upon María Salomé's departure to the lodging of her new wet nurse, the clerk opened a fresh page in a different book, the Book of Exits. Externally similar to the Book of Entries, it registered the exits of children who were to be fed outside and the wages paid to wet nurses. In this way, the Book of Exits functioned as an accounting book, but it also kept information on the child's location. The clerk also updated María Salomé's parchment label (or tag) and wrote in its left corner the page number of the Book of Exits.[57] Along with the child, the parchment would be presented each time the wet nurse came to be paid, in order to avoid cheating, swapping, or loss of children. After just ten days with the wet nurse

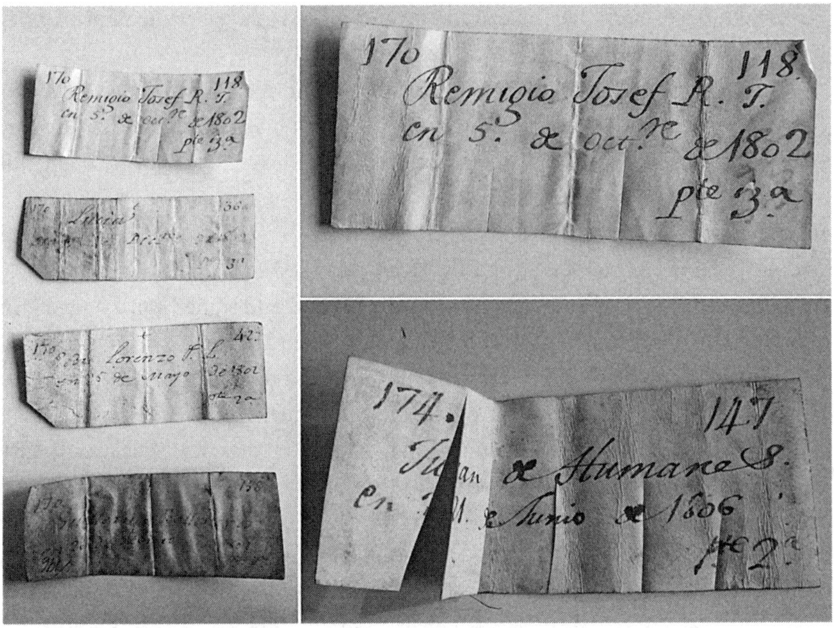

Figure 5.1. Parchments of children received in the Madrid Foundling House in 1802 and 1806. Note the number of the books (170 and 174) and the pages (118 and 147) in which the child was inscribed, and the cuts for the prevention of counterfeits. Caja 7721/7, Archivo Regional de la Comunidad de Madrid, Fondo Diputación, Inclusa.

outside the house, María Salomé was returned. Eventually another external wet nurse was found for her, but she again was shortly returned to the house. Finally, in mid-April, a home was found in the countryside where she stayed for a year. All of these movements were carefully annotated in the Book of Exits.[58]

From a managerial point of view, foundling houses were handled like a business. Children circulated from wet nurse to wet nurse like goods traded from hand to hand; a foundling house might be seen as a company that bought and sold children. Or perhaps it might be viewed as a company that supplied wet nurses to its always in-debt clients: children. Whatever sort of business it might have been, children's register books functioned in foundling houses much as accounting books did for commercial business.[59] Consider, for instance, the method of keeping ledger books for eighteenth-century Spanish companies. Manuals recommended opening a single account for each client, in which all related transactions should be neatly registered. One should write down "all the particulars of each business, all information that would help to handle related matters in the future . . . in a way that in these paragraphs one should find all that is needed for handling the credit or debt."[60] In a similar manner, the clerk noted all the particularities of the "acquisition" of the child: the date on which she was received, who brought her, where she was found or born, what the names of her parents were, and the expenses she incurred.[61]

Hence, children's record books played the economic, social, and moral roles that accounting books had in commercial business. It was assumed that the way one kept and took care of the books reflected not only one's own moral being but also the respectability of the company. Accounting books were purposely written for others to look at with inquisitive eyes. In this sense, "notebooks collected people," as Anke te Heesen has put it.[62] In order to avoid deception, formal ceremonies in which the accounting books were revised and signed were staged.[63] In the Madrid Foundling House, each year a new notebook for registering children was opened and identified by a unique number and the pages of the book were periodically reviewed and signed by the rector of the institution.

Given their religious, economic, moral, and social roles, it is thus meaningful that one of the Junta's first concerns was to look after the books and eventually reimagine the way they functioned. Changing the structure and content of the children's record books conveyed a symbolically powerful message about who ruled and was in charge of the foundling house. And beyond this narrower goal, these changes represented the new role that the Junta envisioned for Spanish upper-class women.

From 1799 until 1802 the Junta tried to work with the system of two books (one for entries and one for exits). The pages of the books for these years appear clean and well structured. This is particularly evident in the Book of Exits number 169 (the one from 1801). Furthermore, the Junta encouraged keeping all the information relating to each child, such as notes of abandonment, certificates of baptism, the parchments of deceased children, and so on, inside the books, tucked into the volume next to the corresponding pages. With their latches and closures, children's record books under the administration of the Junta became more like ordered boxes for keeping precious possessions.[64]

The Junta significantly increased the paperwork carried out in the house. Forms, tables, certificates, and reports were written and filed for ascertaining that children were well tended. Parish priests were enlisted to keep an eye on those already serving as wet nurses in their parishes, while women who wanted to be employed by the foundling house needed certificates of good behavior.[65] Measures for stopping the illegal traffic of parchments were issued.[66] We do not know if it was thanks to the Junta's efficient and effective administration, but in 1802, one year after María Salomé was abandoned, the dire economic times of the family were over and her father, Crisanto de Baza, came to retrieve his daughter, who was successfully located at the home of her wet nurse and brought safely to him.[67]

The calculus of reliable indices of mortality was a constant concern of the Junta de Damas. Monthly reports of the movements of children and tables of mortality rates (in which a substantial decrease of mortality was evident) were sent to the Crown. Medical treatises and journals also reported the decreased mortality rates under the Junta's governance.[68] The Junta also published and commented on these mortality rates in its annual report. From 1799 to 1802 mortality rates were prepared by consulting the two books, entries and exits, but according to the Junta this was a source of mistakes.[69] Finally, in 1802, the Junta decided to establish one unique book, the Book of Entries and Exits of Children (Libro de Entrada y Salida de Criaturas). Book number 169, in which María Salomé was inscribed, was the last edition of the old two-book system. This innovation was publicly announced by the president of the Junta, the Duchess of Osuna, in her annual report: "The two Books of Incomings and Outgoings have been reduced to one, with a noticeable increase in the order and the clarity [of the page] and with the advantage of putting together what is said of every individual."[70]

KNOWING EACH INDIVIDUAL

The new books optimistically reserved three pages for each child, which, if completed, would encompass a sort of schematic biography. As this would have resulted in a book too thick for handling as a single volume, it was divided into three. The Junta provided a page model for the official clerks that specified how the space on each page should be organized. Each piece of information had its place, which made it easier to look up the information for any given child.[71] The tidy, neat, new books were the material representation of the new order that the Junta imposed and they represented its new authority and governing capacity. As it can be seen in figure 5.2, the record is divided in two parts by a line—in the upper part is information about the child, and below is information about their movements with the wet nurses. See also how the page is spatially divided for the different functions that it would perform. The creases for dividing it in four equal quarters (two central for the information, two lateral for signatures and accounting) are still visible.

With the new books, the Junta embarked on more sophisticated statistics. It annotated and quantified the most frequent causes of death, and how mortality rates were distributed by age and sex. In 1811 the general procedure of writing down the cause of death allowed for the tabulation of the most frequently occurring life-threatening illnesses, ordered by age and sex. From 1802 onward we have records of a new notebook, the Book of the Deaths of Children (*Libro de Muertos*), which had the long and narrow format particularly suited for writing down lists.[72] The clerk made monthly summaries of deaths by age and sex, and noted if they occurred in or out of the foundling house (fig. 5.3).

The Junta de Damas also introduced new categories for data collection of statistical interest. In particular, the new category of *inconservable* (literally, "impos-

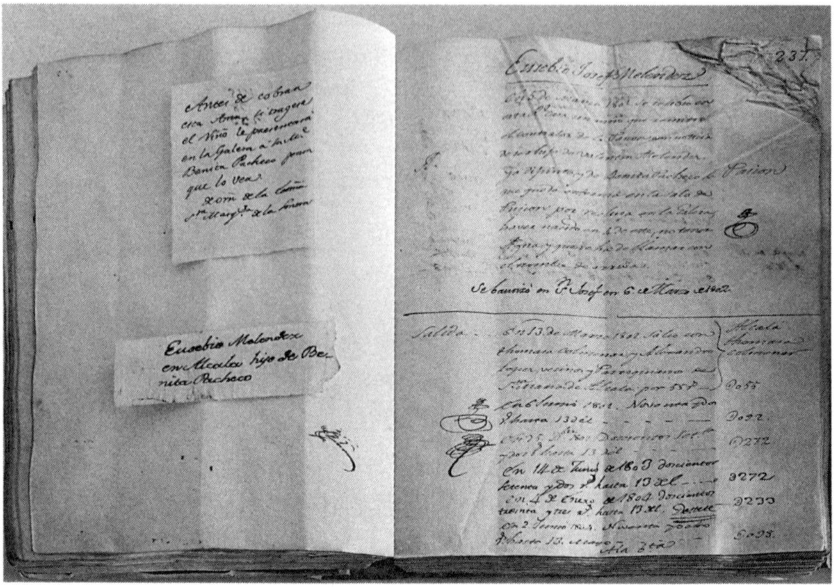

Figure 5.2. A page in the new books of Entries and Exits of Children (Libro de Entradas y Salidas de Criaturas). Note the loose pieces of paper that were kept between the pages. Caja 9227/1, Archivo Regional de la Comunidad de Madrid, Fondo Diputación, Inclusa.

sible to keep alive") required scrutinizing the health condition in which children arrived, not least to absolve the foundling house of responsibility should they die. Further rubrics distinguished between illegitimate and legitimate children, recorded those who came from poor families as well as those who died in the wards of the foundling house or with an external wet nurse, and if the children had been born in the hospital for poor women or at home. The new books thus allowed for a more in-depth assessment of the children's condition, allowing the Junta to more completely take account of their health when they entered and the circumstances of their deaths.

In other words, an attentive look at the children allowed the Junta de Damas to face the problem inherent to any process of quantification: What should be counted? Mortality rates were to be used to evaluate the effectiveness of the Junta's administration, to give a numerical measure of its success. Should one then count the children deemed to be among the "inconservables" along with the healthy ones? What if—as it happened in 1805 and 1806—famine and war brought in a crowd of malnourished children, all of whom perished? Would it be the fault of the Junta?[73] As Ted Porter put it, counting "is senseless unless the objects counted can be mobilized and defended as homogeneous."[74] But as Porter also notes, once numbers were used to assess the effectiveness of the institu-

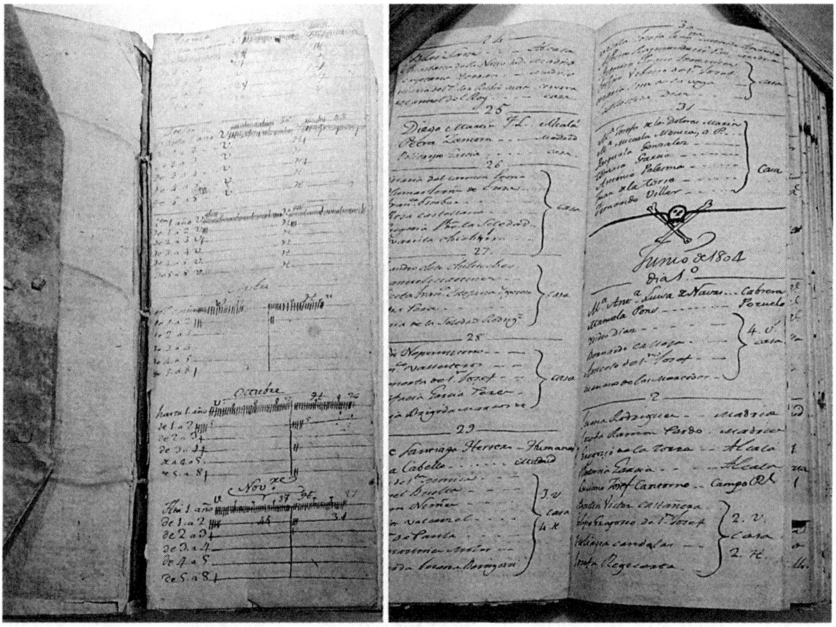

Figure 5.3. The long, narrow Book of the Deaths of Children (Libro de Muertos). Note how children were classified according to sex and age. A mark was put down in order to count the deceased in each group. Caja 10264/2, Archivo Regional de la Comunidad de Madrid, Fondo Diputación, Inclusa.

tions, administrators began to do some "creative" accounting, and new ways of interpreting the numbers appeared.[75]

Paradoxically, then, the new books designed for calculating more accurate mortality rates undermined the rationale on which the logic of mortality rates stood. In the new books, records became compressed stories—as the Duchess of Osuna put it, one had in the same place "all there was to be said about each individual." It is tempting to say that the conception of the child changed, no longer being so evidently a commodity that moved from one point to another, as was the case when their data were registered in two books. No longer were their lives split between two books, but instead all the essential details were kept together in one place, limited and fixed space. And so it was not evident how one could erase the differences between individual children for the purpose of making statistics and general statements.

CONCLUSIONS

Zeroing in on their material characteristics and on how they were written, we see that the children's registry books played different roles—religious, technical, social, and moral. On the one hand, they functioned as parish books did in Catholic churches, dutifully registering the birth of a new Christian. On the other hand, they functioned as lists, indexes, or catalogs of hundreds of children, and therefore as searching tools linking names with actual children's bodies through numbers and labels. They also functioned as the accounting books of the foundling house, registering the expenses that each child generated and localizing where the children were to be found at any given time.

The newly ordered and clarified books were the material representation of the organized, oeconomic skills of learned women, who were used to household accounting. The children's registry books became like ordered boxes where all the documents relevant to the children were kept. But above all, the books provided a means to accurately calculate an "objective" measure of success. Quantification, according to Porter, provides arguments for scientific objectivity.[76] Numbers severed knowledge from the local and the personal, rendering it independent of the producer and the specific circumstances within which it was produced.[77] Numbers can be used as measures of achievement, as rhetorical arguments for demonstrating success or failure. In the case of the Madrid Foundling House, numbers backed women's battle for power.

Yet in changing the material way of inscribing the information—whether by ordering the distributions of spaces on the page or by designing an entirely new book, a female association subverted the prior regimen of accounting. As an association of women, the Junta was not a legitimate actor either in the medical arena or in the administration of public institutions. Therefore, in order to accomplish its goals, it needed to mobilize powerful rhetorical arguments. These were gender and number, and the new children's registry books provided the Junta with both.

In the new books of entries and exits, each child was given her own space of three pages in the book. At least on paper, the children had a sort of home. The new books represented the hope that children could live to fill their three pages. Although few filled their allotted three pages, they were entrusted with them. The board subverted the former accounting regime and set up a new regimen based on caring. It is in this sense that their gynotechnics were put into service for the community.

SIX

Unpacking the Phrenological Toolkit

Knowledge and Identity in Antebellum America

Carla Bittel

Opening Orson and Lorenzo Fowler's 1850 edition of the Illustrated Self-Instructor, one comes face to face with phrenology (fig. 6.1). There, a large, ambiguously gendered head invites readers to see a version of themselves on paper. A profile is mapped into multiple sections, each section containing an illustration of a localized character faculty, or "organ," of the brain. These "symbolical heads," could look masculine, with strong jaws and large foreheads, but the same one could also look feminine, with pronounced eyelashes, rosy cheeks, and pink lips. Coupled with a chart of measurements, completed by a phrenologist, the text served as an interactive guide for touching and assessing one's skull. The chart, the head map, and one's own body—paper and flesh and bone—provided the material means for a cohesive set of epistemic practices that could be exercised and learned by consumers and practitioners, women and men alike.

In the early nineteenth century, phrenology, considered a science of the mind, thrived in the American context, a product of expanding health reform movements and a flourishing consumer culture. Phrenology articulated a relationship between the interior and the exterior, the brain and the skull, the mental and the physical.[1] It described the brain as not one organ, but many, and each

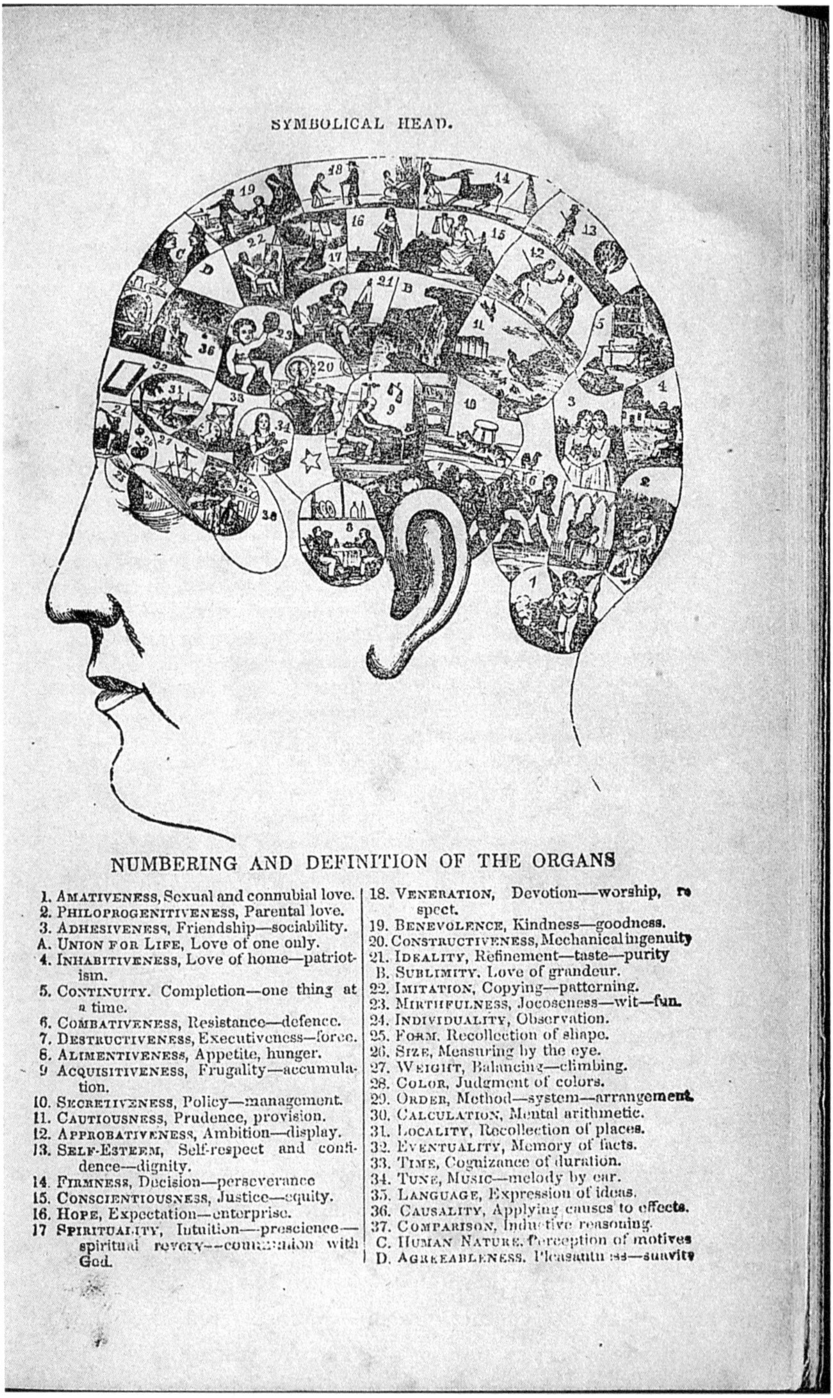

SYMBOLICAL HEAD.

NUMBERING AND DEFINITION OF THE ORGANS

1. Amativeness, Sexual and connubial love.
2. Philoprogenitiveness, Parental love.
3. Adhesiveness, Friendship—sociability.
A. Union for Life, Love of one only.
4. Inhabitiveness, Love of home—patriotism.
5. Continuity. Completion—one thing at a time.
6. Combativeness, Resistance—defence.
7. Destructiveness, Executiveness—force.
8. Alimentiveness, Appetite, hunger.
9. Acquisitiveness, Frugality—accumulation.
10. Secretiveness, Policy—management.
11. Cautiousness, Prudence, provision.
12. Approbativeness, Ambition—display.
13. Self-Esteem, Self-respect and confidence—dignity.
14. Firmness, Decision—perseverance
15. Conscientiousness, Justice—equity.
16. Hope, Expectation—enterprise.
17. Spirituality, Intuition—prescience—spiritual revery—communion with God.
18. Veneration, Devotion—worship, respect.
19. Benevolence, Kindness—goodness.
20. Constructiveness, Mechanical ingenuity
21. Ideality, Refinement—taste—purity
B. Sublimity, Love of grandeur.
22. Imitation, Copying—patterning.
23. Mirthfulness, Jocoseness—wit—fun.
24. Individuality, Observation.
25. Form, Recollection of shape.
26. Size, Measuring by the eye.
27. Weight, Balancing—climbing.
28. Color, Judgment of colors.
29. Order, Method—system—arrangement.
30. Calculation, Mental arithmetic.
31. Locality, Recollection of places.
32. Eventuality, Memory of facts.
33. Time, Cognizance of duration.
34. Tune, Music—melody by ear.
35. Language, Expression of ideas.
36. Causality, Applying causes to effects.
37. Comparison, Inductive reasoning.
C. Human Nature, Perception of motives
D. Agreeableness, Pleasantness—suavity

Figure 6.1. A "symbolical head" from Orson S. Fowler & Lorenzo N. Fowler, *The Illustrated Self-Instructor in Phrenology and Physiology, with One Hundred Engravings, and a Chart of the Character of _____ as given by _____* (New York: Fowlers & Wells, 1850), BF870. F787i 1850, History & Special Collections for the Sciences, UCLA Library Special Collections.

organ represented a particular "faculty," or mental trait, its degree of activity indicated by size and measurement. This way of knowing the brain took hold through "sociomaterial" practices, particularly with paper.[2] As books were abundant and circulated widely, paper charts, dictated analyses, notebooks, lecture notes, broadsides, receipts, stationery, and traced profiles all could be worked with in tandem as a paper system. Printed and plain, the inexpensive, industrialized papers of the nineteenth century facilitated and made the phrenological enterprise.

Importantly, the use of multiple, interdependent paper mediums illustrated and materialized phrenological concepts of gender. Among the American middle class, gender expectations hardened in the antebellum period; codes of industrious manhood and self-sacrificial womanhood proliferated in prescriptive literatures. Phrenology, too, was laden with gender dualism, but the science, its practices, and its paper technologies also allowed for some mixture of masculine and feminine qualities. While men and women were assigned divergent roles in reproduction, they shared the same phrenological organs, and therefore, the same anatomy of the mind. Those organs differed in size and influence; men and women both shared faculties of reason and sentiment, of Destructiveness and Benevolence. If deficient or too abundant in certain categories, one could supposedly cultivate or minimize those faculties. In practice and on paper, while phrenology preached and promoted the differences between the sexes, it also allowed for negotiations of gender identity and common characteristics.

While scholars have long studied phrenology, they are only now truly exploring its material nature,[3] and the centrality of gender within this knowledge system.[4] This chapter brings to light the vital intersection of gender and materiality by examining the central role of paper in phrenological practice. By unpacking the phrenological toolkit, we find paper serving as the primary social and epistemological "interface," or the mechanism of interaction for both producing and using knowledge about one's mind. While the term "interface" did not come into scientific usage until the late nineteenth century, the concept still applies.[5] In 1882 *interface* was defined as "a surface lying between two portions of matter or space, and forming their common boundary."[6] Rachel Plotnick, who has studied electric push buttons, has encouraged historians to examine the role of surfaces as mediators between users and technologies.[7] In this volume we see multiple mediations with paper—for example, Beth Linker's essay vividly illustrates the vital epistemic role of paper in the posture sciences and in producing a poor posture epidemic, making the slouch visible and tangible. With phrenology, paper serves as the material interface between one's human head and phrenological self, as well as one's gender identity.[8]

Paper was the most vital material and marketable element in the phrenological toolkit. Practitioners and consumers alike did buy, sell, and handle human and animal skulls, life and death masks, and phrenological busts. A skull or a bust was no doubt powerful, for observers could see and touch a three-dimensional object with human contours, similar to their own. But paper analogues also captured phrenological organ anatomy; a printed chart or a set of playing cards could do the work of a bust or a skull in compact, portable, shareable, and concealable forms. And, some busts were material hybrids of texts and anatomical models, plaster in form but covered in small, printed paper labels glued onto the locales of corresponding organs. As Jacob Eyferth discusses in his afterword to this volume, paper's perceived lightness and "lack of fleshy physicality" produced particular sets of "affordances." In the case of phrenology, paper afforded the reproduction and dissemination, handling and storing, transcription, and transformation of knowledge.[9] Charts and large foldouts were not simply addenda or bonus plates to a book but a means of integrating users into the text and the phrenological system.[10] As Anke te Heesen has shown, paper technologies like the notebook were machine-like, serving as mechanisms for storing and guiding entries.[11] More than simply surfaces, phrenology's paper platforms were technologies of interactivity that captured and transferred knowledge, and enabled both input and output.

In the pages that follow, we will see how a series of phrenological paper interfaces at times reaffirmed dualistic codes of gender and other times allowed for gender variables. We will also see how phrenological paper tools were technologies of race, forms of interface that hardened social hierarchies, and the privilege of the middle classes and white users.[12] While humans all shared the same phrenological organs, phrenology sorted people into categories, and made normative certain physical and mental characteristics, delineating differences of race, gender, ability, and disability.[13] These ideas not only circulated in books and on charts, they were perpetuated through technological and material interactivity. As we will see, phrenological interfaces could be platforms for the punctuation of differences as well as their negotiation.

PHRENOLOGICAL TOOLS OF THE TRADE

Henry B. Gibbons and Deiadamia Chase had much in common, but likely never met. Both practical phrenologists, they spent the 1840s and 1850s exercising their trade, she from her home in upstate New York, and he on the road.[14] For many years, they built their reputations in order to create and sustain loyal customers and eager audiences. While Gibbons and Chase owned phrenology busts and handled skulls at one time or another, their daily practices depended on paper, and the same paper products. They purchased many of the same books and charts; they used paper to collect knowledge and make knowledge, to study phrenology and sell phrenology. And yet their livelihoods were based on varied gendered forms of constructed authority and identity: his as a gentleman and accurate analyst, hers as a mother and sympathetic healer.

Paper validated phrenological practitioners, regardless of gender, granting them degrees of credibility in a science that was very much contested. Men and women shared epistemological paper tools, when other tools of science and medicine were not easily accessible to all. And while their identities and reputations as practitioners reflected dualistic gender roles, practitioners like Gibbons and Chase shared the same paper toolkits; paper granted them the means to both practice phrenology and negotiate gender role expectations at the same time.

In the summer of 1843, Henry B. Gibbons went on the road and commenced his phrenological work. Gibbons's journey was both a job and a cause, involving both individual remuneration and a social crusade to spread the gospel of phrenology. His travels in 1843 and beyond reveal a great deal about phrenological practices, that they were nothing short of a "campai[g]n."[15] Gibbons was a protégé of Nelson Sizer, who became a lead agent for the firm Fowler and Wells, the most famous American phrenological enterprise. For Gibbons, paper made his practice possible as he moved from town to town. His many trips to Manhattan to purchase phrenological paper products at the Fowler and Wells headquarters demonstrate his reliance on a system of paper.

Gibbons announced his arrival in a new town with paper "notices," or broadsides, which previewed his lectures and described his services. The phrenology head was printed front and center, surrounded by descriptions of Gibbons's cabinet, his years of experience, and his promises of improved self-knowledge. With blank lines for the time and place of his lectures, Gibbons could adapt the same notice as he traveled, inscribing specific information and tailoring it to local conditions. Using paper to fill lecture halls, he often claimed to have a "full house."[16] Gibbons's use of blanks reveals how paper sustained his practice in his very itinerant and fluctuating enterprise.[17]

Gibbons employed multiple methods of paper authentication, trying to prove himself and phrenology as true and credible. Facing skepticism as he entered a new town or met a new audience, Gibbons collected written testimonies from satisfied clients and townspeople, which he used to verify that he was both a skilled analyst and a true gentlemen. Phrenological practitioners were often subject to the same expectations and requirements as "regular" physicians in this period, needing to demonstrate good character.[18] Thus, client testimonials and letters of introduction carried vital information for assuring Gibbons's acceptance. One testimony from Queens stated that he had a "good moral tendency" and was "an excellent practical delineator of character and an amiable man."[19] Sometimes groups of townspeople, often men of standing or title (such as a physician or minister), signed a joint testimony. One in 1845 stated, "We the undersigned having attended your lectures at this place and also your examination on the science of phrenology take *this Method* to express to you our best wishes, believing that your moral rectitude, and instruction, is well calculated to advance and improve Society."[20] "This method," a paper testament, could be carried in a pocket or in a trunk. A letter, penned, signed, and wax sealed, could be handed to and held by others; a stack of letters gave added weight to a practitioner's validity. Further, testimonies could be printed in local newspapers as a public proclamation of his skills.

For Gibbons and Chase, phrenological charts, the inscribed results of client examinations, lay at the heart of their practice.[21] Capturing the analyst's measurements in an orderly fashion, the paper chart was the centerpiece of the phrenological encounter.[22] It structured interactions—between practitioners and clients, clients and their acquaintances, clients and their own bodies. It also packaged personalized behavioral prescriptions. Narrative delineations often accompanied charts or stood on their own; these were descriptive declarations of character traits and recommendations for improvement. But one sheet of paper could serve as a visual representation of the self.[23] For both Gibbons and Chase, selling those sheets of paper sustained their businesses.

While chart designs varied, most combined print and inscription, and standardized categories of assessment with personalized measurements. The chart could be a large, free-standing sheet of paper, a book foldout, or commonly, a smaller table, published at the front or back of a phrenology text, bound and paginated with the book itself. Single-sheet charts were often visually striking, mixing illustrations with information, the symbolical head with lists of phrenological organs, described in brief. The top of the chart usually announced the name of the practitioner and his claims to expertise, along with space for inserting the date and location of the examination. Vacant lines provided a space for inserting the name of the client and sometimes the person's current occupation or a defining characteristic, such as age. The gender of the client was made evident by the name, and for women, marital status was indicated with *Miss* or *Mrs.* A list of faculties, often grouped according to type or location, stood next to blank lines awaiting an inscribed measurement. An area for assessing temperament often followed. Some charts included an explanation of the measurement system, describing the meaning of the assigned numbers and the faculties themselves. The chart was a vital tool of the trade, for it preserved the exam; without a chart as a record, measurements were meaningless.

For table-like charts incorporated into books, the title page identified the practitioner and client, literally incorporating their encounter into the title and content of the text itself. For example, a typical title page of a book from the

1850s by Orson Squire Fowler and his brother, Lorenzo Niles Fowler, would state: *The Illustrated Self-Instructor in Phrenology and Physiology, with One Hundred Engravings, and a Chart of the Character of _____ as given by _______*.[24] A blank chart followed, listing the phrenological temperaments and faculties on the left, accompanied by columns with boxes ready to be checked to indicate size. Each box contained a number, which referenced a page in the book and a corresponding explication of the faculty and its measurement. Boxes were often accompanied by markings such as plus and minus signs, or additional boxes indicating which characteristics to restrain or cultivate. The chart visually sectioned the mind, "boxing" traits to reflect the compartments (or organs) of the brain.

Henry Gibbons combined charts and cranial measurements with paper silhouettes and painting. As he analyzed profiles, silhouette cutting and painting offered a way to preserve an individual face and head for posterity, but also for study, in the physiognomic tradition of Johann Lavater.[25] Gibbons "gave . . . examinations & charts," but also reported to "take," "cut," and "frame" profiles as a separate activity that complemented his other paper practices.[26] He tried to enhance his technique by copying drawings based on the black paper cuttings of Auguste Edouart, the famous French silhouette artist who visited North America.[27] The methods used to create silhouettes varied. In the 1840s Gibbons may have sketched outlines of his clients freehand and used scissors to cut profiles, either by trimming away paper or hollowing out their image; others used silhouette "machines" like the physiognotrace, a device designed to simultaneously trace a shadow in actual and reduced size, as clients sat behind a backlit screen. Silhouettes clearly marked and accentuated gender identity, especially through hair and clothing, but also with facial features coded masculine or feminine; they also marked racial identity, reflecting the artists' rendering of European features. Before the photographic era, silhouettes created a paper shadow of the self that could be "read" according to phrenological principles. Gibbons eventually moved on to paintings, presenting an array of "likenesses" to his audiences and clients, as a transportable phrenological museum. Gibbons did learn to create molds and make busts, but busts still had to be mapped. With printed paper labels cut into small strips and applied with paste, he turned plain plaster into a phrenological head.[28]

Gibbons also used assorted papers and notebooks to manage his enterprise. He drafted and then formalized lecture notes, filling a brown notebook with ideas and transferring them to flat sheets for reading in public. He charted his progress as a practitioner in his twelve diaries, covering almost a twenty-year period, as he used paper to plot his own narrative. He even used paper to envelop specimens of his own hair, cut from particular phrenological regions of his head and stored in his diary.[29] As an itinerant practitioner, paper afforded him a constant home and way to integrate himself into the phrenological project.

For Deiadamia Chase, settled in the village of Eaton in Madison County, upstate New York, paper provided a means for organizing medical knowledge and combining it with phrenological practice. Her biography, *Trials and Triumphs of an Orphan Girl*, by Luna M. Hammond, dramatically narrates Chase's journey to becoming a trusted and well-known practitioner of phrenology, eclectic and botanical medicine, and clairvoyant healing.[30] Chase was not a stranger to the phrenological world, having been featured in the *American Phrenological Journal* for her "practical talent, and strong social and moral powers."[31] Both biographies paint her as extraordinary—a talented female physician, who stood out among her sex. And yet she was often praised for still being one of her sex; she exuded maternal qualities and had the ability to care for her "suffering sisters" whose "delicacy" forbade them from calling a male physician.[32] Consciously promoting her identity as a practitioner "in sympathy with [the] feelings" of patients,[33] Chase combined medicine and phrenology from her home, attracting a local and regional clientele who saw her for related purposes of healing and self-knowledge.[34]

As a phrenologist, Chase relied on materials from Fowler and Wells, using their charts and filling her library with their books. And charting was one of her most remunerative services. Her biography recounts numerous examples of successful analyses and pleased clients.[35] Men and women sought Chase's health advice and her phrenological readings specifically to inform childrearing practices and to influence their choice of marriage partners. Paper was key to the interface for both client and practitioner, as seen in the case of a "Mr. B." who came to see Chase for a phrenological analysis of himself. Once she had completed his chart, "he produced a piece of paper, whereon he had copied numbers from another chart, and which he said belonged to a lady." Mr. B. wanted to hide the identity of the lady at first, for Chase had done her analysis as well, and he wanted an "unbiased" opinion about their compatibility for marriage. For him, the charts, as records and reference points, were paper modes of verifying suitability.[36]

Bound into the end of her biography is the Fowler and Wells pamphlet *Synopsis of Phrenology*; it waits there for an individual analysis to be recorded. This pamphlet could have stood alone, and been used by Henry Gibbons or any other practitioner. But when combined with Chase's very personal story, the attached pages took on new meaning. Her gendered authority, or her roles as a mother and healer, phrenologist and personal advisor, were built right into the texts she sold and used during her examinations. The very personal nature of the volume also mirrored the ways in which her clientele was built from a regional reputation, local networks, and even friends and acquaintances. A surviving copy begins with a poem, and a dedication to "her friend," Miss Emily Griffin, whose reading is included in the back of the text.[37] By merging Chase's own biography with phrenological information, and her own story with a personalized reading, her narrative became a permanent part of the phrenological encounter.

Finally, as a healer, Chase worked from a "catalogue" of "receipts," her own compilation of recipes collected and curated over many years of treating "infirmities of the body as well as the mind."[38] Based on botanical and eclectic remedies, some were compounds guided by "reason," while others she discovered in a clairvoyant state while treating patients.[39] While Chase's original catalog does not survive, Hammond reprinted the recipes as an appendix. There one finds instructions for making powders, poultices, oils, lotions, washes, syrups, and serums.[40] This was her own personalized storehouse of medicinal concoctions, saved, compiled, and organized into a compendium of specialized knowledge, intimately bound together with phrenological practice.

CHARTING GENDER

In November 1838 Dr. Jared Potter Kirtland, a physician, botanist, naturalist, and professor of medicine at the Ohio Medical College in Cincinnati, had his head examined. He received an analysis from phrenologist Robert H. Collyer, who gave Kirtland a copy of his book, *Manual of Phrenology, or the Physiology of the Human Brain*.[41] Opening the cover, Kirtland interfaced with a large, mapped head. It was a foldout that illustrated the phrenological organs, from three vantage points, side, front, and back (see plate 8). He also interacted with a second foldout, containing a map of his own mind, or rather, the results of his examination broken down into a graphic representation (fig. 6.2). Kirtland could see the measured size of his organs, locate them on the mapped symbolical head, feel them on his own head, then read about the meaning of those organs in the interior pages of the book. Paper afforded this tactile interaction, and also his ability to react to those measurements and input information himself. Kirtland penciled interventions in the text, both agreeing with and disputing Collyer's findings.

From Collyer's measurements, we see that Kirtland supposedly had large "domestic propensities," such as Adhesiveness (attachment) and Philoprogenitiveness (love of offspring), usually viewed as more feminine traits. He had more moderate-sized Combativeness and Destructiveness, and a smaller Self-Esteem, which were usually large in men. Interfacing with the text, Kirtland concurred with Collyer that he had a lower Self Esteem, writing: "Have no confidence at all in myself." And, of his small Combativeness, he remarked, "True to the fullest extent." But he rejected his low score for Constructiveness and said, "The least correct—[I] have some mechanical ingenuity—."[42] More intellectual and

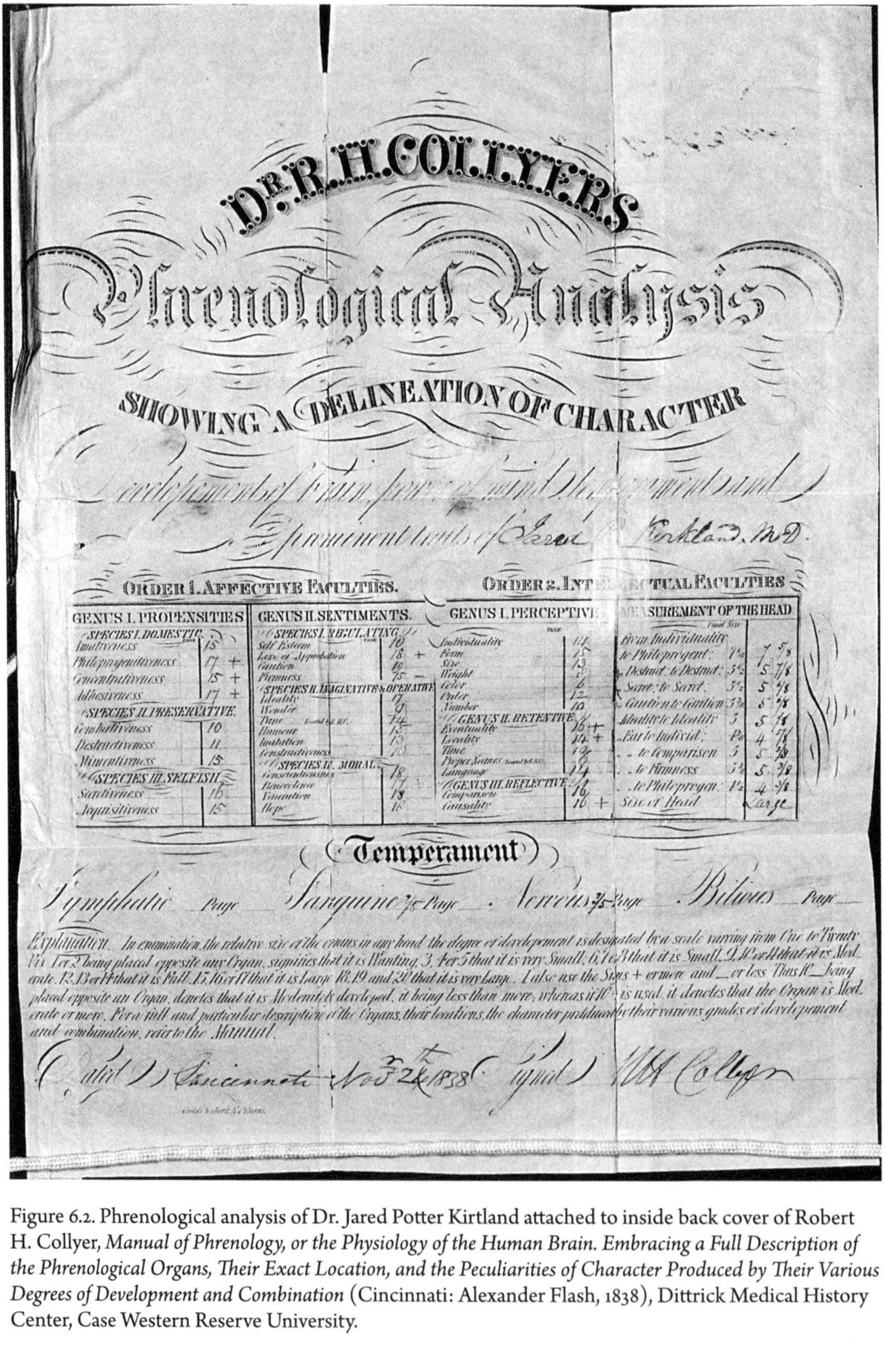

Dr. R. H. Collyer's

Phrenological Analysis

Showing a Delineation of Character

Development of brain, power of mind, disposition and temperament traits of Jared P. Kirtland, M.D.

Order 1. Affective Faculties.

Genus I. Propensities	Page	
Species I. Domestic.		
Amativeness	15	
Philoprogenitiveness	17	+
Concentrativeness	15	+
Adhesiveness	17	+
Species II. Preservative.		
Combativeness	10	
Destructiveness	11	
Alimentiveness	15	
Species III. Selfish.		
Secretiveness	16	
Acquisitiveness	15	

Genus II. Sentiments.	Page	
Species I. Regulating.		
Self Esteem	10	
Love of Approbation	18	+
Caution	19	
Firmness	15	−
Species II. Imaginative & Operative		
Ideality	17	
Wonder	9	
Tune	14	
Humour	15	
Imitation	13	
Constructiveness	13	
Species IV. Moral.		
Conscientiousness	18	
Benevolence	17	
Veneration	13	
Hope	18	

Order 2. Intellectual Faculties

Genus I. Perceptive	Page	
Individuality	14	
Form	15	
Size	13	
Weight	8	
Color	6	
Order	12	
Number	10	
Genus II. Retentive		
Eventuality	16	+
Locality	14	+
Time	10	
Proper Names	6	
Language	14	
Genus III. Reflective		
Comparison	16	
Causality	16	+

Measurement of the Head	Usual Size	
From Individuality to Philoprogent.:	7½	7 5/8
" Destruct.: to Destruct.:	5½	5 7/8
" Secret.: to Secret.:	5½	5 6/8
" Caution to Caution	5¾	5 6/8
Ideality to Ideality	5	5 1/8
" Ear to Individ.:	4½	4 5/8
" " to Comparison	5	5 3/8
" " to Firmness	5½	5 3/8
" " to Philoprogen.:	4½	4 3/8
Size of Head		Large

Temperament

Lymphatic — Page   Sanguine 2/5 Page   Nervous 3/5 Page   Bilious — Page

Explanation. In examination the relative size of the organs in any head the degree of developement is designated by a scale varying from One to Twenty viz 1 or 2 being placed opposite any Organ, signifies that it is Wanting, 3, 4 or 5 that it is very Small, 6, 7 or 8 that it is Small, 9, 10 or 11 that it is Moderate, 12, 13 or 14 that it is Full, 15, 16 or 17 that it is Large, 18, 19 and 20 that it is very Large. I also use the Signs + or more and − or less. Thus 10 − being placed opposite an Organ, denotes that it is Moderately developed, it being less than more, whereas if 10 + is used, it denotes that the Organ is Moderate or more. For a full and particular description of the Organs, their locations, the character produced by their various grades of developement and combination, refer to the Manual.

Dated Cincinnati Nov 24th 1838 Signed R H Collyer

Figure 6.2. Phrenological analysis of Dr. Jared Potter Kirtland attached to inside back cover of Robert H. Collyer, *Manual of Phrenology, or the Physiology of the Human Brain. Embracing a Full Description of the Phrenological Organs, Their Exact Location, and the Peculiarities of Character Produced by Their Various Degrees of Development and Combination* (Cincinnati: Alexander Flash, 1838), Dittrick Medical History Center, Case Western Reserve University.

domestic, and less aggressive and competitive in character, Kirtland's analysis delineates one model of masculinity of the antebellum period.[43] As we will see, paper charts could visually affirm a person's physical alignment with social codes of gender or suggest that they were inclined to thoughts or behaviors in opposition to their sex. But, as charts also reveal, phrenological gender was negotiated and subject to intervention on paper.

Phrenological analysts expected men and women to be more or less pronounced in certain faculties. And with the evaluation of each faculty came the expectation for a certain type of result, differing by sex in size and capacity. As T. Barlow, a colleague of Lorenzo Fowler, stated in 1838, the phrenological "peculiarities of the sexes" were evident: "The male has a larger forehead, a deeper, broader occiput . . . higher in the locality of self-esteem and firmness. The female has greater benevolence, veneration, [and] adhesiveness."[44] Illustrated with male and female profiles, Barlow's idealized characterizations of masculine and feminine heads harkened back to antiquity with neoclassical hairstyles and dress.[45] They also embodied middle-class notions of oppositional sex differences. These phrenological specimens reflected racial ideals as well, as so-called "Anglo" or "Caucasian" facial and cranial characteristics were presented as exemplary while Native American and African features stood in contrast throughout the phrenological literature.

While Barlow and others promoted notions of divergent sex differences, they also described male and female as complementary and harmonious. As William Leach has shown, the antebellum period witnessed a new model of marriage in the white middle classes, one based on compatibility and balance; it was supported by many feminists, as well as men and women health and hygiene reformers.[46] The Fowlers and other phrenologists were among them; they envisioned compatible marriages and equated them with healthy domestic spaces. Covers of early editions of the *American Phrenological Journal* bore the words "home truths for home consumption," welcoming men and women readers, at the intersection of science and domesticity.

Lorenzo Fowler amassed phrenological charts as a form of data collection, which informed many of his later writings and practices. In 1834–1835 Fowler traveled the country and conducted over two hundred analyses. Fowler may have communicated the results to his clients, in exchange for 12½ cents, but they were his research subjects, and he gathered information for future reference and comparisons. Each subject had their own "phrenological card,"[47] a small chart with blank lines for measurements. Among his many subjects were women and men, children, the imprisoned, and the insane.[48] While each printed card was standardized, measurements personalized each one, along with his short descriptions on the opposite page.

Fowler's marked charts, or "cards," were particularly attuned to sex differences and extraordinary female subjects, whose brain size or intellectual qualities surpassed expectations. For example, Miss Burchans had "the largest head, the strongest and most comprehensive mind that I have ever see[n] for a female."[49] He noted Burchans's unusually high score for a woman in Comparison and Causality, the faculties of reasoning. The phrase, "for a female," appears repeatedly, as Fowler seems surprised by the quality of some feminine specimens. For a Miss Briscoe of Mississippi, he wrote: "A very large head for a female, has much strength of feeling and judgement."[50] And Miss Palmer of Kentucky, had "a very singular head for a female, all life, energy, spirit, motion imitation and conversation."[51] Studying several charts, side by side, Fowler used them to create a gendered scale of ability. Women, who it was believed generally had smaller brains, were expected to score lower in the intellectual faculties. So when they scored higher, they challenged expectations.

Tools like the phrenological chart materialized gender, sometimes reinforcing a gender binary and other times reflecting masculine and feminine as overlapping and complementary. Mrs. Thankful B. Paige seemed to fulfill expectations for a woman in her social position. Analyzed by her husband, the Reverend A. Paige, her evaluation may have been shaped by his preconceived notions of her as a good wife and mother. Her Destructiveness and Combativeness were lower, while she scored higher, a 6, for Philoprogenitiveness, or care of offspring, as well as other more feminized traits, such as Benevolence and Veneration.[52] Lorenzo Fowler analyzed a Miss Alverson in 1859, who fulfilled some expectations for women but challenged others. He measured her "domestic group" of faculties as "large," but she also had large Destructiveness and Approbativeness, or vanity, not expected in women.[53]

Husbands and wives, suitors and potential mates were often analyzed alongside one another and on paper. In March 1851 Orson Fowler analyzed Mr. L. M. Ninde and Mrs. B. C. Ninde, a newly married couple from Indiana.[54] A lawyer and his wife, they were soon to be active agents on the Underground Railroad and advocates of women's suffrage.[55] In this dual exam, Fowler measured their crania side by side and incorporated his findings on a single table, using a dash for him and a dot for her.[56] In several categories, both husband and wife were in sync, with both symbols in the same box. She scored higher in domestic propensities, with Parental Love, Adhesiveness, and Inhabitiveness marked as "very large"; this was expected of a woman. But what may have been surprising was that Mr. Ninde and Mrs. Ninde both had large Combativeness (resistant, argumentative, defiant), a faculty usually more pronounced in masculine subjects. In a middle-class culture obsessed with feminine modesty and gentility, Fowler advised Mrs. Ninde to "restrain" her Combativeness. Did Mr. and Mrs. Ninde have a turbulent home? Or, could a good wife and mother be combative, especially one who was an abolitionist and suffragist? By marking boxes, Fowler signaled the degree to which individuals were endowed with attributes considered to be more masculine or more feminine, while illustrating gender variances and common characteristics. Thus, phrenological charts staged a dynamic interface around individual identity, both boxing people in and allowing for variability.

Letters, Transcription, and Knowledge Exchange

Graceanna Lewis, a Quaker writer, abolitionist, and naturalist who later specialized in ornithology, dove deep into phrenology by receiving readings from multiple practitioners and comparing their findings.[57] In October 1842, in a detailed cross-hatched letter, she wrote to her mother and sister, stating that her latest reading "differed so much from that given by Fowler I must tell you about it and you can judge which is the nearest truth." She then proceeded to explain the 1 to 7 numbering scale, and said, "As a chart was given, I will copy from that what parts I think will interest you."[58] But for Lewis, this was no mere act of copying. She had learned the measurement systems employed by phrenologists, as seen in her informed references to size and numerical assessments. She then used transcription and annotations of her chart in private correspondence to document, transfer, affirm, and question the results. Paper, a transportable method for interfacing with phrenology, carried these communications.

Personal letters doubled as phrenological paper tools, and were particularly significant tools for women, who found that familial correspondence was a useful means for touting or challenging practitioners' findings. As James Poskett has shown, letter writing, as a material practice, helped make phrenology a global reform movement.[59] He demonstrates how correspondence captured racial politics and carried the phrenological gospel. As we will see here, letters played a significant role in the American consumer culture of phrenology as well; they also fostered a sense of scientific sociability. Users acquired a phrenological vocabulary and knowledge of the brain's geography. And then in letters, they treated charts and measurements not as conclusive, but as preliminary and open to deliberation. Letters were also extensions of books, as writers deployed phrenological terms, concepts, and even chapters in their personal relations and gendered conflicts.

Graceanna Lewis moved from one paper interface to the next, taking and transcribing her chart into a letter, and with underlines, annotations, and added punctuation, transforming the results. Lewis seemed pleased with the evaluation of her Amativeness (love of the opposite sex), recounting the phrenologist "remarked first that were I to love it would be romantically, intellectually!!" Even more, she delighted in her ability to control the mind, for she underlined "*Con-*

centrativeness full." Lewis also included the analyst's numbers: "Individuality +6 Has a *great* desire to know & investigate. . . . Ideality gives [a] great fondness for *beautiful* paintings *Order +6* ~~very~~ particular, is annoyed by disorder."[60] Here, she edited her own results, with abbreviated phrases, strikethroughs, and use of the first person where she seemed to most agree. Lewis's transcription amounted to a double analysis of her character, by both the phrenologist and herself.

Other correspondents used letters to reference specific phrenological organs and to explain the behavior or character traits of themselves or others. Abolitionist Sallie Holley stayed in the family home of fellow activist Thomas Garrett in 1852. In a letter, she conveyed her admiration for his assisting fugitive slaves, and cited his "Very large benevolence."[61] Susanna Strickland Moodie, a Canadian writer, explained her own qualities and shortcomings to others in letters by referring to specific phrenological organs. "I have not much number 10 with regard to my poetical talents. Nature gave me the mind and the imagination but withheld the power of language to express correctly what I feel." Referring to "number 10," she pointed to the sequence of faculties on phrenological charts.[62] Many writers casually employed the expression "bump of _____," followed by a phrenological organ, to account for a particular personal trait. So familiar with the faculties from paper technologies, they naturally wove phrenological discourse into their letters.

Users learned about phrenological organs via charts and texts, and then applied that knowledge to explain racial differences and what they saw as unusual behavior. Letitia Hargrave, who moved to North America from Scotland, used phrenological vocabulary to describe what she perceived as the failings of her child's servant. She wrote, "Tash's Indian servant is an excellent & willing worker, but the organ of acquisitiveness must be large & philoprogenitiveness absolutely wanting as she hates playing with children & seizes upon every article that poor Tash has."[63] Phrenological texts, packed with typologies of gender and race, led many readers to associate certain phrenological characteristics with particular racial groups and then transfer that knowledge on paper.

As a type of correspondence with the self, diaries also served as a place for the transcription of phrenological examination results. For example, Lucy Chase (no relation to Deiadamia) agreed with and also challenged the analyses of one of the Fowler brothers. She asserted no disagreement with Fowler's rather high marks for a Mr. Dickens, documenting his numerical measurements in her diary: he received "7 Ideality, —Mirth—App—7 Conscientiousness large—Veneration 3. Self esteem small several other organs were numbered 7." With her own reading, Lucy Chase noted frustration. She wrote that Fowler "told me that I must not study, that my brain is too large," and that her strength was physical, not mental. But when she one day needed to work in the family kitchen, she quipped, "I presume Fowler would say that is the place for me."[64] Chase's diary became her private interface with phrenology.

Playing Cards & "Phrenological *Bijou*"

As we have seen, diaries and letters allowed for continual reevaluation, beyond the hands of the practitioner or the walls of his examining room. This also happened via cards and boxes. In 1835 Ackermann & Co. publishers in London released a phrenological volume intended for women. Mrs. L. Miles's *Phrenology and the Moral Influence of Phrenology,* directed "to the female sex," was packaged and presented in a box, labeled on its exterior, "the casket of knowledge."[65] The box carried a set of forty thick paper cards the size of playing cards, inscribed with phrenological information (see plate 9). Resembling an unbound, abridged book, the cards condensed the works of Franz Joseph Gall and Johann Gaspar Spurzheim, the intellectual founders of phrenology, and incorporated "the latest discoveries of the present period." Printed with phrenological images, each card was dedicated to a separate phrenological faculty. Ackermann and Mrs. L. Miles had deconstructed phrenological knowledge only to reconstruct it as a consumer product overtly directed at bourgeois women. They took advantage of paper's affordances to stage a phrenological interface of feminine sensibility.

The casket of knowledge was a book repackaged in a "pleasing and elegant form." It was a small cardboard box, covered in leather, with gilt lettering and decorative designs of flowers and a rising sun, "locked" with a small clasp. The interior of the box was lined with light pink paper; one side served as a title page, illustrated with cherubs and leaves, with the phrase, "Man's Greatest Knowledge is Himself to Know." Pink ribbons attached to the sides could serve as a place marker while one used the cards. A subsequent edition was dedicated to the ultimate feminine user, "her most Gracious Majesty Queen Victoria." As reviewers described it, it was "indeed a splendid Phrenological *Bijou*," "prettily executed," and "a unique and beautiful little work on Phrenology," derived from a "tasteful hand, and that hand a lady's."[66]

Miles's cards created a very tactile interface, as one studied phrenological images and texts in tandem, on the back and front of each card. The small, separate cards allowed for either more private, independent use, or for group study, as cards could be taken from the box, passed, shuffled, reordered, and hidden away. As te Heesen has described, knowledge boxes functioned differently from books, for cards required specific handling. Her study of Enlightenment boxes shows that "the box presents us with the act of taking possession of the world, the subject's ability to reach for something 'always to hand' (Heidegger), which can appear as a whole only when contained within a box."[67] For women users, the box concentrated knowledge of the mind and put it within reach.

The casket also supplied a phrenological head, in paper form, that was tangible and easily transportable. The top card of the box contained an embossed representation of a phrenological bust, sporting a classical, Roman-style helmet with plumage; the phrenological markings here were not on the head itself, but on the helmet. This warrior had long hair and both feminine and masculine facial features. Ackermann saw the top card as Miles's work's greatest advantage, for it enabled phrenological study for women, "altogether superseding the necessity of a bust, and in strict accordance with the principles of the original founders, Gall and Spurzheim."[68]

The identity of Mrs. L. Miles remains somewhat mysterious. In addition to the phrenology casket, she is also associated with a similar box on physiognomy, as well as several poems in early nineteenth-century British literary digests.[69] Her phrenology box was sold in Britain, and was republished in Philadelphia by Carey, Lea, and Blanchard, also in 1835.[70] The American version, though not in a box, represented the same type of condensation of information, and had similar "cards," though in page form. A deconstructed book, made into cards in a box, was once again a book on the American market. And in this version the embossed head was only a flat print, missing the texture, but not the three-dimensional look of the original. The change did not seem to take away from its usefulness. Reviewing it in the *Southern Literary Messenger*, Edgar Allan Poe called it a "beautiful little work" that was "exceedingly neat and convenient" for exercising the most important "uses" of phrenology, "*self-examination and self-knowledge*."[71] Poe did not describe it as a book for women, but other reviewers perpetuated the notion that it was "designed more particularly for ladies."[72]

Either as box or book, the phrenological work of Mrs. L. Miles was packaged for women, in a simplified form, as a source for moral reference. Knowing phrenology, Miles suggested, could only enhance women's roles as guides for children. It could also expand their sphere of knowledge. As one of the title cards suggests, "To the female sex, in particular, this science opens a wide field for the exercise of those quick and perceptive faculties with which they are peculiarly gifted." By implication, women could actually practice phrenology, becoming analysts, not just the objects to be analyzed, since they could hold the cards.

Through this sociomaterial analysis of gender and paper, we can arrive at some new understandings of phrenology as a knowledge system. No longer simply a prescriptive science, phrenology emerges as an epistemic exchange, made possible by material means. Furthermore, phrenological analysis appears as a series of interactions, in which practitioners and mostly middle-class clients served as participants in charting and challenging predetermined categories of character, while reinforcing social hierarchies. Gender differences were made visible and tangible, but were also subject to negotiation. A completed phrenological chart was not the final word but the beginning, as knowledge moved through letters, into diaries, within margins, and on and off bookshelves. In the end, we see how phrenological paper forms—the head map, the chart, the letter, the silhouette, the card—were inseparable from the gendered users who made and remade them.

SEVEN

Keeping Prussia's House in Order

Census Cards, Housewifery, and the State's Data Compilation

Christine von Oertzen

In 1871 the Prussian census bureau introduced a new paper tool, the so-called individual counting card. Central to Prussia's ambitious and far-reaching census reform, the small form was robust and mobile. It was designed to speed up the compilation effort and facilitate statistical complexity, serving at once as a recording and a sorting and counting device.[1] Counting cards consisted of simple-sized heavy-duty paper, cut to twenty-one centimeters in length and twelve centimeters in width.[2] Providing just enough space to record all required enumeration data for one person on a single page, each card weighed merely 7.5 grams—but together, the almost 29 million cards, lists, and envelopes produced for the 1871 census amounted to 375 metric tons of paper.

Mass compilation efforts are primarily understood with reference to the inscriptions on paper, and metaphors such as "floods of data" or "data deluge" emphasize how historical actors struggled with a perceived overload of information.[3] But data have mass, and for the Prussian census the intricacies posed by the quantity of paper were at least equally trying as the amounts of data it carried.[4] The material sequence of census compilation is hence more accurately depicted as a heavy, hard-to-maneuver machine. Once set in motion, the weight and sheer volume of the material, apportioned in wooden boxes containing either five thousand or ten thousand cards, dictated the allocation of spaces and of people attending to the loose paper forms and the inscriptions on each card. The practicalities of census compilation bore similarities to moving troops and heavy weaponry described in Carl von Clausewitz's now legendary strategic analysis of warfare.[5] The Prussian general predicated that far-sighted, detailed planning would most certainly encounter unpredictable frictions, which in turn required steering through fast and deliberate decisions for adaptation and adjustments.

Historians of science have used the notion of paper tools mostly with regard to what happens on paper: as material practices or devices of concretion, which helped to structure theoretical and practical thinking.[6] However, the masses of paper forms to be dealt with in nineteenth-century census compilation make it indispensable to use the concept of paper tools in a three-dimensional, literal sense: the new cards functioned both as carriers of inscription and as material objects. The cards provide a prime example of how inextricably print and writing on the page are interrelated with the material properties of the paper substrate on the one hand and with the everyday power structures among the historical actors tending to the paper work on the other.

Following the sociomaterial approach shared by the contributions to this volume, this chapter explores a peculiar set of actors, workflows, and spaces in which the paper work of the Prussian census bureau took place.[7] The Prussian census bureau made considerable efforts to incorporate homebound spouses and other relatives of their workers into census compilation, nurturing a workforce that required circulating tons of paper carrying irreplaceable information between the bureau and at times hundreds of different homes scattered across and beyond the city limits of Berlin. This laborious procedure was considered essential and valued as necessary to achieve the best possible census results. To understand why this was so, I explore in detail how a growing female at-home workforce was incorporated into the workflow of manual tabulation for the census. I pay special attention to the epistemic, social, and gendered dimensions of the state's paper and data work.

Francesca Bray's application of sociotechnical systems to the concept of gender and the household provides a framework to grasp the complex dynamics between the epistemic, material, and social aspects of nineteenth-century manual census tabulation. With reference to Wiebe Bijker and others, Bray emphasizes that all technical practices are "embedded in, or constituted by, a network of materials, procedures and relationships, . . . a 'seamless web' in which it is never clear a priori and independent of context whether an issue should be treated as technical or social." This approach allows her to draw attention "to the full range of resources upon which a technical activity depends and to highlight the social and gendered identities of and relations between the actors involved."[8] Bray's approach to technology enables me to scrutinize the often-assumed binary between statisticians as experts inventing the tools and directing the process on the one hand and the mere routine or "ancillary" work of the tabulators by investigating the dialectics between design and usage of the technology (i.e., the paper tools) in question. In this context, the housewives' skills become apparent as a simultaneous conglomerate of knowledge and communal practice, a concept that Jacob Eyferth develops in his concluding essay to this volume on rural papermakers in China.[9] The Prussian case provides a case in point that skills deemed necessary for census compilation were specific to gender and estate. By examining the foundation of such skills, I challenge stereotypes of knowledge making in data work as primarily a domain of male expertise.

In contrast to earlier counts, the Prussian census conducted via counting cards accumulated information unbound. However, as mobile and loose as these cards were, all depended on the order in which they were kept. The new movable paper tool, geared to establishing multiple relations between the collected data, turned the compiling effort of the Prussian census into an extensive and also exceptionally delicate operation. Throughout the entire process, orderliness was a prerequisite of success, *conditio sine qua non* for obtaining viable census results. Envelopes, labels, wrapping paper and thread, and the wooden census boxes, as well as lead blocks, served as means to secure the order of the filled-in cards while they were collected, passed on, sorted, and moved about, as no knowledge could be drawn from them if they were mixed up.

Processing data, as historians of science have recently emphasized, is a fundamental and tedious, and by no means a trivial activity in producing knowledge.[10] Nineteenth-century Prussian officials underscored this appraisal in their careful handpicking of workers to be entrusted with the task of turning data into numbers and tables, be it in the bureau itself or at home. They insisted that the sorting and counting of data for the census off-site could only be commissioned to sufficiently educated women keeping what they called an orderly home. This assessment had very specific social, spatial, and political, but also epistemic implications. Orderliness, one of the core values of late nineteenth-century housewifery, was most prominently displayed in a specific space within middle-class homes: the parlor, typically the most presentable room, carefully furnished and decorated.[11] Off-limits for most household members, the function of the parlor, or *gute Stube*, was not to accommodate the families' daily social interactions but to demonstrate orderliness, modest affluence, and social aspiration. The Prussian census bureau regarded orderly housewives and their unused parlors as critical to census compilation. By using the mental skills and household technologies of the parlors' gatekeepers, and by making use of the very space of the parlor itself, the Prussian paper system of census compilation provides us with hands-on evidence of how tightly entangled nineteenth-century state governance, household skills and spaces, and gender norms were.

The first part of this chapter sketches the new paper tool and the paper technology embedded in the Prussian census to render enumeration data movable and readily convertible into numbers, and explains the dialectics of flexibility and order that the new system entailed in material terms. I then turn to the practices of compiling census data. Here I pay special attention to the material challenges and frictions posed by the masses of paper and the solutions pursued to navigate them, first within the census bureau but soon also beyond its premises, as circulating the material to private homes across Berlin was deemed necessary. Mapping the itineraries of the counting cards between the bureau and the workers' dwellings allows me to reconstruct networks of materials, people, and resources set in motion to secure the timely compilation of census results. I then consider both the people and parlors designated for these labors. Here I reconstruct the tasks requested to process the information on the cards as well as the space and skills required to master the material, and how each household managed to incorporate the bureau's fluctuating demand with its own objectives. In the fourth and final part, I assess the broader implications of housewifery skills for Prussian state governance, census taking, and paper work in the parlor, by turning to economic theory and to the published tables that the women helped to produce. While economists valued orderly housewifery as a foundation of the state's well-being, the refined census statistics fueled concerns that industrialization eroded women's "natural" role as wives and mothers, trends deemed especially hazardous for the lower middle class and consequently for the state itself. The practices of Prussia's census compilation offer rich insights into this interplay between the micro- and macropolitics of everyday life and the workings of governance, exposing dynamics of gender as they translated into material activity. Through the medium of the state's paper technology and its skilled at-home applications in the parlors, such entanglements show how the production of knowledge challenged the gendered fabric of the Prussian state.[12]

MAKING DATA MOVE

Turning census information into numbers and tables via counting cards had much in common with older, long-established learned paper technologies.[13] However, the new paper tool employed in the Prussian census was inspired by playing cards.[14] The same was true for the sorting and counting of the data, a process in which cards with matching information were piled into stacks according to a predetermined laying list—just like in matching card games.[15] Prior to counting cards, census information had been collected in bulky enumeration lists and subsequently added together via a process called check-listing, a simple method of counting by strokes of five which yielded entries like this: 𝍸 𝍸 |||. This slow, error-prone, and inflexible technique was prohibitive to composing aggregates. By way of check-listing, entries from enumeration lists were either read aloud to a group of tabulators, each responsible for marking strokes in one particular rubric of spacious, unwieldy interim tables (the so-called synchronous system), or worked through by individual tabulators many times, rubric by rubric (known as the silent system).[16]

The Prussian counting cards replaced both enumeration lists and the cumbersome method of check-listing. Whereas the lists recorded information on each inhabitant in a fixed line-by-line grid, the counting cards contained the complete dataset of one individual all on one page (fig. 7.1). In order to maintain the social and geographical context of the people represented by the cards, they were embedded in a broader set of forms in card format. The respective materials were distributed to and collected from each household in a so-called counting letter. This letter contained counting cards for each member of the household; a "household list card" in which all members present at the day of the count had to be specified by name and position within the household; a card asking for information on absent household members; a double-sided card with instructions to be read beforehand (as indicated on the counting card itself); and, finally, a model counting card showing how the blanks should be filled in.

Census taking in nineteenth-century Prussia became a kind of citizen science project: insofar as possible, the counted were expected to fill in the cards by themselves.[17] The method of self-inscription was put into place for the sake of efficiency, as fewer enumerators had to be deployed to note down the information. More importantly, Prussia's statisticians believed that the act of writing down one's own answers yielded more accurate data. For this very reason, the cards did not offer preprinted options to be ticked or circled. Almost all answers had to be entered as written words on dotted lines, which was considered to elicit more committed and trustworthy responses than checks or circles.[18] The method of self-inscription seems to have yielded satisfying turnouts, albeit at a high price. Cards were filled in with millions of ambiguous entries in different, and often not very dexterous, handwritings, spilling over into the space allotted to other answers and thus cluttering the cards' surface.[19] Deciphering and classifying content thus required a good deal of interpretation and experience.

The card's layout divided its surface in two parts. Eleven questions in the main body established each individual's data record. The card's header asked for locality, street address, and for three numbers to identify each card: a serial number, the number of the counting letter it belonged to, and the number of its respective enumeration district. Through such metadata, every card's individual dataset

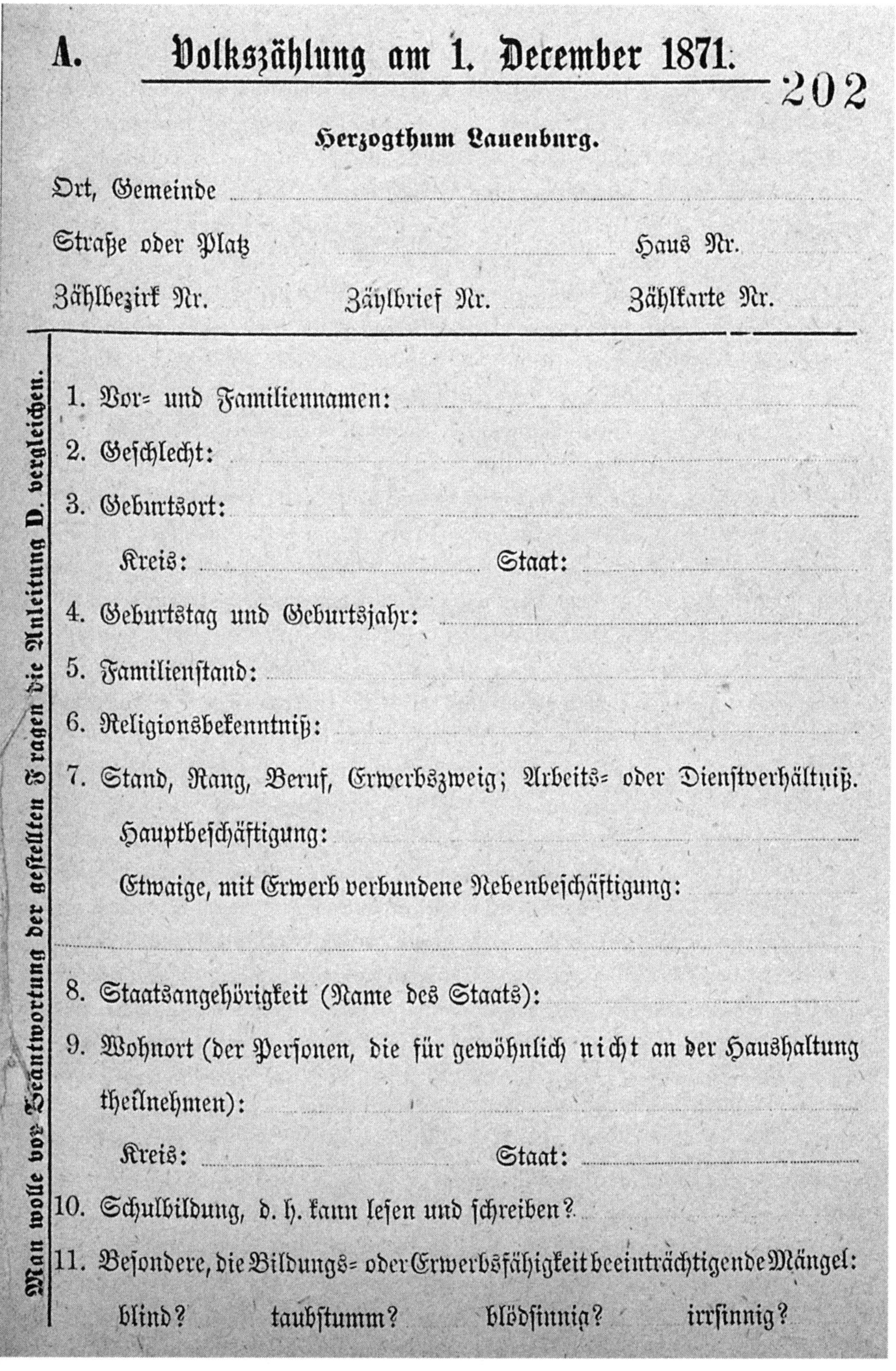

A. **Volkszählung am 1. December 1871.** 202

Herzogthum Lauenburg.

Ort, Gemeinde ....

Straße oder Platz .... Haus Nr.

Zählbezirk Nr. Zählbrief Nr. Zählkarte Nr.

Man wolle vor Beantwortung der gestellten Fragen die Anleitung D. vergleichen.

1. Vor- und Familiennamen: ....
2. Geschlecht: ....
3. Geburtsort: ....
   Kreis: .... Staat: ....
4. Geburtstag und Geburtsjahr: ....
5. Familienstand: ....
6. Religionsbekenntniß: ....
7. Stand, Rang, Beruf, Erwerbszweig; Arbeits- oder Dienstverhältniß.
   Hauptbeschäftigung: ....
   Etwaige, mit Erwerb verbundene Nebenbeschäftigung: ....
8. Staatsangehörigkeit (Name des Staats): ....
9. Wohnort (der Personen, die für gewöhnlich nicht an der Haushaltung theilnehmen): ....
   Kreis: .... Staat: ....
10. Schulbildung, d. h. kann lesen und schreiben? ....
11. Besondere, die Bildungs- oder Erwerbsfähigkeit beeinträchtigende Mängel:
    blind? taubstumm? blödsinnig? irrsinnig?

Figure 7.1. Prussian Counting Card, 1871, GStA PK, I. HA Rep. 77 Ministerium des Innern, Tit. 94 no. 132 vol. 1, p. 202, Secret State Archives Prussian Cultural Heritage, Berlin.

was not only traceable but also verifiable and chained to the individual it represented. At the same time, the metadata linked these "data doubles" inscribed on each page tightly to the material mass of all of the census cards.[20] Each referred to a specific social and geographical context, as the cards were re-collected by enumerators household by household, house by house, street by street, town by town, and district by district, a spatial lineup held together by the numbers on the cards and on each counting letter that structured the order of the material as it was packed into the wooden boxes for its return to Berlin. Wrapping paper, thread, and pen were used to parcel and label bigger units within each enumeration district, so that each box could swiftly be checked for completeness. Hence the cards were movable, albeit in a strictly orderly fashion.

Through the process of enumeration, identical preprinted paper forms sent out in wooden boxes were transformed into millions of distinct and irreplaceable carriers of original and oftentimes hard-to-decipher census information. Returned to the census bureau in Berlin, the boxes filled with this precious bulk functioned as a wandering archive of Prussia's inhabitants' data doubles, mirroring the geography of the state's enumeration districts down to every household and the people in them. Each box kept the state of its flexible papery content stable while being moved around. Within each box, the counting cards were not physically linked but held together through other paper devices and the metadata on the parcels, envelopes, and the cards themselves.

To turn the gathered individual data into numbers and tables, the cards had to be moved and sorted without ripping them out of their social and geographical context. Hence, as mobile and flexible as the cards were, they could not be shuffled at will. Carefully peeled out of the counting letters and separated from their accompanying material, counting cards required multiple rounds of sorting and counting depending on how many combinations of criteria had to be determined. This process followed a strict, predetermined procedure in which each new round of sorting and counting built on the order of the cards from the previous round. As the order of the cards was changed in a step-by-step fashion, so was their location, with wrappers and boxes securing the order of the moving data.

MAPPING ITINERARIES

The introduction of individual counting cards in 1871 went hand in hand with the centralization of the entire compilation process in Berlin. Months before the actual census took place, twenty-nine million cards and envelopes, control lists, and instructions for enumerators were printed and 2,500 wooden boxes built. The delivery of these materials to the census bureau set Prussia's census machinery in motion, transforming the bureau's premises in Berlin into a beehive of activity. As census clerks and their assistants checked and apportioned the paper material into the custom-made wooden boxes, the packing department teemed with packers, porters, and carpenters getting the boxes and their load ready for their journey to the respective enumeration districts.

When the 2,500 boxes with 375 tons' worth of millions of filled-in counting cards and various control lists from all fifty Prussian enumeration districts arrived back in Berlin in early 1872, the bureau's director hired three hundred additional men to handle the material. The bureau's headquarters had recently been expanded to an impressive three-story building, and two large empty rooms on the top floor had been allotted for the upcoming census work.[21] A huge shelf with six thousand pigeonholes, each capable of holding four stacks of up to five thousand counting cards, was installed to accommodate the cards in between sorting rounds. Furniture and working equipment, including custom-made laying tables for checking, sorting, and counting cards; washing pans and towels to ensure that all workers had clean, dry hands; chairs; ladders; petroleum lamps; ink; pens; pencils; rulers; and lead blocks to prevent the sorted piles of cards from getting mixed up were put in place for the newly recruited men.[22] With a sorting plan meticulously dictating each step of the procedure at hand, the bureau had estimated that processing all data on-site would need at maximum 236 days.[23]

In practice, the actual workflow unfolded quite differently from the original plan. The overall structure of the new building proved ill-suited to handle the colossal masses of enumeration material. In order to keep track of the boxes, each one had been assigned a fixed storage space either in the attic or in the basement of the building. Because of the movement to and fro, more room was needed for the sorting and counting of the cards.[24] An adjacent building was rented as a supplementary workspace; additionally, in order to reduce the constant motion, the bureau's director ordered that each box must not be moved more than three times. This directive implied a dramatic change for the tabulation process. Rather than exhausting the material by sorting the cards category by category in multiple successive rounds, the tabulators were to complete their task in three comprehensive counts, each including the compilation of numerous criteria.[25] This change of procedure made the sorting process both more complex and harder to control than the originally planned step-by-step method.

Notwithstanding such modifications, the bulk of sorting and counting the cards, for which so much more space was needed than had been anticipated, was eventually not done on the bureau's premises but distributed as piece labor to be completed in private homes. Home work thus emerged as a key component in the processing of Prussian census data over the following decades, growing in volume considerably each year. By 1895 the census bureau employed one thousand wage workers on-site and about three thousand piece-rate workers at home to compile the commerce, trade, and agricultural statistics of that year, the most comprehensive statistical investigation the agency had ever undertaken. From October 1895 to May 1896, no fewer than 27,760 boxes filled with cards were circulated between the census bureau and the workers' dwellings, transforming Berlin's neighborhoods into a fine-meshed net of actors, destinations, and routes on which the paper cards were moved to be assorted, controlled, and reassorted.[26] Apportioned in the wooden boxes by units of five or ten thousand cards (37.5 kilograms or 75 kilograms, respectively), the census records were circulated by horse carriage (and later by truck) to the workers' homes and their families, along with table forms and meticulous instructions explaining the sorting plan. Each box was sent back and forth up to three times between the bureau and individual apartments.

Archival sources from the 1890s reveal that the main part of the bureau's daily wage earners consisted of men experienced in accounting, banking, insurance, or secretarial work, and of former salesmen, teachers, or farmers.[27] All the men were married, and most of them listed as having additional income through at-home work. Census tabulators were entitled to take two hours' worth of work home each day, but the extra income listed in the bureau's register indicates that the extent of home work actually far exceeded this limit. In a few cases, piece-rate work from home almost doubled the worker's yearly earnings, revealing that while the bureau's male workers performed abstraction work on-site, their wives, children, or extended family members undertook the same or related work at home. Over the years, at-home compiling work was encouraged not only for the bureau's daily wage earners. At peak times, mobilization efforts were explicitly extended to include the spouses of the bureau's office clerks.[28] Additionally, regulations were abrogated that prohibited the bureau's workers to entrust people other than family members with at-home work. As a consequence, many households came to resemble busy office floors. P. Friedrich, one of the bureau's secretaries, provided extra work and income not only for his wife but also two sisters, a brother-in-law, an unemployed salesman, and four unmarried or widowed women from their neighborhood. This group of nine performed such excellent work that the bureau suspended all set limits for maximum income, arguing that at-home abstraction work caused extra expenses: Friedrich's wife had hired a maid to free herself for compilation work.[29]

The extension of the bureau's workforce beyond its own premises fueled the spatial expansion of census work, as figure 7.2 demonstrates.[30] Although by far not all of the workers' addresses could be determined, the forty locations depicted here point to some general patterns.[31] Somewhat clustered in the southwestern outskirts of Berlin's nineteenth-century city limits, most apartments to which home work was delivered were far from close to the bureau. On average, the distance amounted from four to eight kilometers, in some cases a good fifteen

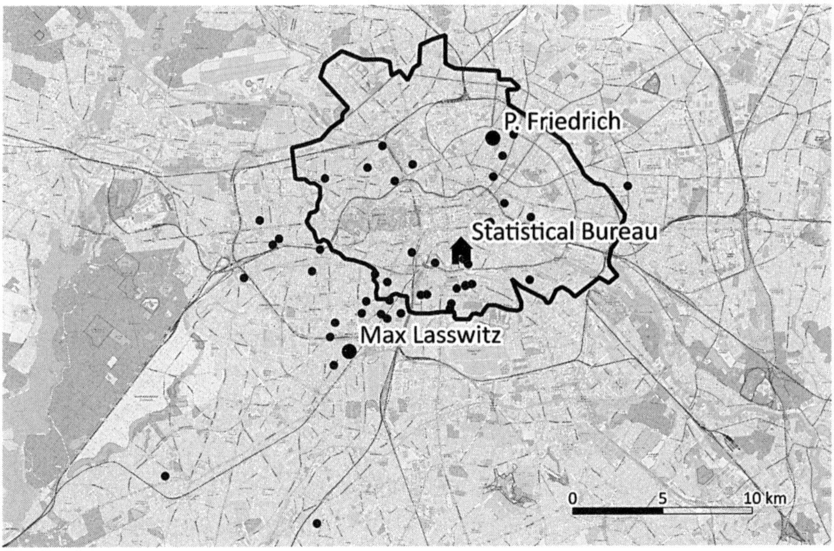

Figure 7.2. Selected census workers' homes, 1890 and 1898, applied to a contemporary city map of Berlin with the 1910 city boundaries included as reference. Laura Selle, Berlin, and © OpenStreetMap contributors.

to twenty. Most apartments were located in the modest middle-class neighborhoods of Tempelhof, Schöneberg, Wilmersdorf, and Charlottenburg, a few also in Prenzlauer Berg. The scope of this web of spaces, hands, and minds engaged in census tabulation gives a sense of the effort required to circulate the material and the challenges to control the handling of the cards, but also of the trust in the efficiency and accuracy of those working off-site. The census bureau targeted particular kinds of workers and their families for the skills and the space they had to offer. The boxes full of census records (see plate 10) were delivered to families, where—as one of the bureau's directors emphasized—wives, widows, daughters, and selected neighbors were able "to handle the precious original data with the necessary care, because they have an orderly home at their disposal."[32] Living in well-ordered circumstances was regarded as a skill, prerequisite to become an at-home census worker, as the orderliness performed in social terms was valued morally, and also epistemically (see plate 11).

TECHNOLOGIES OF ORDERLINESS

As Elaine Leong, Elena Serrano, and Elizabeth Yale explain in this volume, paper work in the home was common since early modern times, as note-taking and accounting figured prominently in the production, accumulation, and orderly preservation of household knowledge (see also plate 12).[33] Judith McGaw points out that such practices corresponded with the sorting or filing of things in the house, as closets, dressers, and pantries became organized idiosyncratically to the different functions of each room. With reference to the nineteenth century, McGaw calls such practices "feminine technologies," as "rules of appropriate order" had become increasingly evident in household management, and had also become more gender-specific as feminine virtues.[34] Census officials emphasized the intellectual endurance resulting from such virtuous technologies. Trustworthy women were lauded as being "meticulous, assiduous, mentally adept, and undiscouraged by the cumbersome tabulation work."[35] In fact, sorting cards may have been described as routine work, but was by no means self-evident or simply mechanical. A close look at each step involved in this technological process exposes what Bray calls "skill, judgment, creative compromise and distributed cognition" from women and men involved at all levels.[36] All workers, women and men alike, were trained for two to three days in the bureau before they were trusted to follow the bureau's instructions at home. Just like the workers toiling on-site, home workers mastered a complex set of tasks: while unpacking each box, they double-checked all material for completeness and then sorted and resorted the cards multiple times according to prescribed criteria. This filing work required interpretation, improvisation, and experience to offset the flaws in the cards' design, operations not unlike what Harry Collins has described as repair work performed by today's software users to overcome deficiencies so that technical devices appear to function well.[37] Not only did these operations entail deciphering millions of handwritings; additionally, ambivalent entries on the cards had to be made sense of and subsumed to the correct rubrics designed for the tables. Auxiliary means such as model tables listing over fifty possible terms with overlapping meanings indicate how difficult it was to classify the entries.

Once the cards were sorted, home workers filled in numbers in rubrics of the respective interim table forms in black ink, drew subtotals and totals, and checked for errors, then repacked the cards in their new order, wrapped and labeled each pack in a specific manner, re-packed the box, and returned all of the material to the census office for control. At peak times, flying revisers circulated throughout the city to check the paper work where it had been done. Hefty wage deductions punished slipshod abstraction work. However, control was first and foremost imposed through family ties via the heads of household employed in the bureau itself. The bureau presupposed that female dependents of the agency's workers and officials "enjoyed their guidance and supervision," as these workers were held responsible for impeccable results of their own as well as of their dependents' work.[38] For that matter, experience counted, with respect to long-term daily wage earners and officials in the bureau, as well as for their wives and relatives engaged in temporary paper work for the state, whenever it was needed.[39]

PAPER IN PARLORS

Few traces permit us to get a glimpse of how these arrangements played out for the workers and their dependents in practice, but the case of Max Lasswitz and his family is exceptionally well documented.[40] In 1896 the thirty-eight-year-old had retired from his position as a stationmaster's assistant in Stettin, a harbor city near the Baltic coast (today Sczeczin in Poland). After fourteen years of service, Lasswitz suffered from ill health. As a civil servant, he received an annual pension of 696 Marks a year, insufficient for him, his wife, and family to live on. The Lasswitz family relocated from Stettin to Berlin, where Lasswitz was quickly hired by the census bureau. The family rented an apartment outside of Berlin in Friedenau's Sponholzstraße No. 46, as marked on the map above.[41] For Lasswitz himself, the commute to his new employer was easy. Designed as a suburban country house colony, Friedenau had been equipped with a train connection to Berlin's city center in 1891. Friedenau was a typical middle-class neighborhood: the owner of Lasswitz's new four-story tenement was a locksmith running a workshop nearby. The building accommodated ten parties: two unmarried women living on pensions, two salesmen, a butcher, an upholsterer, a retired captain, and an editorial journalist with a doctoral degree.[42] All apartments within this building had a kitchen and at least two more rooms, and consequently, maintaining a parlor would have been the norm.

Lasswitz's employment at the statistical bureau turned out to be fortunate for the family.[43] In Stettin, Lasswitz had brought home 2,332 Marks a year, whereas with paper work in Berlin the family was able to make almost twice as much. During the year 1896–1897 Lasswitz earned 3,898 Marks in addition to his pension. With a total income of 4,594 Marks, he stuck out as one of the top earners among the census workers. However, the sickly pensioner himself made only between 80 and 100 Marks each month with wage work in the bureau itself, amounting to a mere 790 Marks that year. The family's main source of income during that time was through home compilation work, fluctuating greatly from month to month. In May 1896 the Lasswitz household raked in 348.09 Marks with piece-work from home, which equaled eighteen hours of daily work. In July, the household reached a peak income of 515.76 Marks corresponding to twenty-seven hours of work each day, a clear sign that just like in the case of P. Friedrich mentioned above, a group of people took to compiling census data. Peaks of work, however, were usually brief. In Lasswitz's case, the income through paper work in August dropped sharply to less than 300 Marks, and in October, the income through home work sank to 100 Marks for an average of 5.4 hours of daily work, before it rose again close to 300 Marks in November and December.[44]

The fluctuating income demonstrates how flexible the system of at-home tabulation proved to be, depending on the bureau's demand as well as the availability of each household member and additional helpers at a given moment. In the case of the relatively young, invalid pensioner Max Lasswitz, the paper work from home provided an opportunity for his wife and probably other (unknown) members of the household to make up for the declining income of their head of household, a fact all the more important as the family was deeply in debt when they arrived in Berlin.

The Lasswitz family was by no means the only one in financial trouble. In general, compensating for the risk of impoverishment inherent in daily wage work for the state was a crucial factor for census workers and their families to generate as much additional income through home work as possible. Numerous petitions to the bureau's directors render evident the lack of social security provided by the state. No matter how many years a worker had supported the bureau's urge for accurate population tables, there was little clemency when knees got arthritic from yearlong errands with heavy paper loads, illness hit, or accidents occurred. At times the few officials who held civil service positions collected money to help their most truthful day toilers in acute emergencies, but destitution loomed large, making tabulation home work a—probably welcome—necessity.[45]

Needing work, and being able to meet the bureau's expectations in spatial terms as well as with regard to middle-class orderliness, however, was a fine line that also long-term daily wage earners and their dependents could fail to walk. Max Hoyer, a retired low-rank officer praised for his diligent and trustworthy performance in the bureau, was never given home work during his yearlong employment due to his domestic circumstances. Maintaining a family of eight, with one child blind, another mentally disabled, and a sickly wife, Hoyer had accumulated horrendous debt. Then he passed away, and his widow Helene appealed to the bureau's director in October 1895 to help her out or provide her with work, claiming that "our misery is now beyond words, not even the most basic apartment, unheated rooms, sick children." But her plea was denied, as was a widow's pension, an entitlement not applicable to daily wage earners' dependents.[46]

Documentation on the Lasswitz household allows us to aquire a clearer sense of how each household functioned as input and output of the state's paper machine. Between February 1896 and March 1897 the Lasswitz home in Friedenau received thirty-two deliveries amounting to 1,491,575 counting cards.[47] The material arrived in four sequences, each of which required quite different attendance. The first eleven shipments kept the household busy from February to June 1896 with preliminary inspections of incoming census records, a process in which each and every counting card had to be read and checked for completeness and obvious mistakes. This is highly revealing, as inspecting cards before the actual count was supposed to always happen in the bureau itself. Hence in practice, house workers were even entrusted with more responsibilities than the bureau would admit in written form. The subsequent deliveries prove that householders had to be highly familiar with the whole process of tabulation, as deliveries required jumping from one round of counting to another, and backward.

Usually the household received a new box of material each time when a finished box was picked up. But pickup and delivery could also overlap if work was delayed, so that two boxes with material from different enumeration districts (and also in two different stages of processing, which the Germans called "exploitation") found themselves in the Lasswitz home, a fact that required extra care. Also, each box sat for at least two weeks in the apartment. The varying time spans between delivery and pickup of each portion of records suggest that the number of people working at a time or the daily time available for census work at home was in constant flux. The patterns emerging here lead me to imagine that the state's paper work may have been treated in the same way as needlework, or even an interrupted card game: I see the cards neatly ordered in packs, tables half filled with numbers, and the sturdy census box open and exposed in otherwise unused parlors, awaiting further attention and thus blending in with other household demands, adjusting to the rhythm of daily chores.

PARLORS AND NUMBERS

An important reason why the Prussian census bureau was so invested in the new paper tool, novel methods of tabulation, and engaging at-home tabulators was to achieve statistically accurate documentation of social change. Population tables and graphs were meant to lay the groundwork for state-sponsored social reform to address the social ills wrought by industrialization. What the numbers brought to light troubled Prussian statisticians, most of whom were engaged in social reform. A fine-grained analysis of Berlin's living conditions, published in 1869 as an early proof of what could be achieved with counting card technology, showed an alarming decline of affordable housing, especially in the segment of two- and three-room apartments, designed for the lower ranks of the middle class.[48] Other signs gaining texture through statistical sophistication (and biases) seemed equally distressing: the rising number of married women depending on employment, soaring divorce rates, and rampant mortality rates among children. With socialism on the rise, Prussian statisticians saw the middle class on the brink of disappearing.[49]

Saving the middle class became a top priority for social reform, best achieved by maintaining the "status quo" with the nuclear family and the orderly housewife as a vanguard, a fact widely reflected in economic theory, especially within the emerging field of historical economics. Ernst Engel, the director of the statistical bureau in Berlin, published a small monograph titled "The Account Book of the Housewife and Its Impact for the National Economy," in which he made such observations explicit.[50] In this treatise, Engel considered the "welcoming order of the household" as the foundation of a happy marriage, but also as a wife's fundamental contribution to the overall economy of the state.[51] Much in line with eighteenth-century concepts of *oeconomy* discussed in other chapters of this volume, Engel, as well as Gustav Schmoller, professor for economics at Berlin University and longtime chairman of the German Economic Association (Verein für Sozialpolitik), elaborated on the economic prominence of housewifery.[52] They defined domestic economy as complementary to economy as such. The functioning of state and society rested on a gendered concept of separate spheres. Whereas the male heads of household should earn enough to support their families, it was married women's first and foremost task to keep house within their own four walls, by fighting the daily struggle against debt, disorder, and the wear and tear of things. As hard and challenging as economists acknowledged this struggle to be, the very nature of housewifery (which daughters should be taught from early on) implied that women performed such cumbersome daily routines with infinite loving attentiveness for their husbands and family, and—by extension—for the state. It was for this very reason that housewives' skills of orderliness could be put to task for state matters.

Complementing the professional expertise of the middle-class male workforce on-site, wives, widows, and unmarried female staff working from home secured the functioning of Prussia's data processing, stirring conflicts of gender and class. The smooth expansion of their on-site workforce to the female gatekeepers of middle-class parlors encouraged statisticians to bypass social policy regulations that required bureaucrats to offer piece-rate employment first and foremost to retired servicemen. Many veterans voiced their anger, feeling betrayed by the bureau's decision to favor middle-class women over men who had risked life and limb for the Prussian state. Forced to justify their procedure, the bureau's directors produced reams of documents, stating repeatedly that efficiency rather than charity must rule the bureau's employment strategies—otherwise, the value of the work done would suffer and its completion delayed, resulting in cost overruns.[53]

For Berlin's census officials, home-based data work done mainly by housewives from the middle class was the most reliable, fastest, and cheapest way to produce numbers and tables for the Prussian state. What made this arrangement necessary and at the same time possible were the masses of paper slips on which the data traveled, creating bonds of trust, subservience, and control between the statistical bureau and countless private homes. Everyday technologies of orderliness rooted in housewifery were deemed so vital for creating the bureau's statistical knowledge that census officials condoned the principle of separate spheres meant to maintain social stability and the well-being of the Prussian state. They recruited their workers' wives to keep Prussia's house in order, trusting the efficiency of their female at-home workforce to the extent that they refused to mechanize their machinery of data power for decades to come.

EIGHT

Tracing Paper, the Posture Sciences, and the Mapping of the Female Body

BETH LINKER

At the turn of the twentieth century, experts in the social and medical sciences worried about rising rates of defective human posture in the United States. Some studies estimated that as many as 80 percent of Americans demonstrated faulty form.[1] In the view of these experts who believed that good organ function and character depended on upright carriage, slouching bodies threatened to bring about an increase in infectious diseases and an overall decline in the physical and mental fitness of the nation. By the turn of the twentieth century, standardized posture assessments became a defining feature of the burgeoning practice of the physical examination in the United States, an essential part of health assessments in schools, the military, industrial work sites, immigration inspection stations, and insurance agencies.

How did certain types of bodily gestures and comportment come to be seen as pathological, and how did it become a known and accepted fact—by experts and the general public—that a majority of Americans suffered from such defects? How did posture scientists distinguish between good and bad posture, and how has that knowledge been disseminated? By focusing on the development of posture inscription technologies in the early twentieth century, this essay explores the role that paper and gender played in the making of the posture epidemic. More specifically, I look at the work of Dr. Clelia D. Mosher (1863–1940), who, in her laboratory work as professor and director of women's physical education at Stanford University, became a forerunner in quantifying human posture.

In 1915, Mosher announced the invention of her "schematograph," a device that captured the human form on a piece of tracing paper (see figs. 8.1 and 8.2).[2] What made Mosher's schematograph unique in the modern posture sciences was its appropriation of rather conventional technology: the device was essentially a camera obscura outfitted with translucent drafting paper. The name of her apparatus, she liked to explain, derived from the Greek *schema*, meaning "form or shape," and *grapho*, "I write." Mosher self-consciously rejected photography—the predominant tool used by a group of mostly male posture experts at the time. She opposed the modern camera for many reasons, the most important of which was that she felt it robbed its female subjects of their privacy and placed too much power into the hands and eyes of the examiner and record keeper. A key difference between the schematograph and photography, then, was the kind of paper involved, which determined whether identifying features of the examinee would be recorded or not. Mosher's paper choice had political and epistemic import; it was a way, as the authors of this volume put it, for her to "negotiate gender and knowledge, and thus, power" in a science that was otherwise dictated by and for men.[3]

Mosher's insistence on a more basic inscription process in many ways confirms Bruno Latour's theory of paper work, for the schematograph allowed for a simplified visualization of the body, and thus could be construed as producing "harder facts" than camera photography.[4] To explain a similar point, Latour uses the example of astronomy. "The photographs of the skies," he writes, "are still much too rich and confusing for the human eye to look at; so a computer and a laser eye have been invented to read the photograph."[5] While Mosher's schematograph long predates computers and laser eyes, Latour's example is apt, for the schematograph similarly simplified the human body. Instead of photography, which rendered a "noisy" image that captured a complex topography of skin and musculature—not to mention individualizing features of the human face—the schematograph produced a clean, simplistic rendering of the human form, reducing it to a singular line on an 8½-by-11-inch piece of paper (fig. 8.2).

By the time of Mosher's death in 1940, photography would ultimately surpass the schematograph as the tool of choice for measuring posture in most schools and clinics nationwide. Mosher's appeal to produce a cleaner graphic rendition of the body and preservation of patient privacy was drowned out by mounting—

Figure 8.1. Dr. Clelia D. Mosher operating her schematograph. SC0011, box 11, folder 5, Stanford University Archives and Special Collections. Courtesy of Stanford University.

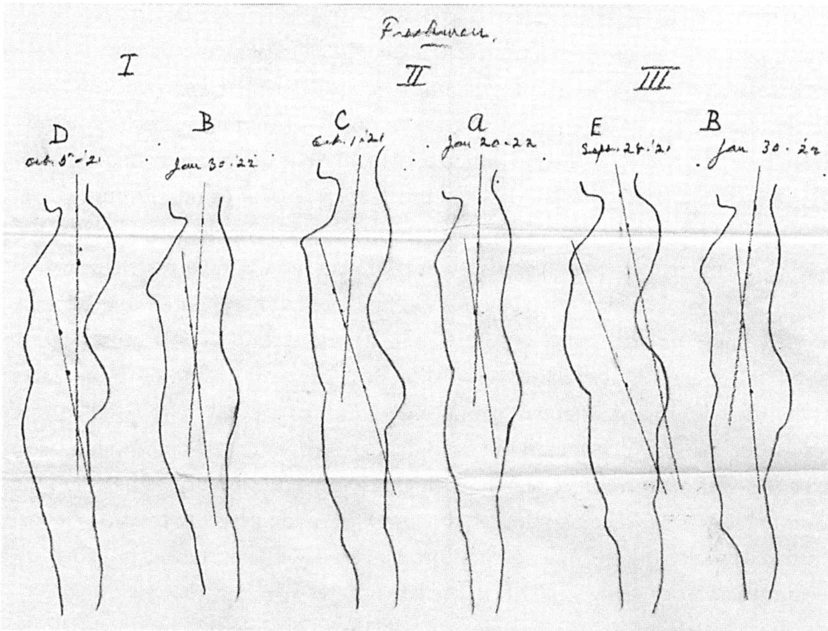

Figure 8.2. Schematographic before/after body sketches of Smith College students' posture; attached to a March 13, 1922, letter of Dr. Joel Goldthwaite to William A. Neilson, Joel Goldthwaite, College Archives, Smith College, MA.

and largely male—concerns for getting rid of the "I" in the schematograph process. Photography came to be seen as producing "harder facts" than other, competing posture-recording technologies, even though concerns for patient privacy dogged the posture sciences up until the end of the twentieth century.

ENGINEERING THE SCHEMATOGRAPH

Unveiling the invention of her schematograph in a 1915 issue of *School and Society,* Mosher insisted that all of the "uneven shoulders and hips, the drooping heads, the winged scapulae, and the flat chests" among American college women needed correction. Mosher was by no means the first physician or scientist to fret about bad posture. Many late nineteenth-century experts in the evolutionary sciences, physiological sciences, and population health had a vested interest in the vertical stance of human beings. Commercial interests in posture also abounded at this time. Popular newspapers and magazines regularly featured articles and advertisements for clothing, furniture, shoes, undergarments, and exercise regimes that could improve and support good posture. Like many of her generation, Mosher saw posture as a window onto both individual health and overall fitness of the human race.

What made Mosher unique among posture experts was her focus on women's health and her desire to collect data that would scientifically establish posture norms specific to female anatomy. Female scientists had "taken over the [posture] methods in use for men," she wrote in 1915, "without inquiring very extensively into the real effects [on women]."[6] Mosher considered herself a "physiological feminist," and as such, devoted most of her career to undermining what she deemed pseudoscience myths of anatomists and physiologists who insisted that women were physiologically more fragile than men.[7] University men such as Edward Clarke, Herbert Spencer, and G. Stanley Hall maintained that because women had smaller brains, monthly menstruation, and weaker musculature, they were inherently debilitated and thus best suited for a life of domesticity where they could be protected from the demands of the public sphere.[8] During her career, Mosher conducted physiological experiments in respiration, muscle strength, and posture, all of which highlighted physical similarities between the sexes. Much of Mosher's comparative work was made possible by the fact that she had a faculty appointment at Stanford University, one of the few private institutions to accept female students in the late nineteenth century.

Mosher's posture work was a piece of her larger physiological feminist agenda. More specifically, she believed that teaching women about good posture—including conducting research so as to establish sound posture norms—would help to normalize menstruation.[9] Slouching posture was blamed for many ills in both men and women at this time—from dyspepsia and constipation to tuberculosis. Certain physiologists and evolutionary anatomists insisted that since *Homo sapiens* was originally accustomed to moving on all fours, in a quadruped position, they suffered from pooling of the blood in the abdominal region, especially when erect.[10] The anxiety about inherent abdominal weakness in human beings easily lent itself to the widely popular concerns about the health of women and their reproductive system, especially among white college women, who, bearing fewer children on average than their mothers, became the object of blame among adherents to the belief in race suicide. While Mosher did not endorse race suicide arguments, she did believe that women had a "very abundant supply of blood in the pelvic organs," an anatomical makeup that "favor[ed] physiological congestion," especially during menstruation.[11] Mosher reasoned that posture health would improve abdominal circulation in women, and thus result in diminution in the duration of their flow and prevent menstrual pain and dysmenorrhea. An advocate of gender equality in the workplace and in higher education, Mosher insisted that women had to physiologically manage their menstruation rather than taking precious sick time from work and school.[12]

In order to better establish this correlative relationship between posture and menstruation for other scientists as well as for American college-bound women across the country, Mosher sought to create a tool that would both legitimate her work as a scientist and prove to her skeptics that posture-focused physical fitness programs improved the health of women. She also needed a device that would help motivate and recruit all of the college-age women to see posture training as necessary to their well-being, and to ultimately overcome the culturally prevalent taboos concerning menstruation and female reproductive health.

Photography was one option. The camera entered the clinic in the mid-nineteenth century, soon after the device's invention. Certain physicians used photography to zero in on particular pathological abnormalities, focusing the lens on certain body parts and—in the case of photomicrography—cells, essentially erasing the individual subject. Or as Daniel M. Fox and James Terry write, in the late nineteenth century "the affliction rather than the patient became the focus of [medical] photographic interest."[13] Other, more holistically minded physicians saw photography as a logical extension of anthropometry and physiognomy.[14] At Harvard, Dr. Dudley Allen Sargent photographed scantily clad students so as to further his work in determining "ideal" bodily proportions, research that resulted in the "average man" and "average women" sculptures that he exhibited at the 1893 World's Columbian Exposition. Sargent and other physicians like him participated in a wider cultural trend in the United States whereby photographs of seminude athletes and illustrations of fig-leaf-adorned men became a visual norm in a host of books and periodicals.[15]

Physicians and advocates of the new public health initiative of the early twentieth century also employed photography to educate the populace about the dangers of germs and the importance of self-care. The American Posture League (APL) was no exception. Founded in 1913, the APL—an association of like-minded physicians, physical educators, efficiency engineers, and social reformers who came together to combat the slouching epidemic—produced one of the more popular posture wall charts of the era using camera photography. Employing a triptych format, the APL reproduced photographic images of women, children, and men who modeled the posture ideal, so that onlookers could learn by example.[16] Each chart included three photographs of the same model; the photograph in the middle—the mean between the other two extremes—demonstrated correct posture, while in the other two the model demonstrated "exaggerated (military) posture" on one end, and "incorrect (slouching) posture" on the other. The APL intended these charts to be hung at home, in schoolrooms, and at the workplace, so as to inspire self-surveillance among—and policing between—onlookers, and to bring about a visual consensus concerning the virtues of good form as well as the vices of unaesthetic slouching.

Mosher had distinct misgivings about using photography in the clinic despite its prevalence and promise. In addition to citing prohibitive costs, especially for female scientists who were perennially underfunded compared to their male counterparts, she believed that photography endangered a woman's "personal reserve."[17] A female scientist who researched indelicate topics such as sexual intercourse and menstruation, Mosher believed that she had clear insights into recording posture of female college students.[18] Nude posture photography, in her view, threatened a woman's privacy, both during the actual exam and in the resultant artifact produced by the camera, a paper object that could be easily reproduced and circulated, landing in the hands of other scientists and the public at large, exposing an unsuspecting subject's flesh.[19] Knowing that many college women would likely feel uncomfortable posing nude, or even scantily clad, for a camera, Mosher developed the schematograph with the goal of creating a more agreeable examination experience, a matter that would lead more women to consent to the exam and also, Mosher argued, lend itself to more accurate data collection.

Having rejected photography, Mosher sought the help of Everett P. Lesley, a professor of mechanical engineering at Stanford and known for his work in airplane propeller design between the two world wars.[20] By collaborating with Lesley, Mosher moved the posture sciences into the direction of cartography, a fact made most evident by her paper choice. While silent on the actual construction and design of the camera obscura part of the schematograph, Mosher made clear that 8½-by-11 inch sheets of "Keuffel and Esser Colonna Tracing Paper" was a defining feature of her device, setting it apart from photography and other camera technologies.

Even though tracing paper could be construed as a feminine tool akin to that of sewing patterns, the schematograph paper—and specifically the Keuffel and Esser brand—would have been seen as a gendered-male tool (see plate 13).[21] Established by two German immigrants in 1867, the Keuffel and Esser Company (K&E) was one of the first American-based companies to specialize in the production and sales of drafting instruments and supplies. Beginning in the 1890s K&E became known for its slide rule, an early calculation device essential to the work of every practicing twentieth-century engineer, at least until the invention of the pocket calculator. Mosher, then, adopted one kind of masculine technology (drafting equipment used by male engineers) to offer an alternative to another competing masculine technology (photography), with the goals of preserving female privacy and advancing the posture sciences (and her status within that specialty).[22]

Mosher's decision to use tracing paper dictated the setup of her entire posture recording system, from the layout of the exam room and the maintenance of paper records to the construction of the schematograph and the training of examiners. Foremost in Mosher's mind was that tracing paper afforded subject privacy, and more specifically it completely obscured the "individuality of the face."[23] But the preservation of privacy occurred at other levels, too. By virtue of the materiality and dimension of the paper, full-body posture tracings could capture only a limited amount of detail, and the face, in proportion to the rest of the body, was less than one-eighth of the entire image being captured. Moreover, because control of image making was literally in the hand of the examiner who held the pencil, facial features could be purposely blurred, crudely outlined, or not included at all.

Mosher believed that preserving subject privacy improved the accuracy of her posture results. With their identities secured, examinees felt "less under observation," she argued.[24] Posture exams necessitated that students undress entirely, or at the very most, don skintight bathing suits or undergarments. Such exposure flouted Victorian-era mores, especially for women of virtue who were expected to maintain their modesty in dress and behavior. Knowing this, Mosher created a laboratory environment that put her subjects "at ease," and thus made it more likely that the examiner would get a more accurate recording of a subject's "natural" posture. If her subjects were still uncomfortable, she offered them the option to pose behind a light screen instead of posing in front of a dark screen, as was often done in camera photography.

Because tracing the human form was a rather simple and uncomplicated process, the exam was relatively quick, taking only a couple of minutes. The scientific benefits of a swift exam were of particular importance in the posture sciences. In addition to making the subject feel less exposed, a speedy exam was necessary in order to avoid subject fatigue and thus faulty data. Dispelling any worries that such speed might lead to sloppy work or inaccuracies, Mosher contended that to make a subject pose any longer than a few minutes would actually jeopardize the scientific process. Her concerns were well founded, for her male counterparts who insisted on the superiority of camera photography frequently complained that, due to slow shutter speeds and other complexities that led to longer exam times, it was often "difficult to capture the exact human profile," and that the camera made these profiles "difficult to repeat exactly, due to the swaying and shifting of the student after he has been posed," thereby leading to blurred bodily lines.[25]

Accordingly, Mosher maintained that her schematograph offered a better image than photography did. Even when posture photographs turned out to be clear, they offered distracting details, such as facial features, undergarments, and skin tone, "noise" that, Mosher held, detracted from the scientific study of body alignment. Additionally, the schematograph permitted the examiner and examinee to occupy entirely separate rooms. Mosher advised placing the schematograph "in a closet . . . with an opening in the wall just large enough for the lens." This method of use, Mosher maintained, had a "distinct advantage," for rather than gazing at the subject through a lens, the schematographer would need to keep her eyes down, to stare at the paper upon which the subject's image was thrown, with full concentration on rendering the correct postural outline.[26]

Not only did the schematograph produce "harder facts," according to Mosher, it also lent itself to better communication of those facts. The resultant tracings could be easily understood and read by, as one colleague put it, the "uninitiated."[27] Embarking on a science that had not yet achieved acceptance among the scientific elite, let alone the wider public, Mosher wanted easily readable images that required little, if any, interpretation. The tracings served this purpose not only in their simplicity of the data presented but also by the ease with which the paper tracings could be shared from person to person and scientist to scientist, without putting privacy and the integrity of scientific fact in jeopardy.

THE SCHEMATOGRAPH IN PRACTICE

Posture examinations at Stanford took place in the school gymnasium that served as Mosher's laboratory. On a designated week of every fall semester, Mosher set up her schematograph equipment, examining approximately twenty-four students per hour, five days of the week. In preparation, she would pencil an axis and vertical line on each piece of paper to more easily create duplicate schematograms at the time of the exam. The extra copies would often be handed to her examinees so that they could continue their posture work at home.[28] In effect, Mosher put a crucial part of the university health record into the hands of the students themselves, which is especially noteworthy since schools that used photography held sole proprietorship over such material.

What students did with their own personal schematograms is difficult to ascertain. One could imagine that, similar to phrenological charts that Carla Bittel investigates in her essay in this volume, schematograms became "objects of sociability" in the home. Lillian C. Drew, director of corrective gymnastics at Columbia University's Teachers College, noted that during the school's annual "posture week," students seemed very "curious to see [their] own 'graph'" and that once the students had their own tracing, they would often show them to friends, drawing comparisons between one another.[29] Drew's report would have cheered Mosher, for posture crusaders like herself hoped that educated women would develop a sense of self-governance concerning their own health and that of their future families and children. The push for mothers to inculcate good posture at home continued steadily even after Mosher's death. During the interwar and post–World War II years, women's and parenting magazines regularly featured articles offering tips on how to get kids motivated to stand upright. One magazine even advised mothers to chart the geography of their children's bodies at home by creating full-body silhouette tracings using proper lighting, white walls, and large sheets of paper.[30]

Compared to the private use (or disuse) of schematograms at home, we know far more about how these records were handled in university gymnasiums and clinics. At Stanford, students viewed their schematograms in consultation with Mosher a week or so after the exam. During these meetings, Mosher would make additional markings, often with a colored pencil, to indicate posture failings, pinpointing the areas where the student needed to improve. In certain cases, Mosher took a schematogram that represented perfect posture and placed it on top of a tracing of an examinee with bad posture, shading in the difference between the two so as to demonstrate more clearly the nature and degree of the subject's posture deficiencies (see plate 14). At other times, Mosher would layer a follow-up schematogram on top of a baseline schematogram of the same individual, creating a "before" and "after" record of a student's progress.[31]

Mosher also mandated that her students view their own schematograms during their monthly flow, when they were to report to the gymnasium for additional physical education instruction and lectures. As a matter of course, students would remove any constricting garments—especially any type of corsetry—and lie in a recumbent position with knees flexed in order to relax the abdominal region. Mosher then instructed these women on how to attain abdominal muscular consciousness through deep breathing exercises, a regime that would become known at other institutions as "the Moshers." While the students attended to their diaphragmatic breathing and muscular control, Mosher lectured them on proper, erect standing, showing them schematograms as well as images of ideal posture. Pictures of Greek statues further emphasized the importance of good abdominal tone to the proper functioning of the respiratory and reproductive system.[32]

The ease with which posture silhouettes could be reproduced and manipulated in terms of their color and size from the original record afforded many possibilities when it came to instruction. Take, for example, a 1926 photograph of a posture clinic at the Massachusetts General Hospital in Boston (see fig. 8.3). Here, two experts, identified as public health nurses, engage their pupils in various methods of posture instruction and measures. Mirror work, as it was called, formed the basis of many posture clinics, for it was believed that in conjunction with hands-on cueing (as is happening in this photograph, left), students could both see and feel proper form for themselves. The respirometer (photo, right) was also a mainstay in posture instruction, for erect carriage was thought to bring about improved lung function and vice versa. Of particular importance, however, is the wall chart hanging in the background. The silhouettes for the chart likely came from student records at a local university. Both good and bad posture

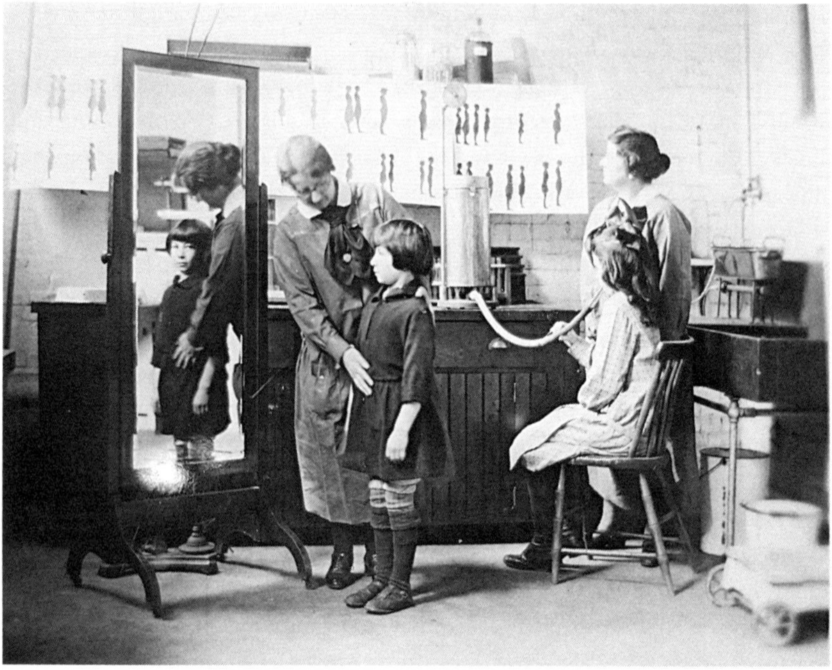

Figure 8.3. A 1926 photograph of a posture clinic in the Department of Orthopedics at the Massachusetts General Hospital in Boston. Note the various technologies of posture measurement and training here, from the respirometer and posture wall charts to mirror work. Of special note is the posture wall chart, silhouettes likely created from either the schematograph or an early model of Fradd's silhouetteograph, showing that no matter the device, paper images could be reproduced in seemingly similar ways. Courtesy of the Subject file Orthopedic Department, Massachusetts General Hospital Archives and Special Collections, Boston.

are represented here, serving as a constant source of visual instruction for what should be mimicked and what should not.

Posture tracings also passed through the hands of college administrators, donors, and trustees. In 1922 Smith College president William Allan Neilson expressed hearty appreciation for a set of student records that Dr. Joel E. Goldthwait, friend and overseer of the college's posture program, sent his way (see fig. 8.2).[33] "The tracings you sent me are extremely interesting," Neilson wrote to Goldthwait. "I shall be glad if I may keep the tracing you sent me," he continued, "to show to people who need to have the point of our physical training explained to them."[34] Goldthwait agreed that the tracings could be very persuasive. "No one would question the fact," he wrote to Neilson, that each of the three students achieved significant gains, for every follow-up tracing demonstrated a "more efficient individual, as well as a more aesthetically perfect creature."[35]

Posture advocates frequently pointed out how their work enhanced human beauty. The promise of physical enhancement held wide cultural appeal in early twentieth-century America, when consumerism, popular eugenics, leisure, and youth culture would all begin to take hold. But the beauty that posture experts promoted was a distinctly different vision from that of fashion houses and social elites. Just before the APL formed, American newspapers and magazines enthusiastically reported on a new fashion trend: the "debutante slouch." Such reports relied on numerous illustrations of woman in a distinctly relaxed pose: slumped back, one foot outturned, hips asymmetrical. These models relied on the relatively unrestricted dress of the day, which allowed for greater freedom of movement.[36]

Combating such media depictions, which they saw as injurious to a woman's health, posture experts hoped to visually train a new generation of white—and eventually also black—middle-class college students in "posture conscious" beauty through the use of schematograms.[37] Posture tracings recorded a woman's physical appearance and thus served as a material object that she could examine carefully—stare at, even. Posture experts hoped that this kind of self-study and self-scrutiny would persuade women that slouching was a vice. Mosher had high hopes for the ability of her technology to visually persuade. "When the girl sees how her defects detract from her good appearance," Mosher wrote, "she is sufficiently interested so that with her cooperation and the aid of her gymnasium work, the correction is only a matter of time."[38]

RECEPTION AND CONTESTATIONS

A year after Mosher announced her invention, the APL heralded it as "one of the most practical and valuable contributions that has been made to corrective and educational work on the subject of posture."[39] Indeed, in a 1916 membership letter the secretary of the APL, Dr. Henry Ling Taylor, reported that the demand for the device from schools, universities, and physicians was overwhelming. By the 1920s and 1930s the schematograph had even found its way into all-male settings, including the navy.

But some experts complained that the device was imprecise and fell short of objective scientific standards. One of the most vocal critics was Norman Fradd of Harvard University.[40] Using Mosher's device for a 1916 study, orthopedic surgeons at Harvard estimated that 80 percent of freshman had faulty posture. In response, the university hired Fradd to tackle the epidemic, outfitting him with state-of-the-art equipment, including a schematograph. From the outset, Fradd expressed dissatisfaction with Mosher's device, complaining that the "hand tracing camera is poor at best and, although used extensively, is inaccurate and wastes much time."[41] To create a more precise and efficient measuring tool, Fradd teamed up with a photographic expert, first from a private Boston company and later with Eastman Kodak, to come up with a method that would improve upon the schematograph's accuracy yet still remain affordable and able to be used without the services of a photographer who needed to oversee printing and developing.[42]

In 1923, Fradd replaced the Mosher schematograph with his own device, which he called the "Silhouetteograph." Eastman Kodak Company loaned Fradd a new type of camera that used sensitized bromide paper instead of film and plates. The advantage of bromide paper was that after only a few minutes in a developer solution and fixing bath, the photographed image could be rendered without any training. With the use of a high-intensity floodlight, making the space between the subject and the camera as dark as possible, the resultant image ended up being, according to Fradd, a "clear sharp silhouette," a white human form on a black background.[43] Fradd believed his device to be superior to Mosher's because the camera eliminated "the personal equation" of the examiner and the handmade tracings.[44]

Fradd made only one copy of the posture record, keeping it for himself. According to the published record, he expressed no intention of duplicating his posture record for his students, and even if he had, bromide negatives were not conducive to reproduction. Given the context in which Fradd was practicing, his choice of technology reveals the degree to which gender dictated posture practices at the time. He did not have to worry, for example, about managing menstruation. Nor did he have to use his posture silhouettes to legitimize the place of men at American universities. Fradd (and his students) enjoyed more political and cultural liberties, and thus his posture assessments served the more singular goal of creating a legitimate science to improve the health of university men.

Moreover, Fradd had his sights set on aggregating large visual datasets so that he could produce standardized models, manikins, and charts. Ultimately, he wished to create a universal system of grading posture, assigning numerical values to each bodily deviation. In so doing, he made the posture sciences seem more esoteric and something that could be conducted only by trained professionals. To be sure, he wanted posture graphics to be easily read by both commoners and experts, for he wished to popularize the posture sciences and make inroads into posture correction as a public health effort. But unlike Mosher, he wanted to store the paper records for his own exclusive use and analysis.

Another essential difference between Fradd's and Mosher's practices lay in gaining student consent to perform such studies. Whereas Mosher remained sensitive to maintaining a female student's "personal reserve," Fradd's description of his posture-recording procedure does not mention concern for student modesty or privacy. Posture exams at Harvard took place after a student's annual physical exam, when students were required to remove all of their clothing. Naked, each student was escorted to the posture recording room where he would be instructed to adopt a "natural" pose, followed by a "best posture" pose, thereby extending the examination time period from what Mosher had prescribed.[45]

To the untrained eye, the posture images produced by Mosher's schematograph and Fradd's silhouetteograph do not appear to be that different. Mosher's device created a white silhouette with a penciled outline; Fradd's similarly produced a solid white silhouette against a black background. In both cases, the posture records were highly manipulable and reproducible. Indeed, some adoptees of Mosher's device created tracings on black silhouette paper (white on the reverse), then cut out the outline and pasted it on white card (see posture wall chart in background of fig. 8.3).[46] Fradd must have engaged in a similar kind of paper handwork, for the composite images that he featured in his classification photographs are the opposite of the actual silhouetteograph images. Silhouettes, then, whether produced originally on bromide paper or tracing paper, were inherently flexible, for through the craft work of shading, cutting, and tracing, the positive and negative spaces could be inverted depending on the end goal of the examiner and what he or she intended the onlooker to notice and see.

Despite the relative similarities and inherent affordability between Mosher's and Fradd's end products, the adherents of each technique emphasized their differences, resulting in a thinly guised, gender-laden contest that revolved around issues of subjectivity and objectivity in science and the paper practices inherent to it. On the whole, male physical educators and physicians questioned the accuracy of Mosher's female-derived and female-oriented technologies to measure posture. In Mosher's system, the instructor created the posture record by handling tracing paper and pencil, and thus had a fair degree of autonomy and control when it came to rendering a posture silhouette. To Fradd and his followers, this degree and kind of human intervention seemed to put the accuracy of posture recording into question.

While Fradd may have rejected Mosher's use of tracing paper, he did not rebuff paper work all together. On the contrary, Fradd embraced different forms of paper and reproduction technologies to advance his own posture knowledge production. With his large dataset of posture photos, Fradd created composites and posture norms that he reproduced in books, academic articles, wall charts, and posters.[47] Fradd's comfort with translating his images into numbers and his use of posters and manikins made his device and work more popularly known than Mosher's. Mosher never produced a posture chart and grading system to accompany her device, for her focus seemed—comparatively speaking—to be more on an individual level, sharing schematograms with each of her students to render physical improvement.

Because of his rather catholic use of various paper systems, Fradd's method of posture recording superseded the schematograph by the time of Mosher's death in 1940. The US Department of Labor and the US Children's Bureau employed Fradd's device and his classification system in their own studies.[48] And with the help of Eastman Kodak, Fradd enjoyed widespread publicity in an array of health care and education journals. For example, in a 1928 issue of the *American Physical Education Review*, Kodak placed a "See Yourself as Others See You" advertisement for Fradd's silhouetteograph, claiming that the device was so affordable that is was easily "within the reach of any progressive school, college, or clinic."[49] Fradd also engaged in self-advertisement—in a 1924 *American Journal of Nursing* column, he wrote to urge the journal's readers to order his wall charts and posture manikins through direct mail.[50]

Because the silhouetteograph was essentially a film camera—and one produced by the country's leading camera retailer—Fradd's method lent itself quite easily to the widespread adoption of photography, so that by midcentury almost every institution—including the US military and women's colleges—created nude posture photographs rather than silhouettes.[51] In a 1943 retrospective study of the posture sciences, two physiologists maintained that the published literature on the topic "would require a monograph of large proportions," including hundreds of thousands of photographs of men, women, and children. When, by the early 1970s, student protests brought an end to the practice of nude posture photography, Mosher's schematograph was little more than a forgotten artifact of American history.

While it is tempting to attribute the demise of Mosher's device to technological determinism—that is, camera technology became ever more affordable and accessible by the 1930s and, in many sciences, became the norm for objectivity—this line of thinking fails to account for Mosher's rather powerful epistemological stance on the posture sciences. Mosher papered over her subject's bodies for compelling moral and epistemic reasons. She had good reasons for believing that hand tracing produced harder facts than camera photography. Moreover, her paper choice reflected her commitment to maintaining the "personal reserve" of the subject being examined over and above concerns about compromising her objectivity, or in Fradd's words, the "personal equation" of the examiner.[52] In her view, a lack of examinee privacy and comfort jeopardized scientific accuracy more than the handwork involved in the act of tracing.

While Mosher's device may have been set aside for an ostensibly more modern and accurate tool, her criticisms of photography would continue to haunt the posture sciences for the remainder of the twentieth century. Already in 1937—as camera photography became the norm, and patient images became less like a silhouette and more like personal photographs—criticisms erupted around the use of this technology in the university setting. One father whose daughter had been photographed at Wayne University brought charges against the university. "Photos of Nude Co-Eds Stir Up Detroit Furor," read newspaper headlines across the country.[53] The Detroit School Board ended up siding with Dr. Irvin W. Sander, who instituted posture photography at Wayne University, despite the fact that other institutions were beginning to feel that nude photos could compromise student privacy.[54] Had Wayne University used Mosher's schematograph, it is conceivable that no such suit would have been brought against the school administration.

As concerns about student consent and privacy grew through the mid- to late twentieth century, the practice of posture photography seemed increasingly out of step with the times. The student and parent outrage over such exams became so great that, by the 1970s, most schools and universities closed down posture labs, no longer requiring a posture assessment as part of a student medical exam. And yet, in many ways, nude posture photography had always been a questionable enterprise from the outset. Ellen Kelly, supervisor of physical education in Williamsport, Pennsylvania, articulated the perennial fear faced by all posture experts in the post–World War II era. Although Kelly had been successful in instituting nude posture photography of all elementary, junior, and senior high school students in her district, she admitted that the "superintendents and teachers may hesitate to initiate posture photographing because of public opinion." Kelly's advice was to her colleagues was to persevere while also "avoiding publicity," especially when examinations were taking place.[55]

The demise of the posture sciences, then, was due not to a waning belief in the correlative importance of erect stance to good health; indeed, one can still find the logic behind the posture epidemic—bad posture equals bad health—at work in today's professional scientific journals and popular health guides.[56] Students who protested posture exams in the mid-twentieth century did so because they found the *method* of collecting and recording such data reprehensible. When this important aspect of the history of the posture sciences is understood and appreciated, it makes the story of Mosher and her schematograph all the more compelling. At the very least, it allows one to consider that an alternative historical path that might have been taken, one in which Mosher's peers had adopted her paper tracing technology and thus avoided nude posture photograph scandals altogether.

COLOR PLATES

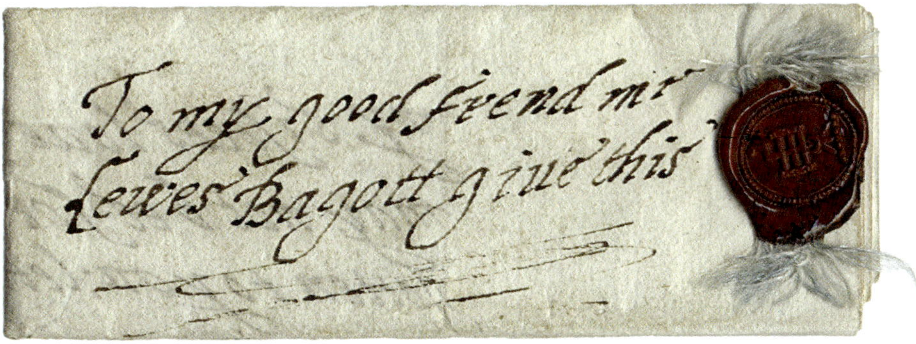

Plate 1a. A pleated letter secured with light blue embroidery floss wrapped around the long edge, with a wax seal on the front and back of the packet. This letter has been refolded after the floss was cut in order to demonstrate what the superscription side of the packet originally looked like. Autograph letter from Jane Skipwith to Lewis Bagot, circa 1610, MS L.a.852, Folger Shakespeare Library, Washington, DC.

Plate 1b. Detail from a full-page image of the blank panel (the side without the address) of a pleated letter. The letter was secured with pink silk embroidery floss originally wrapped around the fore-edge and the long edge of the packet (now cut) and sealed on both sides. Autograph letter signed from Lady Jane Ashley to Walter Bagot, circa 1609, MS L.a.19, Folger Shakespeare Library, Washington, DC.

N. 2

A Mon tresher
Monsieur le Duc
D'Anjou

Queen Eliz: to ye Duke
D'Anjou. Orig.

Plate 3. Detail of autograph letter signed from John Donne to his brother-in-law Sir Robert More, August 10, 1614. The letter was written on Venice paper with gilt edges. MS L.b.539, Folger Shakespeare Library, Washington, DC.

◄ Plate 2. Address leaf of a silk-flossed autograph letter, in French, signed from Queen Elizabeth to the Duke of Anjou, 1581, Cecil Papers 135/10, Hatfield House. The paper, with an eagle watermark most similar to Eagle.268.1 (gravell.org), was probably produced in Frankfurt. Reproduced with permission of the Marquess of Salisbury, Hatfield House.

►► Plates 4 and 5. Two autograph letters with decorative borders signed from Catherine de Bourbon to Queen Elizabeth, ca. 1595 (Cecil Papers 147/64 and Cecil Papers 147/62, Hatfield House). Reproduced with permission of the Marquess of Salisbury, Hatfield House.

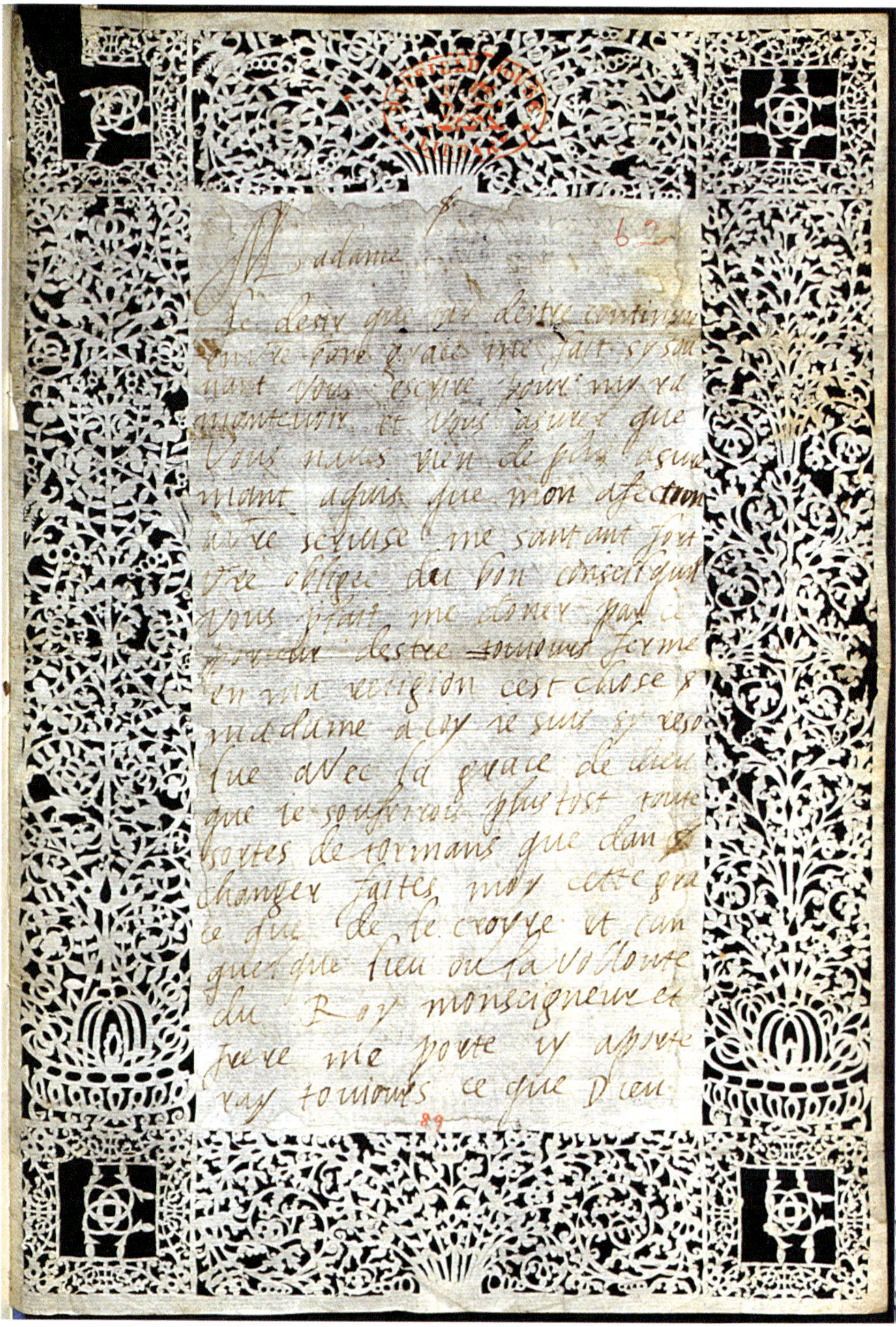

62

Madame
Le desir que iay destre continue
en vre bonne graces me fait sy sou
vant vous escrire pour my ra
mantevoir et vous asurer que
vous naves rien de plus asure
mant aquis que mon afection
a vre servise me santant fort
vre obligee du bon conseil quil
vous plait me doner par ce
porteur destre tousiours ferme
en ma religion cest chose &
madame a quoy ie suis sy reso
lue avec la grace de dieu
que ie soufrirois plustost toute
sortes de tormans que dan
changer faites moy cette gra
ce que de le croyre et can
quelque lieu ou la vollonte
du Roy monseigneur et
pere me porte iy aporte
ray tousiours ce que Dieu

89

8 64

Madame Jay
ette prie par le frere de se
porteur de vous suplier
come ie fais treshumblemant
de luy acorder la charge
quil vous avoit pleu doner
a chateau martin ie vous
asure madame quil est tres
home de bien et quil vous
rrandra fort fidelle service
favorisesle en ma faveur
et croyes que vous en
seres aussy bien servye
que vous saves ette mal
de lautre que ma [illegible]da
tion luy serve et me
tesmougnes en cela que
vous me faites lhoneur
de maymer faisant pour

91

RIDICULOUS TASTE OR THE LADIES ABSURDITY.

Pub.d accord.g to Act of Parll.t July 13th: 1771 by Wady No. 80 Strand. & R. Sayer at the Golden Buck Fleet Street.

Zweytes Muster

Erster Versuch
mit
Tannenzapfen.

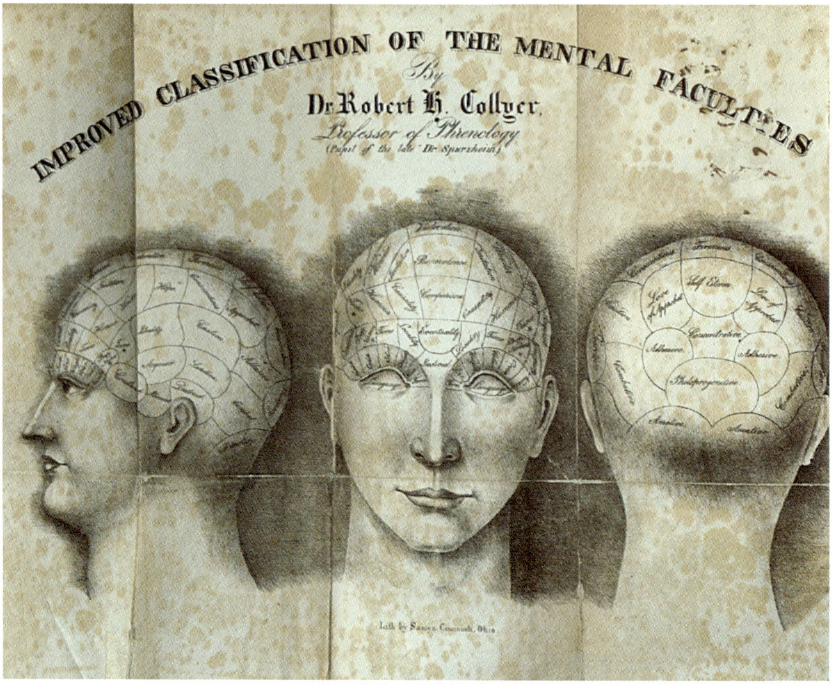

Plate 8. Foldout engraving of a phrenological head in three views, inside front cover of Robert H. Collyer, *Manual of Phrenology, or the Physiology of the Human Brain. Embracing a Full Description of the Phrenological Organs, Their Exact Location, and the Peculiarities of Character Produced by their Various Degrees of Development and Combination* (Cincinnati: Alexander Flash, 1838), Dittrick Medical History Center, Case Western Reserve University.

◀◀ **Plate 6.** Curling with French brown paper. Ladies' increasingly large hairstyles were ridiculed in the later eighteenth century, as were the supposedly effeminate hairdressers who created them. Caricature by Matthew Darly, "Ridiculous Taste or the Ladies Absurdity," 1771, hand-colored engraving. © Trustees of the British Museum, UK.

◀◀ **Plate 7.** "Second Sample, First Trial with Pine Cones" from Jacob Christian Schäffer's paper trials to replace linen pulp, published in his *Neue Versuche und Muster das Pflanzenreich zum Papiermachen und anderen Sachen wirthschaftsnützlich zu gebrauchen* (Regensburg, 1767). Realizing that the paper resulting from this trial was too gray and thick to serve as writing paper, Schäffer suggested that it could be used to satisfy the contemporary craze for wallpaper. He even appears to have hired a painter to decorate the border of each sample made with this material, as all of the surviving exemplars feature a different pattern. 8 TECHN II, 2176:3 RARA, courtesy of the Niedersächsische Staats- und Universitätsbibliothek Göttingen.

Plate 9. The "Casket of Knowledge," including a top card with embossed profile of a phrenological head, and corresponding cards. Mrs. L. Miles, *Phrenology and the Moral Influence of Phrenology: Arranged on 40 Cards, Illustrative of the System* (London: Ackermann and Co., 1835), BF870.P577 1835, History and Special Collections for the Sciences, UCLA Library Special Collections.

Plate 10. Prussian census box as used for the 1871 census, reconstructed to size by Norbert Massuthe, Berlin (2016). Photo by Laura Selle. The larger box contained 70 kilograms of enumeration material (10,000 counting cards), the smaller one 35 kilograms (5,000 cards). Thousands of boxes like these were transported across Berlin for at-home tabulation.

Plate 11. Data processing for nineteenth-century Prussian census statistics was mainly done by housewives in their orderly parlors. Eduard Niese, *Blick in die Wohnstube eines Mauermeisters* [Parlor of a master mason], 1891, watercolor, Inv.Nr. EB 1964,231. Courtesy of City Museum for the History of Hamburg, Germany.

Coffe.
Zucker
Teh

Plate 13. An early K&E catalog advertising tracing paper next to an image of a drafting room, a space depicted as exclusively male. *Partners in Creating: The First Century of K+E* [Keuffel & Esser] *1867/1967*, ed. William Franklin (1967), 7. Courtesy of Hoboken Historical Museum, Online Collections Database.

◂◂ Plate 12 (a–c). Homemaker writing in her account book. Porcelain figure by Johann Joachim Kändler, Meissen, ca. 1756. © Stéphane Piera/Musée Cognacq-Jay/Roger-Viollet.

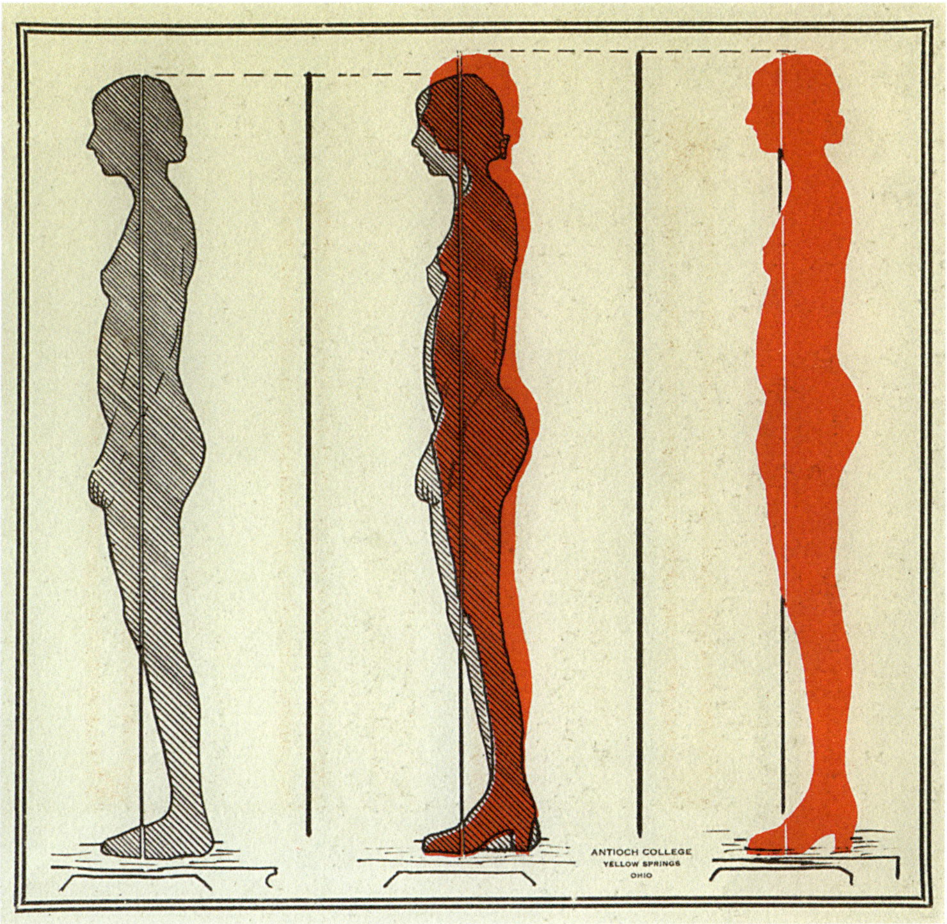

Plate 14. Antioch College, *The Effects of Modern Shoes upon Proper Body Mechanics, 1924–1931* (Antioch College, 1931), figure 12.

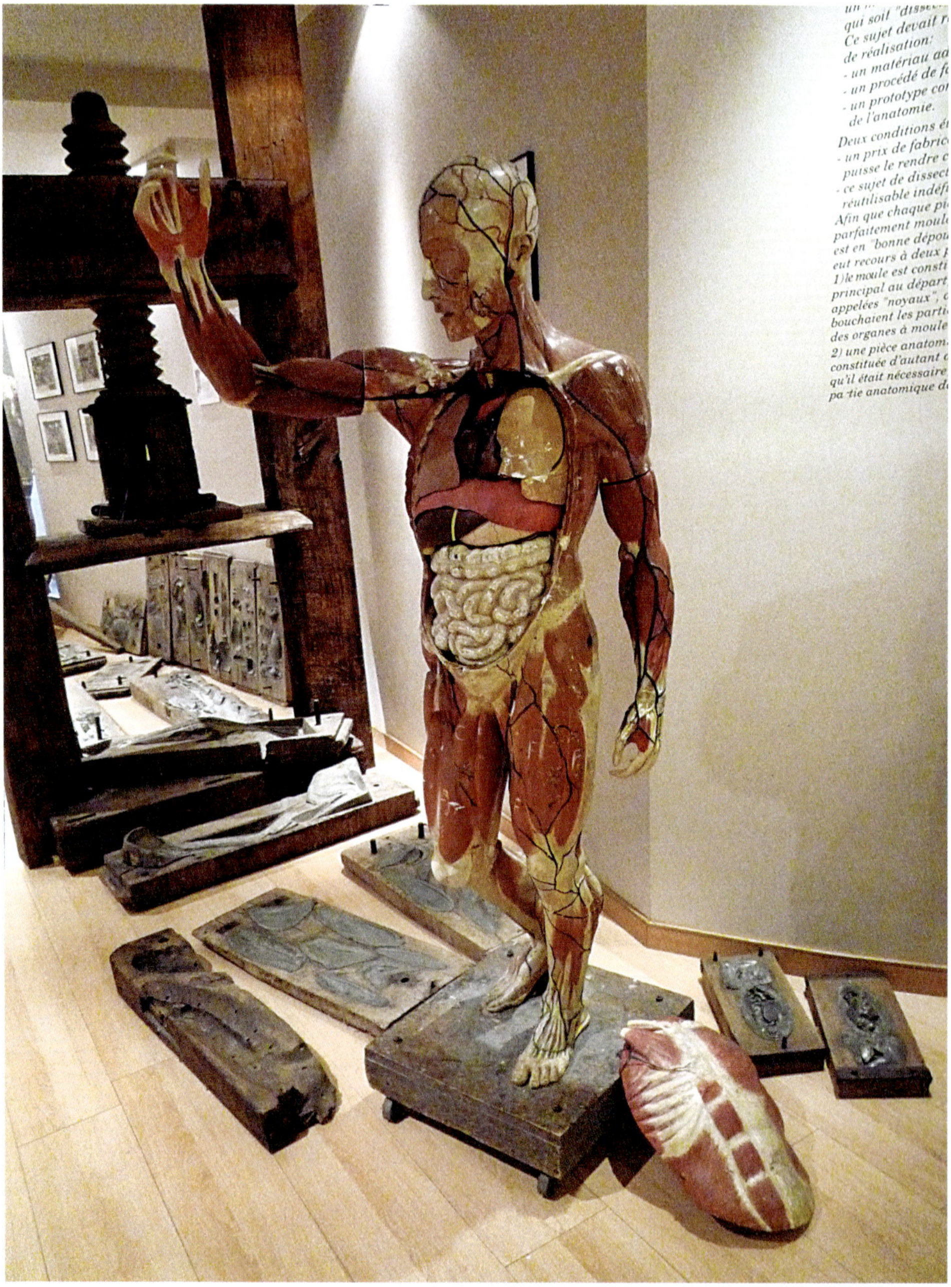

Ce sujet devait r
de réalisation:
- un matériau a
- un procédé de f
- un prototype co
de l'anatomie.
Deux conditions é
- un prix de fabric
puisse le rendre c
- ce sujet de dissect
réutilisable indéf
Afin que chaque pi
parfaitement moul
est en "bonne dépou
eut recours à deux p
1)le moule est consti
principal au départ
appelées "noyaux", c
bouchaient les parti
des organes à moule
2) une pièce anatom
constituée d'autant
qu'il était nécessaire
pa tie anatomique d

Plate 16. These paperboard spinning wheels model the US population in 1930. The top wheel represents data on mortality among white women. Abigail Balfour, Dan Bouk, Erin Burke, and Christy Mills reconstructed these wheels from descriptions in the writings of biomathematician Alfred Lotka. Photo by Dan Bouk.

◀ Plate 15. Molds used in model production of Auzoux's papier-mâché factory and a model of the male body. Musée de l'Ecorché, Le Neubourg. Photo by Anna Maerker.

Plate 17. Household paper workshop in Jiajiang, China, ca. 1990. Photo by Jacob Eyferth.

PART III

Knowledge, Power, and the Everyday

NINE

A Letter Is a Paper House

Home, Family, and Natural Knowledge

ELIZABETH YALE

When the naturalist and cleric John Ray died in his home in Black Notley, Essex, in January 1705, his widow, Margaret Ray, would have washed his body and dressed it in a woolen shift, perhaps with a bit of black embroidery at cuff and collar. She tied the shift at the feet and then wrapped the corpse in a woolen winding sheet (not linen, as required by a parliamentary statute that bolstered the English woolen industry and reserved linen for papermaking).[1] John's body, by his own request, was nailed into a plain wooden coffin so that his corpse could not be seen during his funeral service.[2] Afterward, his body was buried in the parish churchyard, even though the parish minister offered him the chancel, a place of greater honor. Margaret may have distributed funeral tokens (such as gloves or hatbands) and invited the neighborhood to share in ale and cake to mark John's passing and celebrate his life.[3] John, adhering to an old-fashioned charity, set aside a bit of money for distribution to the poor of the parish following his death. Aside from a further small bequest to Trinity College Cambridge for the purchase of books for the library, he left his worldly estate, including his books, papers, and correspondence, to Margaret.[4]

In early modern England, the investigation of nature, like death, was a domestic activity.[5] In this chapter I link the two, investigating scientific practices in the Ray household and among their wider correspondence before and after Ray's death. Family members and colleagues used paper technologies—letters—to develop both natural knowledge and gendered representations of the labor and relationships through which that knowledge was constructed. I argue that a consideration of the Ray household leads us to a more expansive view of what counts as scientific labor in the early modern household.[6] In examining Margaret Ray's role in the domestic spaces in which naturalists met, conversed, shared observations, and conducted experiments, one feature of scientific labor emerges in particular: female family members' roles in providing learned hospitality. Following the paper trail after John's death, through Margaret's disposition of his paper effects, her textual labor comes to light as well.[7]

In recent years, historians of science have opened up the home in early modern Europe as a site for producing natural and medical knowledge. Early contributions in this field present scientific workspaces as distinctly masculine spaces within an early modern English household otherwise dominated by women. In her study of John and Jane Dee's household, Deborah Harkness excavates a world in which a wife's duties conflicted with her husband's work: from Jane's perspective, John's alchemical, astrological, and magical natural philosophy brought disrepute upon their household. Natural philosophy rested uneasily in the home.[8] In Steven Shapin's description of the Boyle household, Katherine Ranelagh may have hosted natural philosophers at her table, but the door that closed off Boyle's study separated the work of hospitality from the work of natural philosophy.[9]

Subsequently, scholars have illuminated a fuller picture of household experimental, natural philosophical, and healing labor, one that incorporates men's and women's roles and the exchanges and collaborations between them.[10] Yet we lack a full account of women's social and textual labors in the history of science. In the Ray household, women's labor and presence were integral to the provision of scientific hospitality, the relational foundation of collaborative scientific work.[11] Women's roles in creating spaces for natural philosophical endeavor were reflected in their representation in male naturalists' correspondence. In their letters, male naturalists used references to the women and children in their lives (though not servants) to build, on paper, a social space in which to conduct natural historical conversation by letter.[12]

While a patriarch lived, his family's involvement in his work was often obscured. But in negotiating the paper work around a husband's death, widows drew these collaborations out into the open. Margaret Ray was not alone among natural philosophical and medical widows in her engagement with her husband's textual legacy. Alice Culpeper sold her husband Nicholas's unpublished medical writings to the printer Peter Cole and used her husband's name to market and sell "potable gold," a patent medicine.[13] Emma Willughby footed the bill for engraved plates for her husband, Francis Willughby's, posthumous natural historical works, edited by Ray.[14] Margaret Flamsteed, who had worked with her husband, Astronomer Royal John Flamsteed, at Greenwich Observatory, published his *Atlas Coelestis*, promoting it against the incomplete version published by Edmund Halley in 1712.[15]

When we seek to understand the labor of experimental science, natural history, and medicine in the early modern world, we should be alert to the archival traces that indicate the many ways in which women labored in scientific households. This is important through the twentieth century (and beyond), as the essays in this volume show. Women led and implemented the reproduction surveys Dan Bouk studies, yet male investigators (and the archive itself) erased or subordinated their work. Within this context, male naturalists' papers and books—the material traces that form today's scientific archives—were a distinctive legacy that widows capitalized on to ensure economic survival, promote their own careers, and secure their husbands' posthumous reputations. Male colleagues, in turn, through their negotiations over papers, produced their own gendered social relations and ways of being.[16]

THE INVESTIGATION OF NATURE AS FAMILY LABOR

John Ray met Margaret Oakley in the household of Francis Willughby, where she served as governess; they married after Willughby's death. Initially, after Willughby's death, Margaret "taught Lord *Middleton* his letters, and to read *English*," while John prepared Latin and Greek instructional materials.[17] In 1679, following his mother's death, John and Margaret Ray settled at Dewlands, the family home in Black Notley, Essex.[18] Even as Ray's health declined, their household was a lively center for natural historical research, a meeting point for British naturalists, via correspondence and in person. Regular visitors included apothecary Samuel Dale and physician Benjamin Allen. From farther afield came Hans Sloane, antiquary John Aubrey, and William Derham. Through correspondence and travel, Ray's connections extended across Britain and into wider European and even global worlds.[19]

Though the Rays lived on a restricted income—an annual bequest from Willughby's will was their primary source of money—with a house of at least eight rooms, they were, by the standards of the day, relatively "house-rich" (fig. 9.1).[20] Dewlands was a two-story, L-shaped house surrounded by gardens and an orchard. The long arm of the *L* (from left to right, facing the house), was divided into parlor, hall, lobby, and kitchen with bedchambers above. Behind the parlor, a brewhouse, with Ray's study above it, made the short arm of the *L*. Ray's books, papers, and natural collections, which included living organisms, were

Figure 9.1. A woman standing on the threshold of Dewlands, the Ray home, Black Notley, Essex. John Ray, *The Correspondence of John Ray*, ed. Edwin Lankester (London: Printed for the Ray Society, 1848), facing 459.

distributed in various "cupboards" and chambers, primarily on the house's upper floors.[21]

Natural historical labor was also distributed around the house. Guests were likely entertained in the first-floor hall or parlor.[22] In the hall, the most public space, guests enjoyed food and drink. In the parlor, the head of household and his wife would rest, but they might also entertain guests.[23] Ray also "worked" in these spaces. Though he wrote upstairs, amid his books, specimens, and papers, he opened and read letters by candlelight as company chattered around him in the evenings.[24] In wintertime, when cold worsened his leg pains, he rested frequently by the fire, perhaps in the hall.[25] Death, too, happened in the hall. Ray's mother died "in the Hall-Chamber, about Three of the Clock in the Afternoon," on March 15, 1678.[26]

In the 1690s the family's main natural historical focus was the investigation of insects. Ray provided intellectual stimulus for this work, examining and classifying insects, with a particular focus on moths and butterflies. To create a systematic account of insects, he collated his own notes with those of his deceased friend Francis Willughby. The result of this work was posthumously published in 1710 as *Historia insectorum*.

The family's home was a living laboratory of insect metamorphosis. Ray's daughters brought him butterflies and moths. In 1696, his daughter Mary, age twelve, captured butterflies for him—one as it flew around in their garden on an early June evening.[27] Margaret trapped two male moths in an upper room in the house after they entered an open window, attracted (John theorized) by the scent given off by a caged female moth just emerging from its cocoon.[28] On another occasion, John recorded a moth that Margaret discovered and brought to him.[29] Ray's brief references condense his family's expertise and familiarity with insects, as they identified and trapped creatures that would be new additions to his collection.

To understand the role that Margaret Ray and her daughters played in creating spaces for scientific conversation, let us consider a visit that the elderly John Aubrey made in summer 1692. Aubrey sought Ray's opinion on treatises he hoped to publish on natural history and antiquities, shared interests of theirs. The visit also gave Aubrey an opportunity to meet Ray's neighbors with similar interests, including Samuel Dale, who assisted Ray in his research and sorted Ray's natural historical papers after his death, and the family physician, Benjamin Allen.[30]

How might the Ray household have provided for its scientific guests? As Amanda Flather has reconstructed in her study of home and household, using Essex archival records, although the use of domestic space for socializing varied depending on household status, wealth, seasons, and the occupations of its inhabitants, the lady of the house maintained overall responsibility for the home, controlling the keys, and thus access, to the house and its various rooms.[31] Domestic sociability tended to be the province of married couples. It was mixed-gender, with women and men inviting friends, relations, and associates to socialize informally and formally. People met in the home for many reasons: to celebrate life cycle events, discuss business (such as unresolved debts) or neighborhood matters, provide mutual or charitable aid, and conduct religious meetings.[32] Domestic socializing was an important avenue along which married women maintained neighborly relations.

When Aubrey visited, Margaret Ray and her daughters saw to his physical comforts.[33] They, as well as any female servants the family employed, would have found Aubrey a bed. Given the size of the home and the number of available beds, perhaps Aubrey shared a bed with another member of the household, even a servant. On the other hand, given the overall composition of the family—one man plus five women—perhaps the Rays' daughters bunked up together in order to free a bed for Aubrey. The women would have provided for the gatherings of local virtuosi inspired by Aubrey's visit. Typically, men would have socialized after supper at the alehouse, while women stayed home.[34] However, given John's limited mobility, Ray, Dale, Allen, and Aubrey more likely enjoyed each other's company at Dewlands. Perhaps Mary ran to Allen and Dale to invite them for an evening's socializing, while Margaret provided refreshments.

In such meetings, naturalists wove the net of personal connections through which they advanced their studies. Following Aubrey's visit, which established an acquaintance between Aubrey and Allen, Ray drew upon Allen to answer Aubrey's questions about mineral springs in Epsom. Allen planned to collect fossils on an upcoming visit to the springs; Ray promised Aubrey an account of his finds.[35] Though Ray mediated the connection, Aubrey's personal familiarity with Allen thickened the social bond along which information traveled.

In these gatherings, the women did not stand silently by, pouring small beer.[36] Aubrey and the Ray women exchanged gifts and promises of "service" that confirm Margaret and her daughters' participation in the household's learned work and the conversations his visit engendered. When he visited, Aubrey brought the Ray women a microscope—probably a small glass lens—that could be used to observe insects' sexual organs.[37] Such observations were the only sure way to distinguish males and females of a species; Maria Sybilla Merian may have used a microscope in this way.[38] John Ray, skeptical of the value of microscopes, used them infrequently.[39] But as a gift to the Ray women, it suggests Aubrey was attuned to Margaret and her daughters' participation in the household's scientific work.

The visit wove the Ray women into the collaborative net of service in which Allen and Dale participated. In a letter thanking Aubrey for his visit, Ray wrote that Dale and his own "Wife and young girls" "return their service" to him, a customary promise to be useful to one's correspondent, if at all possible.[40] Though John mediated the exchange, Margaret and her daughters could return their service to Aubrey in concrete ways. Ray wrote, including both the Ray women and Dale, "we shall be careful to enquire concerning ancient Coyns or Medals found hereabouts."[41] Notice of such finds would have been valuable contributions to Aubrey's antiquarian research.[42] John, given his physical condition, was relatively fixed in the household, but Margaret, her daughters, and Dale would have been out and about, on market rounds and errands and providing health and charity assistance around the neighborhood.[43] They, more than John, would have had opportunities to ask others about rare antiquarian finds.

A LETTER IS A PAPER HOUSE

In the previous section we reconstructed the four-dimensional social space of the home (three dimensions plus time) from its three-dimensional projection in correspondence (two dimensions on paper, plus time, articulated through a series of letters). Margaret maintained the household in which she and John entertained scientific guests. Through her labor—the work of hand, mind, and body—Margaret and her daughters fashioned the spaces in which scientific friendships were created and sustained. Let us consider now the reverse operation: the projection of the household onto a sheet of paper.

In writing letters, correspondents represented the four-dimensional social space of the home onto the three-dimensional series of letters. Ray, and other male naturalists working in households, framed their letters with references to wives and children, opening and closing their paper conversations with brief domestic reports and inquiries after their correspondents' families. In between, they dissected natural historical observations and specimens, the latest publishing news, and colleagues' activities in male natural philosophical spaces, including the Royal Society and London's coffeehouses. Male naturalists bounded the space of paper conversation with references to each other's families. If a letter was a paper house, wives and children stood at the threshold, as with the woman at the door of the Rays' Essex home (fig. 9.1). Naturalists' representations of family in their letters demarcated the social space in which they conversed.

It was not only learned male correspondents who operated in this fashion: in this volume, Heather Wolfe shows us how the "decorated borders" of well-off Elizabethan women's letters recalled the use of the "decorative arts in domestic spaces of large country houses." Further, Simon Werrett and Elaine Leong, in their essays, show us how both women and men did similar things with paper: wrapping medicines and foods, for example, to filter and preserve them. Yet these practices were (and are, by some modern historians of science) valued or denigrated according to the gender identity of the practitioners or their gendered context. Learned early modern men were not unique in using the home to frame their letters, but it is important to remember that they used this framing to mediate their social relations with each other. We are often tempted to disregard philosophers' domestic settings as irrelevant to natural philosophical work, in part because of their association with women.

In a series of letters written between John Ray and the physician Martin Lister in the early 1670s, Ray asked after Lister's pregnant wife, hoping that she was safely delivered and the child well.[44] Ray, at this point, was unmarried, living in the home of Francis Willughby, his natural historical collaborator and friend.[45] Lister opened and closed letters with reference to Willughby's health and friendship. Lister created the social space of the letter with references to Willughby, placing him in the position of Ray's "wife."[46] In the interval between Willughby's death and Ray's marriage, Lister opened and closed with inquiries after "the Good Lady and sweet Children" and "your excellent Lady," meaning Emma Willughby and her children.[47]

In his correspondence with Ray, Hans Sloane used gifts to build his relationship with Ray's household. Sloane sent presents of sugar, explicitly to Margaret, not John, as John wrote in thanks to Sloane in May 1692:

> Munday last I received your kind Letter attended with a rich Present of sugar to my Wife: They were both very gratefull & acceptable; only the latter was too great & inadequate to any merit of mine to be received without some shame; as well the quality as quantity concurring to render it valuable. You have so highly pleased & obliged my Wife that she is much in commendation of your generosity, & returns you her humble service & hearty thanks; wishing that you were heer to partake of some of the effects of your kindnesse.[48]

Sugar loaves were a common, but still prestigious, food gift in early modern England.[49] But Sloane's gift, though given for John Ray's sake, was to Ray's wife. And Ray disclaimed his own worthiness in the face of such a rich present—he was unable to receive it without shame. This self-effacing politeness is typical of Ray's letters, but it suggests some discomfort here: Sloane pushed the bounds of propriety to develop a connection with Margaret. Margaret's response, offering Sloane sweetmeats baked with the sugar, confirms the bond his gift established between them.[50]

We will return, in a moment, to Ray's shame. Let us note, for now, that Sloane's correspondence is shot through with gifts, not only food but also recipes and health advice.[51] He keyed his presents to his correspondents' status. While Margaret Ray received plain sugar, the wealthy Cassandra Willughby (Francis Willughby's daughter) received a recipe for "cashew sugar." As Lisa Smith has noted, cashews were a "rare and imported food considered to be high status"; the gift signaled Willughby and Sloane's "mutual interest in botanical matters."[52] Plain sugar was no less welcome to Margaret. Though increasingly available as English planters imported more enslaved Africans to the West Indies and brought more land under cultivation, it was still a relatively expensive treat that Sloane likely had special access to, given his wife's Jamaican plantations.[53] In these gifts, we see how male naturalists represented and engaged each other's households in their letters, creating a domestic social space for scientific correspondence.

THE RELICT AND THE RELICS

In the previous two sections, I examined how gender and knowledge were crafted in the home and its paper projection, the letter. As in many other essays in this volume, "following the paper" has revealed unexpected connections between men and women, between masculine and feminine paper practices. Now I turn to the expression of gender relations, and the role of the domestic in knowledge production, through the posthumous negotiations over Ray's papers and books.

Working with Samuel Dale, Margaret broke her husband's paper estate into four categories: Francis Willughby's papers on insects, which John had borrowed from Francis's son, Thomas Willughby, and used in the preparation of *Historia insectorum*; papers and treatises from which short extracts might be drawn for publication in *Philosophical Transactions*; Ray's correspondence; and his library of printed books. With diverse ends in mind, but within a shared framework of charity and mutual obligation, Sloane, Derham, Dale, and Thomas Willughby negotiated with Margaret for access to this material.

Soon after John's death, Sloane initiated the conversation with a letter to Margaret probing her on the disposition of her husband's writings and books.[54] In response, she fixed first on the annual bequest from Francis Willughby's will, writing that she hoped that Thomas Willughby, out of consideration for her circumstances, would be willing to make one last payment.[55] With three teenage daughters, and little income besides the stipend, Margaret wrote, "My circumstances must be but straight."[56] She linked the bequest to the return of Francis Willughby's insect manuscript. Should she approach Thomas directly, she asked Sloane, to ask him to extend the bequest and to let him know that she would transmit his father's papers to him securely?[57] After years of cultivating relationships with the Rays and the Willughbys, Sloane, apparently, advised against it, instead positioning himself as intermediary. Perhaps he had fostered bonds of obligation between himself and Margaret knowing that she might one day control access to John's papers. Perhaps he felt grateful to her, the wife whose skilled care—which included daily dressing Ray's leg wounds—had kept her invalid husband alive, making possible his continued engagement in natural history, in his later years.[58]

Sloane interceded ably with Thomas Willughby. Writing to Sloane, Willughby expressed sorrow at John Ray's death.[59] In fall 1706 Willughby sent twenty pounds, asking that Sloane obtain for him "my Fathers manuscript of Insects."[60] Margaret Ray wrote repeatedly to Sloane to remind him to tell Willughby that she held his father's papers. And she returned them only after Willughby had sent the money (fig. 9.2).[61] When Ray sent the insect papers to Willughby, Sloane independently sent a further present of twenty pounds, transmitting it, at Ray's request, through London bookseller Samuel Smith, the publisher of many of her husband's books.[62] Ray held back the papers as a bargaining chip in what was, for her, a negotiation, though she did not couch it in those terms.

While she delicately managed the transfer of Willughby's insect manuscript, Margaret readily made John's papers—"my husbands papers"—over to his colleagues.[63] Dale, she wrote, would give Sloane an account of the manuscripts (excluding "some letters") and "publish what he finds fitt."[64] Dale selected some extracts and wrote an account of John's archive for *Philosophical Transactions*,

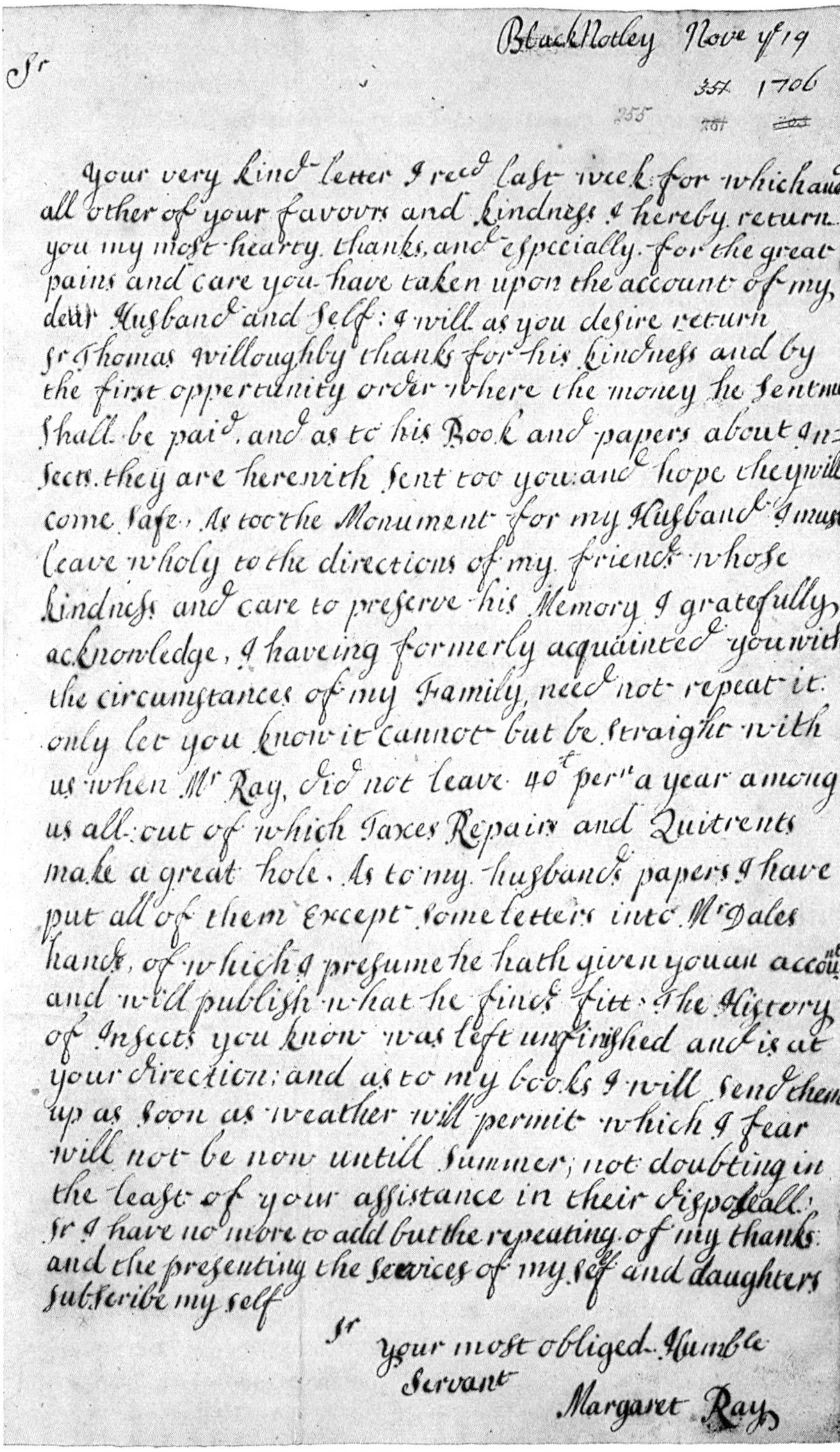

Black Notley Novr ye 19

Sr 1706

Your very kind letter I recd last week for which and all other of your favours and kindness I hereby return you my most hearty thanks, and especially for the great pains and care you have taken upon the account of my dear Husband and Self: I will as you desire return Sr Thomas Willoughby thanks for his kindness and by the first oppertunity order where the money he sent me shall be paid, and as to his Book and papers about Insects they are herewith sent too you: and hope they will come safe. As too the Monument for my Husband I must leave wholy to the directions of my friends whose kindness and care to preserve his Memory I gratefully acknowledge, I haveing formerly acquainted you with the circumstances of my Family, need not repeat it. only let you know it cannot but be straight with us when Mr Ray did not leave 40l per an a year among us all: out of which Taxes Repairs and Quitrents make a great hole. As to my husbands papers I have put all of them except some letters into Mr Dales hands, of which I presume he hath given you an accompt and will publish what he finds fitt: The History of Insects you know was left unfinished and is at your direction; and as to my books I will send them up as soon as weather will permit which I fear will not be now untill Summer; not doubting in the least of your assistance in their disposeall: Sr I have no more to add but the repeating of my thanks and the presenting the services of my self and daughters subscribe my self

Sr your most obliged Humble Servant

Margaret Ray

Figure 9.2. Margaret Ray to Sloane, November 19, 1706, Sloane MS 4040, f. 255r © British Library Board.

which Sloane edited.[65] Dale also advocated for Margaret, pressing Sloane to intervene with Thomas Willughby and to give her the surplus that remained after John's friends had paid to erect a memorial stone in the Black Notley churchyard, her due as the widow.[66]

In her letters with Sloane, Margaret classed John's library of printed books as "my books," fencing the books off from John's other paper remains.[67] Dale affirmed this claim in his letters, referring to "her books."[68] Margaret defended the library against raiders: an irate James Petiver wrote to Dale after she refused to give him her copy of a book by the French botanist Charles Plumier.[69] Petiver, though he would "nott dispute with the Widdow of a Person for whom I had such a value," believed the book was rightfully his, as he had gotten it "from Dr. Hotton now Botanick Professor att Leyden for Mr Ray's Use."[70] Like Willughby and Sloane, Petiver spoke of the respect due to the widow of an esteemed colleague. Yet he also not so subtly questioned her ownership, asserting that a book given to Ray had, in fact, only been on loan.

The papers and the books belonged in two different economic regimes. Whereas only Sloane and Willughby were willing to exchange money for the papers, and that exchange was couched in terms of gift and obligation, the books, a collection of some 1,350 volumes, had many potential buyers. Ray's library consisted of learned books of divinity, medicine, philology, history, and natural philosophy. A number of copies were of interest to collectors who sought fine editions from the early history of print ("Celebrated Editions of Aldus, Stephanus, and Elziver") and elegant modern books (large paper printings and gilt spines).[71] Margaret, with Dale, negotiated for the highest possible price for the library. They coordinated letters to Sloane, who met face-to-face with London booksellers. Sloane and Dale located a buyer willing to spend eighty pounds.

But Ray refused to relinquish "her books" for less than one hundred pounds.[72] Sloane, via Dale, pressed Ray to act quickly and take the lower price, but to no avail. Trusting her own judgment, she could not, Dale wrote, "be convinced that the keeping them will be a detriment."[73] She obtained an independent valuation confirming her assessment.[74] Dale asked Sloane to seek a higher price, in accordance with her wishes. Just as Dale acted as Ray's representative with Sloane, Dale asked the more socially prominent, centrally located Sloane to serve as his representative in London. Ultimately, bookseller Thomas Ballard purchased the library, and it was sold at auction at the Black-Boy Coffee house in Ave-Mary Lane a little over one year later, March 11, 1707/8.[75]

How might we understand the transactions between Margaret Ray, Sloane, Dale, and Thomas Willughby? Ray, in withholding the insects manuscript until Willughby and Sloane paid her as promised, stressed her family's needs, though she did so with deference and respect. Willughby cast the transaction in terms of obligation and charity to the widow of a man he respected.[76] However, he used this language with some ambivalence. Shortly after John's death, he wrote to Sloane, "I am very sorry Mr Ray has left his family in so ill a condition, every body I believe had a great value for him and knowing my father had so particular one I have allways paid him 12£ a year more than I was oblig'd to, and I shall still be willing to doe what you ask of me if there is reasonable occasion in charity to the Widow to doe it."[77] Later that year, he wrote to Sloane to reiterate his willingness to help the Rays.[78] In this letter, Thomas Willughby proclaimed his respect for Ray's memory while noting that "on some accounts" he owed him very little. He subtly questioned Ray's masculinity by highlighting his failure to provide for his family, even suggesting that money collected to build him a monument (and not just the overplus) would be better given to the family instead.

It is difficult to imagine Thomas Willughby giving away, much less selling, his father's papers and collections. In the late 1680s Cassandra and Thomas Willughby, then in their late teens, undertook their own domestic memorial operations for their father. In the process of rebuilding Wollaton Hall, the fire-ravaged Willughby family seat, Cassandra and Thomas transferred their father's collections to their new home, where they cataloged and ordered them, last touched by John Ray.[79] They cleaned and labeled the objects, including "medals . . . dryed birds, dishes, insects, shells, seeds, minerals and plants," "which we were fain to do ourselves," Cassandra Willughby wrote, "fearing the servants might make mistakes, and pull such tender curiosities to pieces."[80]

With Cassandra and Thomas in mind, let us think again about Ray's shame, born from the web of gifts, mutual obligation, and failures to reciprocate binding the Rays, Sloane, and the Willughbys. Ray used humility and gratitude to finesse the shame born from his dependence.[81] Yet the fact remained that Ray was unable to properly reciprocate the kindnesses of others—Sloane's sugar, Willughby's financial support. In the years after Francis Willughby's death, Ray (and his biographers) cast their relationship as devoted friendship.[82] But Willughby's surviving family—and Josiah Child, Emma Willughby's second husband—resented Ray's financial dependence and his control over the estate as one of its executors.[83] They cast him as Francis's servant, a perception rooted in the work Ray performed for Willughby, including helping to order family records and resolve property disputes.[84] Marrying Margaret, whose family appears to have served the Willughbys over multiple generations, can have only heightened the difference in rank the Willughbys perceived between Ray and themselves.[85]

Dueling versions of Ray and Willughby's relationship—were they friends? Or servant and master?—suggest that "scientific collaboration," with its emotional, intellectual, and economic dimensions, fit uneasily into available models for masculine relationships. As in Gabriella Szalay's essay in this volume, competing, "multiple masculinities" were at work, as individuals defined masculinity through different class and professional markers. For her part, Margaret Ray, in withholding Willughby's insect manuscript and pressing Sloane for a higher valuation for "her" library, risked edging over the limits of "reasonable" charity. On the other hand, Samuel Dale clearly supported her. As a poor widow seeking to secure her family's survival, Ray's role was delicate, her success subject to male supporters' perceptions of her deservingness.

"MY HUSBAND'S LETTERS" BECOME *PHILOSOPHICAL LETTERS*

We turn again to John Ray's letters, edited and published posthumously by the cleric William Derham, Royal Society fellow and chaplain to the Prince of Wales. In the years following Ray's death, Derham took possession of letters in the Ray household. Borrowing further from Ray's colleagues, especially Sloane, he organized, edited, and published the correspondence as *Philosophical Letters* (1718). As editor, Derham typically omitted content having to do with family business, nonscientific services naturalists performed for each other (such as Ray's assistance with Willughby's legal affairs), and gossip. He lightly crossed through these sections with red pencil, suggesting that he loaned the letters directly to the printer as fair copy. He also used them, with reminiscences collected from Ray's associates, to write a life of Ray. Derham shaped Ray's memory as a pious, humble man, a religious conformist and a far-seeing naturalist. Projecting Ray's religious conformity required careful editing and writing, as it was well known that Ray resigned his Trinity College Cambridge fellowship in 1662 rather than take the oaths required by the Act of Uniformity.

Derham edited away "private Business, Complements, &c., except now and then a Clause, that may be of use to Mr. Ray's, or some other learned Man's Character, or that may shew their learned Projects, or give some Account of their Labours."[86] He dropped from "philosophical letters" the social pleasantries that correspondents exchanged at the beginnings and ends of their letters, as well as gossip and day-to-day business in the bodies of their letters. This left behind natural historical observations. In some cases, Derham struck out entire letters, such as the one in which John gave Sloane his and Margaret's thanks for the present of sugar. He eliminated Ray's medical consultations with Lister and Sloane—discarding entirely letters in which Ray, early in his marriage, consulted with Lister about his impotence. With his red pencil, Derham cut Ray's friendly letters down into something more like the public dispatches printed in *Philosophical Transactions*, the journal of the Royal Society.[87]

Yet Derham depended on the home, and the women in it, to secure the authority and authenticity of *Philosophical Letters*—of special concern because it was posthumously published.[88] In the preface to *Philosophical Letters,* he thanks Ray's male colleagues, including Sloane, Dale, Lister, and Tancred Robinson, for making Ray's correspondence available. But he gives greatest thanks to Mrs. Ray and her daughters, who allowed him to take away letters remaining in their hands. For letters written by Ray's "Best friend and Patron," Francis Willughby, he thanks Thomas and Cassandra Willughby.[89] Derham's life of Ray also seems to reflect conversations with Margaret Ray—through her, he likely learned that she taught Francis Willughby's eldest son to read and write English.[90] Derham relied upon the naturalists' female relations to provide material—and authenticating provenance—for *Philosophical Letters,* yet edited away their role in domestic and epistolary knowledge production.

So, what is wrong with this, if anything? Derham edited the letters, he said, to save his readers time and money (excess copy meant excess paper, which meant a higher per-unit cost for each book), but also out of an admirable, if unsurprising, respect for John Ray's privacy. We might compare, in Mary Terrall's telling, how the academician René Antoine Ferchault de Réaumur, out of a concern for the artist Hélène Dumoustier's privacy, anonymized her contributions in his published work.[91] Dumoustier herself preferred her privacy preserved. Many of us would hope (I believe) to have our privacy respected as Derham respected Ray's (though perhaps not as Réaumur preserved Dumoustier's). Indeed, there is nothing wrong with it. Derham's editing was not an act of betrayal. But it did have certain effects on our historical understanding of science. Derham constructed a paper divide that did not fully exist in John and Margaret Ray's world.

Under Derham's editorship, the home, as a mixed-gender space, was private. As a suffering body, the body was private. But the male mind, engaged in natural history, was public no matter where it was. If a series of letters is a three-dimensional projection of the four-dimensional domestic space in which a scientific family functioned, then Derham effaced the mixed-gender social and working spaces—the whole house, except, possibly, Ray's study—from the building plan. In *Philosophical Letters,* letters jumped from study to study, point to point, man to man: the man's location in a mixed-gender space, the people with whom he surrounded himself, didn't much matter.

As a space in which the Rays jointly worked and entertained local and visiting naturalists, their home was not entirely dissimilar from all-male "public" spaces like the Royal Society meeting rooms. In both places, male natural philosophers met to converse, coproduce and witness experiments, and read aloud letters written to them by their correspondents. In both places, access was restricted to known and trusted visitors. Unlike the home, however, the Royal Society meeting rooms were all-male; they were also given detailed (though idealized) public representation as spaces of scientific knowledge production in publications such as Thomas Sprat's *History of the Royal Society of London*.[92] Derham assimilated the home to these male spaces. The early modern naturalist's household was a site of mixed-gender natural inquiry and neighborly conversation, linked to the wider world through letters and travel. It was neither public nor private, according to our modern understanding of those terms. *Philosophical Letters*, however, rendered the domestic invisible, piling up papers in that modern boundary between private and public, masculine and feminine.[93]

TEN

Family Notebooks, Mnemotechnics, and the Rational Education of Margaret Monro

Matthew Daniel Eddy

Sometime in 1738 the eleven-year-old Margaret Monro (1727–1802) received a letter from her father. Though she spent most of her early years at home, she was most likely attending a boarding school located outside Edinburgh in the Lowlands of Scotland at this time. Written in a congenial and familiar manner, her father's letter outlined a plan that, if executed diligently, would improve her reason, advance her virtue, and garner the highest esteem from her acquaintances and relations. The plan was to send her a series of letters on female conduct that explained how to make rational decisions in a variety of domestic, social, and commercial contexts.

In addition to reading her father's letters, Margaret copied them out. As he explained in his initial letter, the act of writing served jointly to strengthen her mind and to preserve useful knowledge for future reference. More specifically, the "art of writing" was "an assistant to memory" and it required her to write for at least "one hour per day."[1] The result would be a single manuscript that they could edit together at home. Margaret's father viewed the manuscript as a collaborative project, one in which they were both participating in creating a shared artifact of learning.

Not only did the manuscript preserve a collection of observations and facts, it also communicated *how* to organize, how to visually and conceptually order knowledge on, across, and with paper in a way that helped Margaret think linearly and categorically and, by extension, more efficiently. In making the manuscript herself, Margaret was effectively learning how to treat paper as a material and learning tool through which she gained a specialized collection of writing techniques that she could use in the future to organize a vast amount of knowledge in one handmade manuscript.

Margaret's father was Alexander Monro *Primus* (1697–1767), the University of Edinburgh's professor of medicine and one of the most celebrated anatomists of the Enlightenment.[2] Her mother was Isabella Monro (née Macdonald, 1694–1774), the accomplished daughter of a Highland clan chief. Both parents were highly literate. Primus in particular had written or edited hundreds of manuscripts during his career and his ability to extend his own mind on paper was substantial. A key facilitator of this extension was an advanced form of scribal mnemotechnics; that is, the ability to select and inscribe information into a memorable arrangement of words and space on the page and across a codex.[3] For him, the letters on conduct represented an opportunity to bequeath his mnemotechnic knowledge to Margaret, his beloved only daughter.

True to her father's wishes, Margaret assiduously enacted his graphic suggestions, thereby creating a beautifully written, 386-page manuscript, a notebook of sorts, that she finished as a twelve-year-old in 1739. Titled *Essay on Female Conduct,* it was later bound in red leather and is housed today in the National Library of Scotland.[4] The manuscript is a noteworthy historical document because it shows how a girl learned writing techniques that were perceived as being rational in a manner that taught her how to filter and order a sprawling topic like female conduct on and through multiple forms of paper. Not only does it afford insight into how a girl of her social status learned to use pen and paper, it also serves as a salient example of intergenerational knowledge in transit, or, as I like to put it, knowledge in motion.[5]

In this chapter, I reconstruct the sociomaterial process that unfolded when Margaret set out to design, create, and adjust the physical and visual components of her exquisite multimedia artifact. In following this path, I wish to shed light on how young female learners actively acquired classifying, compositing, and correcting skills as they made manuscript books—codices—in early modern domestic settings. Paper, as Elizabeth Yale shows in this volume, mediated the epistemologies of the household and the study of nature in the early modern home.[6] Similarly, my aim here is to show that as a sociomaterial device, Margaret Monro's codex was a boundary object that inhabited space between scholarly and domestic knowledge management practices.

Figure 10.1. A scholarly female figure, most likely the goddess Isis, stands in front of the pyramids at Giza. She points to inscriptions on her table as she speaks to an architect holding calipers. "Antiquities of Egypt," in Charles Rollin, *The Ancient History of the Egyptians, Carthaginians, Assyrians, Babylonians, Medes and Persians, Macedonians, and Grecians*, vol. 1 (London: J. and F. Rivington . . . , 1768), facing 2.

HOUSEHOLDS, PAPER PRACTICES, AND FEMALE EDUCATION

From the late seventeenth century there was a Scottish tradition of cultivating the rational capabilities of girls. The late eighteenth-century novelist and philosopher Elizabeth Hamilton, for instance, noted that her grandfather "wisely endeavoured" to make his daughters "rational" and that "so successful were his efforts directed that even the third generation had reason to bless him."[7] As revealed in Hamilton's biography, her family participated in a movement in which a rising standard of female education enabled women to actively exercise power through the tools provided by literacy.[8] As readers, eighteenth-century Scottish women engaged with the cutting-edge ideas of their time. As correspondents, they participated in the Republic of Letters. As notekeepers, they formulated ideas and managed household business accounts. By the latter part of the century these skills were employed by leading Scottish authors such as Hamilton, Joanne Bailey, and Mary Somerville, who published philosophical, pedagogical, and literary works that challenged the masculine structures that guided the church, state, and the market.[9]

But though we know what educated Scottish women were capable of writing and reading, we know surprisingly little about how they learned to use paper tools to organize their thoughts before they published books or wrote letters. As the indices in Scottish commonplace books reveal, these tools were certainly used by unmarried women. The power of the handwritten index within domestic settings even made special appearances in popular novels such as Elizabeth Hamilton's *Memoirs of Modern Philosophers*. In one of the scenes of the novel, Bridgetina Botherim, one of three young female characters, attempts to argue the dashing Henry Sydney "into love." She attempts to convince him with "a speech which had long been conned, twice written over in a fair hand" and was arranged on paper as headings that were "a sort of index taken of the contents." Her efforts to order ("con") her argument through writing and rewriting it on paper were effectively acts in which she categorized her thoughts into headings or an ordered index.[10]

How did Scottish girls learn to create a tool like Botherim's index? How did girls who possessed indexical skills use them to plan, organize, or understand the many codices that were present in early modern households? Despite the presence of many kinds of paper-based tools that surrounded literate Scottish women at home, it is unclear how younger female learners acquired the ability to shape, manipulate, and order paper in a way that rendered it a material conducive to managing a large amount of information in a knowledge system.

In order to know more about how Scottish women used paper, we need to study how girls, before they reached adulthood, learned to use it as a material form of representation to which ideas could be attached.[11] Here Monro and her codex become a very helpful case study. Like many Scottish girls who grew up in merchant and professional households at this time, Margaret lived in a diverse and fascinating sociomaterial world that operated on and across paper.[12] Her own family communicated with a variety of handwritten memoranda slips, invitation cards, inventory sheets, receipts, letters, and notebooks. Likewise, the collections housed in the National Library of Scotland readily indicate that it was normal for eighteenth-century Scottish girls of professional and gentry households to be surrounded by many paper-based forms of communication. But though we know that such manuscripts existed, their role within the Scottish educational context is only just being discovered.

Establishing how girls learned to use manuscripts is a historiographic challenge. The history of early modern female education for girls growing up in middle-class households has traditionally used printed sources. But unless it is absolutely clear that a specific source was read by a specific learner (and it often is not), histories based on this kind of material can tell us only what girls might have learned. The sources that speak most clearly to how girls learned to use paper-based forms of organization are the manuscripts that they themselves made. Yet aside from the collections of wealthy families, the manuscript ephemera of female household learning is hard to trace in early modern Britain because the material is often not preserved or adequately cataloged. In cases where material has been preserved for middle-class or gentry families, it is often that of boys. Even when it does exist, it is difficult to determine how it was used.

A case in point is the educational box created by Scottish botanist and author Lady Charlotte Murray (1754–1808). It contains the handwritten collection of mnemonic card games, volvelles, game boards, indices, and botanical drawings that she made with pen and paper for her nieces and nephews during the 1780s.[13] Like the children's print encyclopedias published by Johann Siegmund Stoy in Germany, it was a pedagogical system on paper.[14] But like these kinds of printed works, evidence of existence is not necessarily evidence of usage. Did, for example, the girls in the Murray household use the paper devices differently? Aside from the possibility that the children might have made part of the vovelle with Lady Charlotte, there is little evidence of whether her nieces or nephews used the system, and, if they did, how they used it. If we are to more fully understand how girls developed scribal and rational skills with and through paper, evidence of making and using manuscripts is important.

The significance of learning through making a manuscript in a domestic setting was influenced by the value that households attached to paper-based communication and its applications. The Monro household was progressive and promoted the practical and intellectual uses of manuscripts for both the male and female members. Margaret's parents wished for her to recognize the value of weighing "evidence and probability," and to understand the use of argument, logic, and analogy.[15] One of the outcomes of working with pen and paper was to facilitate these abilities in a way that strengthened her sense of judgment; that is, the quintessential attribute of reason as it was understood at the time. Possessing judgment empowered her to use reason to nurture and select behavioral characteristics in a way that enabled her to accomplish the duties and responsibilities that awaited her when she became a woman.[16]

Knowing how to make and interface with manuscript culture in the foregoing manner promoted the development of what Katharine Glover has called the "female mind," which actively engaged with the Enlightenment ideals of equality and sociability.[17] Monro's father encouraged this kind of thinking. He held that the powers of girls' minds were equal to those of boys, a view that reveals that he fell within a coterie of British men whose progressive views of the female mind constituted a form of nascent feminism.[18] He advocated a jointly empirical and progressive notion of femininity that treated the mental capacities of girls and boys as being the same.[19] When compared to the patriarchal views of womanhood advocated in Scottish learning contexts by the conservative Calvinist establishment at the time, the relationship between femininity and the paper world of learning in the Monro household was reformist.[20] Overall, girls working with pen and paper to craft manuscript books learned that women possessed the ability to shape and exert their mental capacities in a variety of important familial, social, religious, and educational contexts.

EDUCATING A CODEXER

The ample presence of paper-based forms of communication presents a number of stimulating possibilities regarding how Monro learned to use them and to appreciate their potential within the wider world of literate culture that would surround her when she left home. When it comes to understanding how she learned to work with manuscripts, her educational context and the material form of her codex provide a helpful starting point.

Monro's father was educated as an apprentice by his own father, which might explain why he was so keen to play a central role in educating all of his children.[21] He taught Margaret (1727–1802) at home alongside her brothers, John (1725–1789), Donald (1728–1802), and Alexander (1733–1817). Prior to attending a private school, Monro learned Latin, "arithmetick, geography and some other Parts of Education" directly from her father. She studied advanced Latin with private tutors, a practice followed by other girls in Europe at the time.[22] Monro was present when the family read books together, as was popular for many literate families living in early modern Britain. The Monros read books such as Charles

Rollin's *Ancient History* on religion and history, which contained illustrations that depicted highly literate women (fig. 10.1).[23]

At boarding school, Monro took lessons from private writing masters. At some point, she also received personal instruction in bookkeeping, a subject that was taught to Scottish schoolgirls from the late seventeenth century forward.[24] Over the course of her education at home and at school, Monro excelled in all subjects, leading her father to call her a "scholar," a term usually reserved for university graduates.[25] He valued her views and sometimes asked for her opinion or feedback on what he had written. Her brothers held her intellect in high regard as well, including her in conversations about the nature of reason.[26]

Monro's family placed a high value on "Readyness and Correctness in Writing."[27] Evidence of her early efforts to learn this skill no longer seems to exist. But we know that other students her age, male and female, learned school subjects through copying, ciphering, translating, and composing their lessons across different kinds of loose-leaf, bifoliated, and bound paper.[28] In addition to learning the facts of arithmetic, grammar, Latin, geography, and accounting, Monro's early education taught her how to select, negotiate, manipulate, and format paper so that she would manage knowledge in a way that was easier to access and remember.

Other aspects of Monro's education inculcated paper-based skills as well. Translating Latin sentences, for instance, was a process that usually took place over multiple sheets of paper. The process began with a rough translation scribbled on scrap paper. A better version was then copied onto a loose leaf of octavo- or quarto-sized sheet and submitted to a tutor. The corrected version was copied with other revised sentences onto an octavo-, quarto-, or folio-sized bifolia, which was then stacked for storage. Sometimes translations were bound into a sewn or leather-bound manuscript. In the Monro household, for example, Margaret's brothers wrote their own Latin compositions and copied them into a communal book called a *calendarium*.[29] Margaret's father made manuscripts as gifts for his sons as well.[30]

The overarching picture that we have of Monro's education is that it took place in a world surrounded by manuscripts. But, in the end, knowing how to make a sophisticated artifact like her codex required further forms of training. The combined skills of manipulating paper with ink and other forms of folding, stacking, and cutting were essential for anyone at the time seeking to design manuscripts that doubled as information management devices. It was precisely these kinds of skills that Monro's father wanted her to further develop by making a manuscript book of her own.

The subject matter of female conduct conformed to a subgenre of literature for adolescents called the parental monitor, which took the form of a parent (or sometimes an older sibling) giving advice to a young learner in a familiar manner on everything from courtship to comportment. Monitors were printed, but there was a Scottish tradition in which fathers wrote a manuscript book for their children to read and contemplate at home.[31] It was for this reason that Monro's father proposed making a manuscript book together. He viewed the codex as a collaborative project, one in which they were both participating in creating a shared artifact of learning: "If these Papers remain in the form of Letters which I am to send you them in, they will soon be lost. You surely wou'd regret being deprived of such a Mark of your Father's tender Fondness. Wou'd not you rather bestow the same hour which you employ with your writing Master when in Town copying my Letters on Paper that is not folded and has large Margines with every other Page blank, that it may be bound and may allow Corrections that you and I make in [it]?"[32] The project took her well over a year, which means that much of it was carried out at home. By 1739 Monro had finished her beautifully written, 386-page codex. The manuscript provides a wealth of information about the paper-based skills that she had learned and the ways in which it functioned as a learning device within her domestic environment.

Consider the evidence presented in the page exhibited in figure 10.2. The sentences are straight with even spaces occurring between them. The script is clear. The main headings at the top are centered and are typographically differentiated. Though its margins were later reduced when it was bound, it can still be seen that they were wide and proportionate. The editorial correction near the bottom is appropriately positioned so that it can be more easily read. All of these features reveal that Monro was adept at manipulating a sheet of paper into a well-crafted page. The foregoing components are present on the first page and remain through the entirety of the codex, which means that, even before she began to make it, she already possessed a number of fundamental paper skills.

When one looks closer at the skills evinced in the page's layout, these, once again, disclose Monro's previous experiences with manipulating the open field provided by a blank sheet of paper. Unlike the ruled paper used by children today, the paper used in the early modern period was blank. Monro, like all early modern children, had to learn how to grid the paper with graphite lines so that she could write straight lines of text. The advanced writing skills presented in her codex were learned at the time by copying practice lines over and over on loose-leaf scrap paper. Once a sentence was mastered, it was then copied onto a small, rectangular slip of loose-leaf paper. When students finished their penmanship lessons, they gathered, ordered, and bound (sewed) the slips together into a small book.[33]

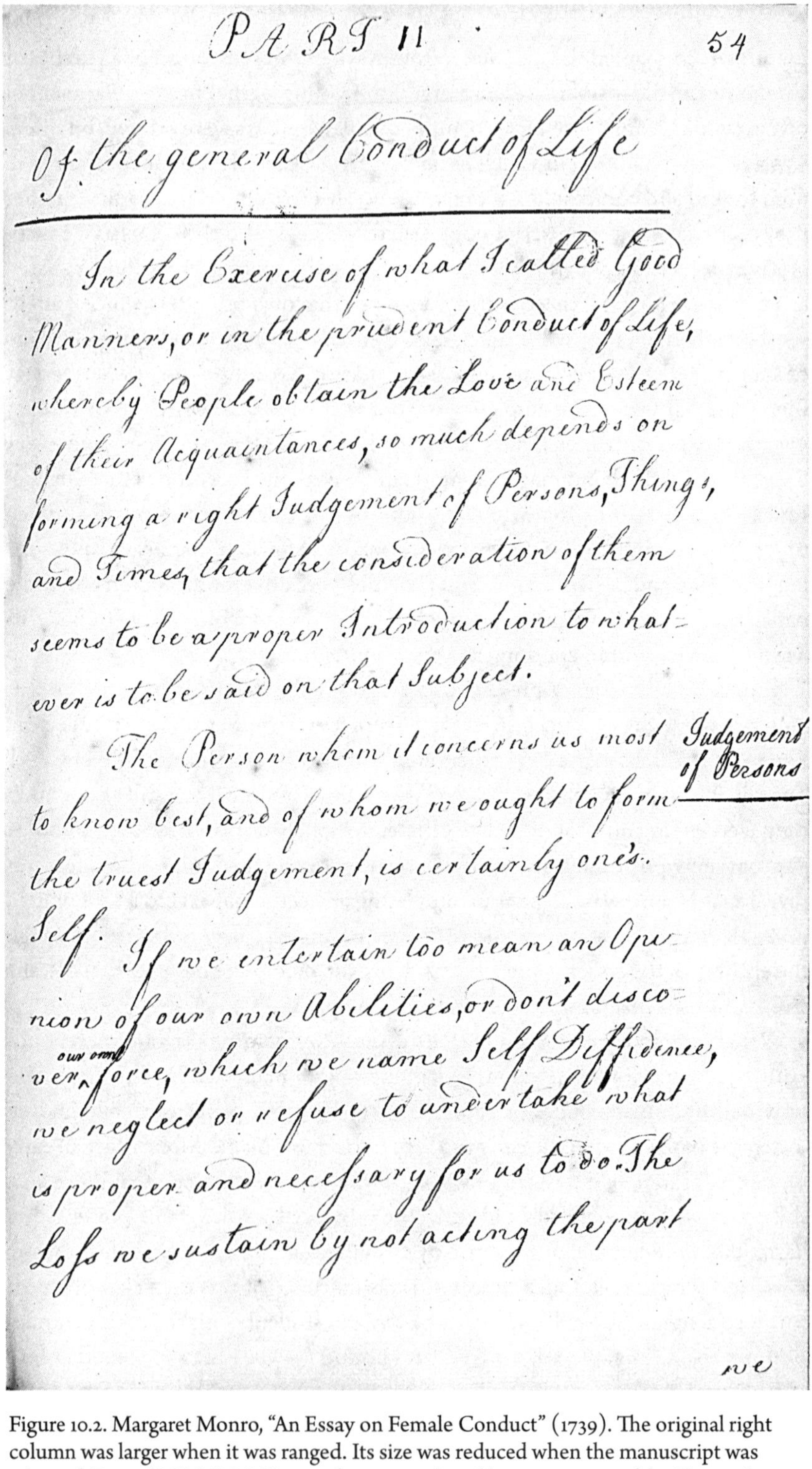

PART II 54

Of the general Conduct of Life

In the Exercise of what I called Good Manners, or in the prudent Conduct of Life, whereby People obtain the Love and Esteem of their Acquaintances, so much depends on forming a right Judgement of Persons, Things, and Times, that the consideration of them seems to be a proper Introduction to what=ever is to be said on that Subject.

Judgement of Persons

The Person whom it concerns us most to know best, and of whom we ought to form the truest Judgement, is certainly one's Self. If we entertain too mean an Opi=nion of our own Abilities, or don't disco=ver our own force, which we name Self Diffidence, we neglect or refuse to undertake what is proper and necessary for us to do. The Loss we sustain by not acting the part

we

Figure 10.2. Margaret Monro, "An Essay on Female Conduct" (1739). The original right column was larger when it was ranged. Its size was reduced when the manuscript was rebound. Bound MS, NLS MS 6659, f. 54, reproduced by permission of the National Library of Scotland.

The process of copying sentences across slips, scraps, and sized sheets, though simple to modern eyes, taught young students like Monro how to match the form and layout of a piece of paper to the kind of information that they wished to preserve and access. In the case of Monro's manuscript, the skill of selecting paper for her manuscript was put to good use. As mentioned above, her father asked her to leave "large Margines." He did this so that she could insert paratextual apparatus and other scholarly annotations at a future date. To accommodate this request, she elected to use loose-leaf folio-sized pages; that is, she matched an appropriately sized piece of paper to the layout she needed to implement in her codex.

CLASSIFYING

The Monros wanted their daughter to acquire a keen sense of judgment, which, as most conduct manuals indicated, served as the basis of moral decision making. Judgment was based on a variety of capabilities, one of the most important of which was the ability to categorize and classify facts and observations. Within Scottish educational settings, ordering words, and, hence, ideas, into multiple categories was considered a logical act if it was done according to a clearly articulated organizing principle.[34] As we will soon see, Monro's codex is organized according to categories and subcategories that were ordered according to the principles of moral conduct.

The surest way for school-aged children to recognize, replicate, and recollect ordered categories was to write them over and over across multiple sheets of paper. Here paper served as a medium through which the skills of logic and, by extension, judgment were learned. Schoolchildren living at the time that Monro made her manuscript implicitly learned to replicate such principles when they kept school notebooks because masters and parents encouraged them to categorize the prose according to headings that served as categorical labels. Students organized arithmetic notebooks, for instance, under the headings of "addition," "subtraction," etc., with the principle being, roughly, the main facts and operations one needed to know about the subject.[35]

Monro's headings were sorted in the same way. Whereas those made in a school classroom were mostly created by their teachers, Monro's were prescribed by her father. When she copied them, she was learning how to think logically through writing out a system of categories that were neatly organized on a piece of folio-sized paper. The categories were given as the main headings she used to label the codex's seven parts. The principle of selection was thematic in that it sorted the information according to the kinds of relationships that required moral conduct. The main headings and thematic principles were as follows:

Margaret's Categorical Headings

CATEGORICAL HEADING	THEMATIC PRINCIPLE
Part I: Of the Education of Girls	Margaret Herself
Part II: Of the General Conduct of Life	Margaret Herself
Part III: Of the Commerce with Men	Husband
Part IV: Of the Government of Servants	Servants
Part V: Of the Management of Children	Children
Part VI: A Summary of Religion	Clergy
Part VII: Of the Origine of Government	State Officials

The parts corresponded to the different kinds of people, including herself, with whom she would most likely interact when she became an adult.

But the sophistication of the categories that Monro replicated in the pages of her manuscript went far beyond those used in most school notebooks. This was because her categories were further divided into numerous subcategories, each of which was arranged according to a subprinciple. Accordingly, each was given its own subheading. The principle of arrangement was again thematic in that all the subheadings were examples of the topic stated in the main heading. Some of the subheadings (subcategories) of part II, "Of the General Conduct of Life," for instance, were virtues or vices that Monro could embrace, reject, or moderate.[36] The subheadings were:

Table 10.1 Subheadings of General Conduct[i]

Judgement of Persons	Envy	Sobriety
Value of Things	Too Much Good Nature	Visiting
Right Time of Acting	Modesty	Air & Manner
Religion	Chastity	Singularities
Honesty	Bashfullness	Conversation
Good Nature	Humility & Pride	Talking
Anger	Vanity	Wit
Surlyness	Complaisance	Laughing
Dureness	Moderation	Diversions
Revenge	Generosity	Companions
Cruelty	Oeconomy	Friends
Peevishness	Imitation	Regard to Character
Petting	Affection	

[i] I have taken the subheadings directly from MS 6658, ff. 14–83. They are also given in Monro, *Professor's Daughter*, 23–48.

On occasion the codex displays alphanumeric subheadings written in the running text of the paragraph as well.[37]

The logical structure of the headings and subheadings in her codex reveals that Monro was mastering the same kind of classification skills that male university students learned when they kept lecture notes. Indeed, as a university professor, her father knew that his colleagues employed headings in the same way to structure their lecture notes into mnemonically expedient categories. University students then used the headings as organizational technologies to order their own notes and to collate hundreds of sheets of paper to create a full set of student lecture notebooks. Like the headings in Monro's codex, the headings of their notebooks doubled as categories of a system arranged according to a thematic principle (which usually related to the subject matter of the course).[38]

Monro's father and his university colleagues taught their students that a good classification system was predicated on the use of a consistent set of upper *and* lower categories. Following the work of Carl Linnaeus, some professors turned to logic and used the upper categories of "class" and "order" and the lower categories of "genus" and "species" to order their lectures. But there were of course other options. Like the main headings in Monro's manuscript, some professors used "part" for upper categories, while others used "chapter" and "section."[39] Her father encouraged Monro to order her manuscript book according to hierarchical categories because he felt that the act of writing helped all thinkers, female and male, to develop and exercise the ability to make rational decisions and to consider how such choices could be best implemented in *all* the significant relationships that they might experience. In Monro's case, she was learning to group the information in her manuscript under headings and subheadings, which was an implicit act of classification that helped her to see how to order a knowledge system in and across a codex.

COMPOSITING

By making the various components of a codex herself, Monro was effectively treating paper as a material learning tool through which she gained a specialized collection of skills that she could use in the future to organize a vast amount of knowledge inside one handmade manuscript. One of the most important abilities gained through this learning process was compositing; that is to say, the ability to format a blank sheet of paper into an accessible layout.

During the eighteenth century, one of the words used for scribal compositing was "ranging." The wider meaning of the verb *to range* referred to the ability to order things by putting them in a line or dividing them into categories. When used in reference to the act of writing and sewing, "ranging" referred to a more specific act of linearization or categorization that took place when the hand worked with the eye to visually align ink or thread across the flat, square surface of a sheet of paper or cloth.[40]

When used in reference to the columns stretched across a manuscript ledger page, "ranging" captured the process of writing in a straight line, particularly within a row or a column. A good example of this usage occurs in William Perry's textbook *The Man of Business*. A successful academy master in Edinburgh, Perry described a ledger's columns as places "wherein the several articles belonging to every different person or subject, which are dispersed in the journal, are collected together and ranged in their natural order."[41] In other words, "ranging" was an act of visually ordering words and space on paper. Like the compositors who laid out print, Monro was a compositor who ranged script.

Monro's father believed that "the Act of ranging" provided a way for her to apply a "distinct method and order" in contexts that required facts or data to be ordered in a visually accessible fashion.[42] To this goal, he recommended that she divide the page into two columns. Monro implemented a layout that effectively transformed the faces of her pages into matrices. She filled the left column with modules that contained the main headings and prose. This column was the largest, taking up around two-thirds of the original page. Its heading and running text operated as a figure and ground gestalt image in that it capitalized on the visual affordances of contrast and symmetry.[43] She left the right column blank so that she could fill it later with subheadings and annotations (see fig. 10.2).

Making a well-formed, highly spatialized word matrix was not something that came naturally to many children in Scotland at this time.[44] When considering the level of precision evinced in the layout of Monro's codex, it is worth investigating whether she was extending several scribal skills that she might have already possessed. Here the context of paper-based household management skills can serve as a guide. The use of columns to divide the space of a page, for example, was a common technique used in bookkeeping and, as mentioned above, Monro was already familiar with keeping accounts when she began her codex.

Monro's father believed that "the art of ranging" was an essential skill for those who wanted to organize "[ledger] books in the most distinct method and order."[45] At the time, accounts were written according to the double-entry method, in which different columns were used for debts and credits. Bookkeeping students were accordingly asked to create ledger pages that consisted of at least two columns. Some students even acquired highly sophisticated layouts that involved plotting multiple columns on a single page. Others treated the ledger as a master document into which they copied information taken from paper slips and other codices such as memoranda and waste books.[46]

Monro most likely would have witnessed the transfer of information into the columns of a basic ledger at home, as ledgers were used in the Monro household. At an informal level, family members and servants were asked to provide receipts, which were most likely slips of paper, for what they had spent. Her mother, Isabella, oversaw the collection of these receipts into the "family book" of accounts. At the very least, Margaret would have witnessed her writing entries in it. It is likely, however, that Isabella showed Margaret how to write the entries herself.[47] But the skills required to make entries in a column were a bit different from those necessary to make the column itself. Margaret's codex shows that she clearly knew how to do this before she began making it.

Girls studying bookkeeping in Scotland during the eighteenth century were taught to range by repeatedly drawing columns across blank sheets of paper. Acquiring this skill was an important part of learning how to visually structure an accounts book. Remnants of this process occur in the notebook kept by the young Barbara Walker (née Montgomery) when she was learning bookkeeping in 1787. Before she wrote any figures, she drew columns across all of the eighty pages (recto and verso) of the new, blank, and bound paper book into which she would be writing her practice accounts. In the end, she needed only thirty of the pages to complete her exercises; the rest remained unused until she decided to employ them again after she was married.[48]

As evinced in every page of Monro's codex, she used a lead pen to divide each page by drawing a line down its blank face. The result was two columns into which she could plot the prose and headings of the script. The lead pen was a graphic design tool included in many writing cases sold during the eighteenth century. The purpose of the pen was not to make a graphite or ink mark; it was used instead to make an impression. Since the paper of the day was softer than today's, because it was linen-based, the dull, pointed metal tip of a lead pen was used to impress faint lines that helped to demarcate columns or to make geometric schemas over which diagrams were drawn. Mastering a lead pen to structure a page undoubtedly came to Monro through executing columns in a manner similar to that practiced by the young Walker. Because accounting columns were deployed to sort numbers according to two basic themes, debits and credits, Monro was familiar with the role they could play as a visual sorting tool.

Dividing a sheet of paper into columns inculcated what Jonathan Gibson has called "significant space," the notion that different places on the page carry different meanings that are transferred to the words written into them.[49] In a simple ledger of two columns, the space of one column represented debits and the space of the other represented credit. The space was labeled with headings that communicated the meaning assigned to them. Sometimes this was as simple as creating a heading for "Debit" and another for "Credit." In the case of Monro's codex, her father asked her to treat the main headings and subheadings of his conduct classification system (as outlined above) as labels for the left columns. The main headings were placed in the left column. They signaled a category of "higher" information—"The General Conduct of Life," for instance. The headings of the right column signaled a "lower" category of information such as "Courage" or

"Modesty." In working with page after page of paper, Monro was appropriating the skills required to create a paper-based mnemotechnic artifact, an object in which the form and meaning overlapped in a way that organized information into an easily accessible format.

Ranging a double-columned matrix just under four hundred times in her codex probably dwarfed the number of times that she had ranged an account book. With every new sheet of paper, she was further solidifying skills that enabled her to make an interface device in which visual order and verbal order were superimposed into one artifact. Through this process, every page of the manuscript was ordered effectively by two systems, one visual and the other verbal. Wrapped around each other, these systems shaped the structure of the page that Margaret then used as a platform for correcting the script.

CORRECTING

The fabric and layout of Monro's codex presents evidence that helps us understand how she changed elements of her father's manuscript through making intentional omissions, insertions, and annotations. Much like the users of phrenology described by Carla Bittel in this volume, Margaret made edits and added emphasis to reflect her own knowledge.[50] All of these transformed her codex into an object that made more sense to her and was easier for her to use in the future. From the perspective of a codex maker, the changes were in fact corrections.

Through interfacing with her father's manuscript, Monro deployed ink as a significant tool to manipulate how knowledge could be organized across the many pages of a codex. She made the most noticeable changes at the paratextual level, as she came to better utilize the layout that she had rendered on the page. Within this customized space, she resolved what (and what not) to underscore, indent, capitalize, italicize, etc. Each of these decisions influenced the presentation of words differently and affected how she interacted with each page while she was writing it.

Indenting, for instance, injected blank space that helped her skim the paragraphs in a way that made it quicker to move through the codex. There are several pages in Monro's manuscript where she broke her father's longish paragraphs in two and placed a new indention at the beginning of the new paragraph. Her father seems to have encouraged the development of this editorial skill. Throughout early parts of his manuscript he wrote marginalia that suggested how she might split several of his long paragraphs. In these cases, he wrote "a new line" in the margin and an *X* at the end of the sentence where Monro might want to make the break.[51] She implemented most of her father's suggestions, which enabled her to edit her text in her own way.

Monro introduced new forms of underscoring as well. Underscoring drew attention to keywords and headings, making the script even more accessible. Thin and thick lines denoted lesser and greater forms of emphasis. Drawing these lines required her to understand the thickness of the paper and the miscibility of the no fewer than three different kinds of ink that she used. With this knowledge, she made thin lines by pressing softly and changing the angle of her nib (or quill). She made thick lines by pressing harder, moving slower, and re-angling her nib.

Monro's manipulation of the manuscript reveals that she understood how to craft ink on paper in a way that communicated things that she wanted to emphasize. Her thick underscored lines, most of which do not occur in her father's script, are placed strategically on words that she wanted to highlight. As can be seen in figure 10.2, she added a thick line to separate the main headings from the script's prose. Her father did not use this technique and, consequently, his main headings are harder to spot.[52]

Another informative example of Monro's underscoring occurs in the subheadings written in the margins. She used thick strokes to underline all of them, a practice that can be seen in the "Judgement of Persons" subheading in the right column of the page depicted in figure 10.2. Though her father wrote his subheadings in this column as well, his were formatted differently. He added some and demarcated others with boxes drawn around them. But Monro's underscoring was her own; here, and for the running headings, she shaped the script in her way.

The most impressive example of Monro's paratextual design skill is displayed in her three-page table of contents, which she made and then bound into the front of the codex.[53] Without this crucial piece of apparatus, the content of the codex would have been difficult for her to access quickly. Like all tables of contents, hers was an index, a list of well-selected and ordered entries. The completed version provided a way for her to overview the contents of the entire manuscript at a glance. Her father's manuscript does not contain an index, nor does it give instruction on how to make one. Unlike the layout of the other pages, Monro may have designed its columns, indentations, and typography on her own. Her index further impressed the thematic order instantiated in the headings and subheadings. Since it was done after Monro had finished the codex, the act of creating it reinforced the modes of organization she had witnessed while writing the main body of the script. Space prevents us from fully examining the other paratexts, such as catchwords and running heads, that Monro added to her script. But it will suffice to say that all of them offer evidence that can be used to examine how she actively thought about the meaning vis-à-vis to the design of her pages.

In addition to making paratextual changes, Monro omitted and added to the group of subheadings that her father wrote to classify and label the subsections. His script contains fewer subheadings than Monro's. It is possible that the first version of the script that he gave her contained hardly any at all. Quite possibly,

Monro invented and inserted a large number of them by herself. In several cases, as for instance in part II, this is abundantly clear. Here, she omitted "Surlyness" and "Petting" and instead introduced "Experience." Such changes indicate that she categorized and recategorized the script as she went along. This activity was facilitated by the fact that she had formatted the margins of the page as wide columns into which she could write her own subheadings.[54]

Monro's additions and revisions to the subheadings are present from the very start of the manuscript. Though she faithfully copied the running text of the first five paragraphs of part I, she did not elect to use the original subheading that her father suggested for this part of the script. He had only included one, "Mercating," which flagged a passing reference in the second paragraph to bookkeeping. Monro, however, omitted this subheading in her codex and inserted new subheadings for "Reading" and "Writing"; that is, topics discussed in the third and fifth paragraphs.[55]

In many respects Monro's headings more accurately represented the larger themes discussed in the opening sections of the codex, demonstrating how she sharpened her skills of categorization. Her sense of categorizing allowed her to see that the "Mercating" subheading was better suited for a later section of the script that more thoroughly addressed the larger importance of household management and the necessity "for a Girl to know how to manage rightly the Things that are purchased." Monro accordingly placed the "Mercating" subheading in a more appropriate place that occurred later in the codex, revealing that she understood how to categorize knowledge thematically and how to design the script in a way that was clearly laid out on and across paper.[56]

Margaret Monro was the daughter of highly literate proponents of the Scottish Enlightenment. I have shown that she exerted power over paper through making a beautiful codex. Notably, her understanding of paper-based forms of knowledge creation and management included both scholarly and domestic forms of manuscript usage. Through classifying, compositing, and correcting the form and fabric of the manuscript, Margaret Monro came to build a knowledge system on and around paper. Her codex provides evidence of what she was learning in real time and how she internalized various skills that operated in conjunction with the acquisition and exercise of reason.

ELEVEN

Papier-Mâché Anatomical Models

The Making of Reform and Empire in Nineteenth-Century France and Beyond

Anna Maerker

Sexuality and reproduction have long been central features of anatomical models and representations in two and three dimensions. Early modern models in wood and ivory presented pairs of male and female bodies side-by-side. Pregnancy, in particular, was a frequent subject: the flap anatomies of early modern Europe featured pregnant women, as did most wooden and waxen models of female bodies. Scholars have paid much attention to the gendered aspects of these objects, and have argued that they have often played key roles in articulating and affirming culturally and historically specific gender roles, despite their supposed allegiance to an objective, timeless truth.[1] Ludmilla Jordanova's classic study *Sexual Visions*, for instance, analyzes the aesthetic features of the celebrated anatomical wax models of eighteenth-century Italy.[2] She shows that despite their medical accuracy and anatomical detail these models were simultaneously deeply indebted to contemporary assumptions about the roles of men and women in society. Female models frequently were posed as "anatomical Venuses"—recumbent and passive, with beautiful faces and skin, adorned by long hair and decorative details such as pearl necklaces. In many cases, these female models were pregnant, and spectators could discover a small fetus hidden underneath the removable cover of the torso. Thus, female bodies were equated with passivity, reproduction, beauty, and sexual allure. Male bodies, in contrast, were often depicted in upright, active poses. Signifiers of beauty such as skin, face, and hair were not present—instead the focus of male models tended to be on musculature or the nervous system. Wax modelers thus represented anatomical facts, but simultaneously their representations were shaped by shared perceptions of gender roles, and by artistic genres and conventions.[3]

The present chapter extends the analysis of anatomical models and gender by approaching these objects as technologies, as material things that "produce people, and relationships between them."[4] The chapter argues that we need to go beyond analyses of models' visual appearance, and recognize how the materiality of models shapes human interaction with and perception of them in crucial ways, from enabling different practices of production and use to widening access by increasing durability or decreasing price levels. A focus on this "sociomateriality" of models reveals how the gendered identities of model makers and model users are implicated in the *practices* of making, distributing, and using models.[5] I use the cases of some prominent nineteenth-century paper-based anatomical modeling enterprises to ask how practices of model making and model use have supported, challenged, or obscured gendered perceptions of model makers and users, and what role the material properties of these three-dimensional paper anatomies played in gendering the people who interacted with them.[6] Paper, in this case, needs to be considered as more than a carrier of texts and images. A sociomaterial analysis of these models reveals how they contributed to articulations of gender in ways that were significantly more complex than a straightforward representation of binary stereotypes.

MODELS, MATERIALITY, AND GENDER BEFORE 1800

The famous anatomical wax models of eighteenth-century Italy had already given rise to complex configurations of gender and materiality. Here, gendered associations of the material wax could be used both to strengthen modelers' claims to authority and to call them into question. In the case of the female anatomical wax modeler Anna Morandi (1714–1774), her gender and her work with wax were articulated by her eighteenth-century contemporaries as evidence of a privileged access to anatomy. Lucia Dacome has drawn attention to the ways in which Morandi's femininity, and especially her status as a mother, were used to argue for her superior understanding of the formation of the human body: "Wax-modelling resonated with long-term views of generation that regarded offspring as the result of the impression of soft matter."[7] However, wax and its gendered connotations not only were used to support a female modeler's claim to mastery and verisimilitude but could also be employed negatively to cast doubt on the abilities of (male) model makers and users. In late eighteenth-century Vienna, Italian anatomical waxes had been introduced as teaching tools for the new surgicomedical academy Josephinum by Emperor Joseph II and his surgeon Brambilla. They were criticized vociferously by medical professionals and satirists who opposed the emperor's reforms more generally.[8] Critics highlighted the models' similarity to frivolous playthings such as dolls, which were frequently made in wax. In this instance, the models were lambasted as expensive, but ultimately useless: nothing "but a toy for children and for persons who have no notion of science whatsoever, who regard such things as the child sees the doll."[9] Students and teachers who used models were ridiculed as immature and unmanly, as "beardless boys" and "hermaphrodites."[10] A key feature of this criticism was the materiality of the models: wax models, critics argued, were fragile and thus not suitable for hands-on interaction, which many medical practitioners and educators considered central to medical education.

PUBLIC UTILITY AND STATE SUPPORT

Arguably, this gendering critique of the Josephinum models was so aggressive precisely because models were widely praised as groundbreaking innovations that might even render traditional dissection-based anatomical instruction obsolete, and make knowledge of the inside of the body more accessible to laypeople.[11] Viennese critics notwithstanding, the models' claims to utility were widely accepted. The international fame of wax model collections, especially those of Bologna and Florence, inspired reformers around Europe to exploit models' educational potential more fully.

A particularly strong commitment to the use of models for medical education came from the French government under Napoleon. The future emperor had seen the celebrated collection of anatomical waxes in Florence during the French occupation of the 1790s. Medical professionals also highlighted the waxes' potential for professional education and public enlightenment in published accounts.[12] Two government decrees of 1805 and 1806 requested anatomical cabinets for every *école de santé*, and the creation of an anatomical modeling school in Rouen. The new school was to produce wax models for use in medical schools, for the Museum of Natural History in Paris, to make pathological pieces for medical schools and large civil and military hospitals, for obstetrical and surgical instruction of midwives, and for examining *officiers de santé*. Due to administrative problems the school was closed down in 1814.[13] Despite its relatively short life span, however, its products made their way into prestigious anatomical collections such as the museum of the Faculty of Medicine in Paris, where they continued to elicit praise and to inspire future generations of anatomical modelers.[14]

EXPERIMENTS WITH PAPER AND PAPIER-MÂCHÉ

Experiments with different materials were central to modelers' attempts at improvement in the service of public utility. The move to new materials changed models' gendered associations and uses. Critics of the celebrated anatomical waxes had pointed out that such models were both fragile and expensive: wax was a costly material and individually sculpted models required high levels of skill and time-consuming labor. In order to address either issue, modelers resorted to a wide range of materials. In Florence, for instance, the director of the Museum of Physics and Natural History, Felice Fontana, had become disappointed by his wax models' lack of robustness, and he tried to produce a decomposable model in wood instead. Midwives and obstetricians, in particular, frequently used combinations of more robust materials such as wood, leather, and textiles to produce models that allowed for hands-on interaction, such as the French royal midwife Madame du Coudray's ingenious "obstetric phantoms," or "birthing machines"; or models that afforded visibility, such as the Bologna professor of obstetrics Giovanni Antonio Galli's glass model of a womb.[15]

A development that would profoundly shape model production and use in the nineteenth century was the move to papier-mâché. In the early years of the century, the Norman physician Jean-François Ameline (1763–1835) succeeded in producing a small series of life-sized models by modeling papier-mâché directly onto real skeletons. Ameline's progress is typical of the development of anatomical models through tinkering in ways that responded directly to present requirements, and it highlights the tension between models that imitate the *haptic* qualities of the body, and those that imitate their *visual* appearance.[16] Ameline began modeling as a student in the 1780s by attaching ribbons and pieces of fabric to real bones in an attempt to understand the placement and connections of the muscles. It was through these activities, he later argued, that he became a competent surgeon.[17] Ameline took up modeling again when he was appointed teacher of anatomy at the new École Secondaire de Médecine in Caen in 1808. His first efforts were "small sacs of leather filled, and shaped like muscles." Plaited threads served to represent arteries and nerves; screws held everything in place. Ameline did not consider these pieces perfect, despite "many years of labor," but reported that they helped students remember details that had escaped their memory. The prototypes' advantages and disadvantages were closely tied to the choice of materials and their qualities. While they "presented to the touch a kind of suppleness which related to that of the imitated organs," this very flexibility simultaneously prevented Ameline from replicating "certain protrusions and contours without which one cannot have a proper understanding of nature." Thus he began to work with different materials, and finally arrived at a model whose muscles were modeled in a "solid cardboard" on real bones.[18] Much as the Chinese used paper for a wide range of uses beyond writing and drawing, nineteenth-century French men and women used varieties of paper and papier-mâché for a variety of applications, from doll making and small domestic items such as trays and inkstands to larger pieces of furniture and architectural ornamentation, and this seems not to have carried particular gendered connotations to the same degree as wax.[19] At least, Ameline kept quiet about potential resemblances to toys so as not to jeopardize models' acceptability to medical students and teachers.

Other paper technologies were Ameline's most important point of reference when he put forward his arguments for the utility of his hybrid papier-mâché-and-bone models.[20] He explicitly compared his teaching tools to anatomy manuals—cheap, small-scale handbooks lacking illustrations that were used to accompany dissections—and claimed that his models were more effective, as they "always offered *images*."[21] The problem of communicating the shape and connections of organs was considered crucial by many medical instructors. Two-dimensional illustrations, Ameline argued, could never fully convey these, not even the layered illustrations recently developed by some of his colleagues.[22] To Ameline, the advantages of his modeled pieces were especially obvious in the description of muscles. On the cadaver, these would lose shape as soon as they had been detached in the act of dissection, and could not be put back into place. The solution, therefore, was to model each muscle individually as it presented itself in situ in the living body, and then to make them detachable and reattachable. Having prepared by studying muscles in this way, Ameline claimed, his students were able to "prepare muscles with the greatest ease in dissection." His choice of material also allowed for the introduction of didactic features such as painting different nerves in different colors.[23]

COMMERCIALIZING ANATOMICAL MODELS IN PAPIER-MÂCHÉ

Ameline presented this model of detachable muscles to the Société de Médecine de Caen in 1816, and to the Academy of Medicine in Paris in 1819. While his models were praised for their accuracy and utility by academicians and colleagues, they required long hours of individual sculpting work, and were not suitable for serial production.[24] The design approach that ultimately succeeded in turning anatomical models into commercial products was developed by a medical student, Louis Thomas Jérôme Auzoux (1797–1880).

Auzoux was a farmer's son from Normandy who became a medical student in Paris in 1816.[25] He began experimenting with the production of anatomical models almost as soon as he began his medical studies, just as Ameline had done. However, by that point the French government had a widely publicized interest in promoting model production, and Auzoux was encouraged to go public with his efforts early to solicit support. In September 1822 he presented his first papier-mâché anatomical models to the Academy of Medicine. These models of a foot, leg, thigh, and pelvis were still mounted on real human bone, like Ameline's.[26] The academy praised his work, but also put forward recommendations for further improvement. In 1825 Auzoux received 1,500 francs (roughly 1.5 times the annual salary of a high-ranking civil servant) from the Ministry of the Interior to produce a life-sized human model for the Conservatoire des Arts et Métiers.[27]

Ameline's and Auzoux's modeling efforts in papier-mâché were produced by the male anatomists themselves, and intended for a male audience of medical students and teachers. Not surprisingly, then, early accounts of the models and their makers did not explicitly gender the models or their use. In the two modelers' descriptions and the reports of their contemporaries there was no indication of a gendered understanding of the models or their materials. Similarly, academicians who commented on the early efforts of Ameline and Auzoux took it for granted that their models would be used by male students and teachers; they did not discuss the models' potential use by women. Several reports put forward the argument that artificial anatomies such as Ameline's and Auzoux's could make the study of the body more palatable to those who were otherwise disgusted by dissection. Historical analyses of dissection frequently highlight its role in socializing medical students into their profession, and as opportunities to demonstrate one's masculinity in the homosocial environment of the medical profession in the later nineteenth century.[28] However, in the early decades of the century, reluctance to engage with the dead body was accepted as a common response among students that did not reflect badly on their masculinity.[29]

In model making and model use the papier-mâché bodies were part of an array of objects, especially other paper technologies. Academy reports suggest that Auzoux's prototypes were based on the study of individual cadavers.[30] Documentation from the development of later zoological models indicates that Auzoux supplemented this study of the corpse itself with illustrations he collected from textbooks and scientific publications, and with his own drawings.[31] He also consulted authoritative scholars on specific subjects such as the horse and the May beetle by correspondence. These consultations came full circle: once the new model was ready, Auzoux proudly highlighted his collaboration with experts in his marketing material.

Model production was a highly complex process that required a large number of techniques and materials.[32] Central to the models was the use of a form of papier-mâché. Auzoux developed his own secret recipe for a mixture he called "la terre," a mix of paper, starch, hemp fibers, chalk, and ground cork, which gave the material some elasticity. Hollow forms were made by pasting layers of paper strips into metal molds (fig. 11.1); the resulting shapes were then sewn together.

Figure 11.1. Molds used in model production at Auzoux's factory. Musée de l'Ecorché, Le Neubourg. Photo by Anna Maerker.

For solid parts the forms were filled with "la terre." Larger models such as the life-sized humans and the horse also included an internal iron support. Other detail was added using different materials: glass for the eyes, transparent membranes for wings, metal wires wrapped in colored paper strips for veins and arteries. The finished parts were painted in layers of color and a protective varnish, and two labeling systems were attached: one to identify details, the other a system of numbers accompanied by hands that indicated the correct order of "dissection." All parts were assembled and held together by metal pins or hook-and-eye mechanisms.

The first life-sized model of a human male of 1825 contained 66 ordinal numbers and 356 details; the model of 1830 contained 139 ordinal numbers and 1,700 details; both cost 3,000 francs. Academicians urged Auzoux to develop smaller, more affordable models as well, and in 1839 Auzoux presented a human model of three and a half feet at 1,000 francs, and two very small models of eighty-two centimeters (at 500 francs) and fifty-five centimeters (250 francs).[33] Other early products included magnified models of organs such as the eye, the larynx, the heart, and the brain, as well as a life-sized female model, organs of generation, and a series illustrating pregnancy. Auzoux also began to branch out to produce models of animals that illustrated different zoological classes, and comparative series of physiological functions such as digestion and respiration (see plate 15).

Like the eighteenth-century wax modelers, Auzoux gave a nod to well-known works of art in his model design. In his case, the life-sized male and female models echoed the poses of two celebrated sculptures, the *Venus de Medici* and the so-called *Capitoline Antinous*. However, one might argue that his choice of subject, a portrait of the young lover of Emperor Hadrian, called into question the masculinity of his male figure. With his downward gaze and soft features, the young man appears removed from the masculine ideals of action and strength embodied in the eighteenth-century waxes. In addition, in the 1840s the male model lost its most eye-catching marker of sex. The penis, which had been prominent on early models, was left out of later copies, as Auzoux began to target lay audiences and schools as potential customers. Overall, Auzoux's main concern seems to have been to adopt artistic references that would be recognized by a wide audience—in the nineteenth century, the *Capitoline Antinous* and the *Venus de Medici* were frequently reproduced in a range of sizes and materials for customers from diverse backgrounds. Auzoux's female figure also departed from earlier aesthetic conventions in crucial ways. While it paid lip service to female modesty and beauty by adopting the pose of the Medicean Venus, it was devoid of markers of female beauty that so fascinated the admirers of eighteenth-century "anatomical Venuses" such as real hair and the appearance of soft skin. Auzoux's females were as upright, skinless, and bald as their male counterparts. Like them, they depicted all functional aspects of the body in equal measure, including muscles and the nervous system.

Model use was shaped by a few central physical features of the objects: their labeling systems, coloration, and "dissectability," but also the array of other paper technologies that accompanied the model itself. A crucial element was the way in which the models opened up: "coupes ingénieuses" very different from those performed in an actual dissection.[34] This enabled full detachability as well as the visibility of the interior of organs. Auzoux's original claim that work with his models could replace work with corpses in medical education was reinforced by the introduction of a steel instrument, a kind of spatula, which was meant to be used in the "dissection" of the model much like a scalpel. However, in practice this tool is not necessary to take the model apart. More important for a productive engagement with the model were its accompanying paper technologies, especially the synoptic tables that, together with the labels on the models themselves, facilitated the identification of details and the order of "dissection." In addition, Auzoux wrote a textbook in an accessible language, using examples and comparisons familiar from domestic life.[35]

According to accounts by Auzoux and his supporters, an understanding of the body's form and function was generated by working with his papier-mâché models because they allowed repeated revision, and an interaction with the model that combined the actions of the eye and the hand, such that one could "fix [the objects] in one's memory, correct one's errors, compare objects among each other, and consider them together with matters of anatomical surgery and medical operations."[36] This form of learning could be autodidactic; the presence of a teacher was not necessary, as authoritative knowledge was fully embodied in the model and its supporting materials.

THE MODEL FACTORY AND INTERNATIONAL MARKETING

In the 1830s Auzoux founded a factory in his small hometown in Normandy. At the factory, men, women, and children performed paper work in several ways: they produced the paper paste from Auzoux's secret recipe, molded paper shapes, labeled details, and wrapped paper strips around metal wire. Auzoux also brought popular textbooks into the factory, which workers were to use to teach themselves gymnastics and choral singing.

After moving away from his initial ambition to replace dissection in medical teaching with his human models, Auzoux aligned himself with the new public health movement. Activists of the movement argued that healthy citizens were central to the achievement of social progress, and that public education was the key to health.[37] Auzoux endorsed these principles with his chosen motto, *Nosce te ipsum* (know thyself), the traditional credo of anatomists that had been adopted in the nineteenth century by many reform movements. Key to the reformers' approach was the assumption that the body was a machine that could be manipulated and optimized by its owner.[38] Traces of this mechanistic view of the body were already evident in the first edition of Auzoux's textbook of 1839, which contained frequent references to the body as "la machine animale."[39] As Auzoux increasingly targeted the market of general public education, this image of the body was made even more explicit in the second edition of 1858, which included, for instance, comparisons of the body to a steam engine.[40] Much like the adopters of photography for posture improvement analyzed by Beth Linker in this volume, Auzoux claimed that his new technology could provide great benefits to public health if it were adopted widely, and the company marketed its products increasingly to schools, universities, and the military.

Like other paper technologies in this volume—plasters, paper curls, registration tags for babies, posture recordings—the papier-mâché models as teaching tools were intimately connected to the body. Physical engagement with the models was central to their power as teaching tools, and thus the first people to benefit from the objects' educational potential were those who built them: the men, women, and children employed at Auzoux's factory. Like many contemporary industrialists, Auzoux provided paternalistic care for his workers, from physical exercise and disciplinary measures to the institution of saving plans.[41] A core feature of his factory was the provision of anatomical instruction: workers learned about anatomy and physiology from the models themselves. To publicize the high level of anatomical knowledge among his workers, Auzoux occasionally staged public examinations of his employees as "model students," model users who, thanks to the artificial anatomies, had acquired a substantial amount of anatomical knowledge. In Paris and in the Norman town of Evreux, Auzoux put both male and female workers on display responding to questioning by professors of anatomy.[42] Students and doctors witnessed how these "peasant anatomists" answered questions about human anatomy, physiology, and generation.[43] Such displays, audiences recognized, served two aims: they highlighted the fact that the models embodied sound anatomical knowledge, and they were proof that, with the help of Auzoux's papier-mâché bodies, even unschooled men, women, and children from the Norman province could acquire detailed knowledge of the human body.

Beyond such public displays, an array of paper objects also contributed crucially to the successful marketing of the new teaching tools, facilitated by the introduction of new machines for faster paper production in the first third of the nineteenth century.[44] Marketing efforts with a wider national and international reach included the frequent catalogs (distributed in French and English, and later in Japanese by a franchise opening in the 1880s), pamphlets, and printed form letters in which Auzoux announced his latest product developments, and that he blanket-mailed to universities. Thanks to its new combination of materials, the finished product was considerably sturdier than the traditional anatomical waxes, and it could be transported over long distances relatively safely. The papier-mâché bodies also largely escaped the corruption through heat and moisture that made dissection in tropical climates difficult. This made the models particularly suitable for medical education in the colonies, as commentators had observed since the presentation of Auzoux's earliest prototypes.[45] Models were exported as far as India, Brazil, Australia, and the United States.

MODEL STUDENTS

Successful learning of anatomy and physiology with the aid of the papier-mâché models was showcased by Auzoux in the case of his workers to publicize the models' power as teaching tools. Such educational success could also be used by (male and female) students themselves for professional advancement, and be instrumental in narratives of civilization and progress.

In the 1830s the viceroy of Egypt Mehmet Ali ordered the introduction of hospitals and medical schools following the European model, and he recruited the French doctor Antoine Clot to supervise the establishment of these new institutions.[46] According to Clot's reports, the local population's resistance to using corpses for medical study posed a major challenge to his work.[47] In Clot's accounts, paper tools such as Auzoux's models served as intermediary steps on the way to European-style public enlightenment. In addition to seeking toleration from local religious authorities, Clot reported the use of anatomical models and animal corpses to accustom students to working with human bodies: "Bit by bit the students overcame all prejudice and repugnance, and became convinced of the indispensable necessity of the study of anatomy." In their turn, the students served to change societal attitudes—to act as "so many apostles destined to spread the light of knowledge in the midst of a people still enslaved by prejudice and ignorance" until the public became "completely accustomed to the idea of dissection."[48] Thus, models worked in the setting prepared by Clot, and in turn prepared the local community to make corpses work as well.

As families were reluctant to allow their daughters to enroll in the new school of midwifery, in 1832 Clot resorted to buying female slaves to train as midwives.[49] A particularly gifted student soon emerged from this group of trainees. The young Abyssinian Fatme showed "a thirst for knowledge" and "persevering industry" that enabled her "to commence, poor slave as she was, a new era of civilization."[50] She was invited to give anatomical lessons at the palace to the pasha's daughter, using "several anatomical pieces." Upon leaving the palace, the pasha's daughter accorded Fatme the honorific *effendi* (learned), a distinction usually reserved for men.[51] Much like female phrenologists described by Bittel and lady reformers described by Serrano in this volume, in Fatme's case, too, her competent engagement with paper tools was used to support her claim to expertise. A practitioner of "great dignity and scholarship," she was eventually elevated to the post of director of the women's division of the hospital at Esbekie.[52] As Auzoux had staged exams of his factory workers to demonstrate the power of his teaching tools, so Clot arranged for public demonstrations of anatomy and examinations of his scholarly slaves to European visitors who concluded that Clot's efforts had rescued young "half-savages" such as Fatme and her fellow slaves from destitution and superstition, and turned them into "agents of civilization."[53] Clot himself publicized the success of his trainees as evidence of Africans' intelligence.[54]

While the models thus contributed to emancipatory claims that "civilization" could transcend boundaries of gender and race, the new medical institutions' imposition of Western attitudes to the body simultaneously contributed to enforcing the gendering and control of bodies. As Mervat Hatem has pointed out, the graduates of the school of midwives were instrumentalized to police female fertility, and in particular they "were to play a key role in the surveillance and criminalisation of abortion."[55] Thus, in Egypt the papier-mâché tools served to articulate both the boundary between male and female and between the "civilized" and "uncivilized."

EMANCIPATORY POTENTIAL

The reformist potential of the papier-mâché models was exploited not only in institutional settings such as the new state-sponsored medical schools in Egypt but also by independent lecturers in the United States who adopted Auzoux's models early.[56] Many itinerant public speakers advocated public education on health and the body for general audiences as a way to overthrow the "medical monopoly" of physicians, and to make "every man his own doctor."[57] Not surprisingly, such activities met with strong resistance from physicians who accused lecturers of quackery and licentiousness. A prominent example of a public lecturer who combined entertaining public events with a rhetoric of public emancipation from doctors' authority was the English-born Frederick Hollick, who traveled to the United States in the early 1840s and whose lectures focused strongly on matters of sexuality and reproduction, for both male and female audiences.[58] Hollick used Auzoux's models of male and female bodies to illustrate his outspoken lectures. In 1846 he was put on trial in Philadelphia for his explicit language, and for the use of "certain indecent and immoral figures and models" in his lectures.[59] Hollick was quickly acquitted, but he had to balance his emancipatory rhetoric with a modicum of respectability to maintain both his reputation as a fearless reformer and his appeal to his largely middle-class audience. As in the case of Auzoux's own marketing efforts, Hollick drew on an array of paper tools to support his activities. He was the prolific author of short, cheap books on the topics of his lectures, especially on sexuality and reproduction, which were frequently reprinted.[60] His books and advertisements publicized the Auzoux models as respectable teaching tools "of the most complete and costly kind."[61] Audience support was crucial to Hollick's success, as it had been in the case of Bittel's beleaguered phrenologists. Women, in particular, campaigned for Hollick's exoneration—and by association for their own respectability. In Philadelphia, female demonstrators argued that it was a "duty they owed the public and themselves" to defend the lecturer against anonymous charges of indecency made by "interested jackanapes."[62] They used the local press as well as public demonstrations to make their case.[63] Hollick's champions in Philadelphia relied on the free and copious circulation of paper in their campaign by instrumentalizing the local press. They also defended Hollick's own paper tools, especially the papier-mâché models. While some reports claimed that "many have fainted away at first view of Hollick's manikins," both male and female members of the audience used open letters to affirm the respectability and educational value of the models, arguing that "the Manikin, or Artificial Anatomy, by which he illustrated his subject, is a most wonderful machine. . . . It is made of papier-mâché, and represents the human body with admirable perfection."[64]

Just as Auzoux presented the body as a "machine animale," Hollick's supporters frequently stressed the machine-like nature of the models, and in particular their capacity for being dismantled thanks to the sturdy paper paste. One newspaper report praised the "artificial female figure" as "one of the most admirable pieces of mechanism that we have seen."[65] Another similarly described "the human figure—a French model, made, we believe, of papier mâché" as "beautifully constructed, and every trifling organ is not only an accurate counterpart of nature, but can be taken apart, opened examined, &c., with an ease that renders the study as perfect as an actual dissection, without the desagremens that extend a scrutiny of the real subject."[66] Users' positive testimony, published in local newspapers, was appropriated by Hollick himself, who reprinted audience letters in his own publications. In the preface to his *Outlines of Anatomy and Physiology* (1846), published in Philadelphia shortly after the trial for indecent libel (a now-obsolete legal term comparable to "public obscenity"), Hollick highlighted that public support was vital to his project of "attempting to destroy the existing monopoly of scientific knowledge," and he dedicated this work to "all those ladies and gentlemen who, with generous zeal, came to my assistance when I was subjected to an unscrupulous and illegal persecution" (fig. 11.2).[67]

Other women not only publicly proclaimed their support for the itinerant medical lecturer but actively adopted similar methods to further the cause of women's education and emancipation. They turned arguments about the necessity to overcome doctors' monopolies into arguments for female physical autonomy. One prominent example was the American activist for women's rights Paulina Wright Davis (1813–1876), who spent four years traveling and lecturing on human anatomy and physiology in the 1840s, using a life-sized Auzoux model of the female body.[68] Just as wax modeler Anna Morandi had used her identity as a mother to argue for her privileged understanding of the formation of flesh, Davis used motherhood to legitimize women's access to anatomical and medical knowledge. She argued that basic medical education was crucial for the improve-

APPENDIX. 285

WRITING DESK AND GOLD PEN PRESENTED TO DR. H. BY ONE OF HIS LADY CLASSES.

DR. HOLLICK—Dear Sir: The members of your class, desiring the gratification of offering you some testimonial of their personal regard, and grateful appreciation of the benefits which you are conferring upon them and their sex generally, respectfully request your acceptance of the accompanying writing desk.

Were it necessary, we might repeat our assurances that your services to humanity will be, by us, long and gratefully remembered. The women of this generation have reason to rejoice that, by your efforts, a new and extensive field of information has been opened to them, whence they may derive treasures of knowledge, of immense importance to themselves and their posterity, hitherto concealed within professional enclosures.

Wishing you health and happiness, we beg leave to subscribe ourselves,

Truly your Friends,
Signed on behalf of the class by,
M. G.
O. W. B.

Phila. March 20, 1845.

Figure 11.2. Expressions of gratitude from Hollick's female supporters, reproduced in F. Hollick, *The Diseases of Woman* (New York: Burgess, Stringer & Townsend, 1850), 285.

ment of public health since "women, with or without qualification, do decide what shall be done, and who shall do it, when their families are ill." Women bore the responsibility for domestic care and for employing a competent family physician, and knowledge of the human frame was the only way to improve their choice.

But education for women in medical and sexual matters was not just a practical measure to improve family health. Women, Davis asserted, had "a right to know everything that concerns our life and happiness."[69] Her ultimate goal was utopian: to restore humanity to its original perfection at the moment of creation. When he was made from "dust of [*sic*] the ground," man was "a model of physical perfection in his entire organisation." While this perfection was lost after the Fall, in Davis's utopian vision knowledge of the body would enable postlapsarian man to bring the body "back to something like its primevial [*sic*] strength and beauty." This enhancement would go well beyond the body, affecting the mind and ultimately bringing about a better society.[70] Paper tools such as "books written in an easy familiar style" could help women achieve their goal. "Models and diagrams," in particular, were highlighted by Davis as "aids not to be despised in this study."[71] Male and female supporters of her "heroic enterprise" praised her abilities as a lecturer, her "delicacy" and "graceful union of simplicity with intellectual superiority," as well as "that wonderful machine, the *modelle du femme* [*sic*]."[72] For Davis, perhaps, Auzoux's models represented her imaginary, prelapsarian humanity: strong and sturdy, physically perfect, and innocently naked.

Established analyses of anatomical models frequently highlight how such purportedly objective representations of male and female bodies articulate gender stereotypes. However, it is crucial to understand such models not merely as representations but as technologies. Paying attention to the materiality and the social context of models' production, distribution, and use shows how such practices have shaped and challenged gendered perceptions of model makers and users. Motherhood, for instance, could be used to claim privileged understanding of the formation of the flesh (in the case of modeler Anna Morandi), and also to justify female access to medical knowledge (for women's rights campaigner Paulina Wright Davis). Thus, while models could serve to reinforce traditional gender roles (as in the case of the aesthetic conventions of the anatomical waxes), they could also be used to subvert or change them through practices of use. In some cases, empowerment and discipline went hand in hand: at the new Egyptian school of midwifery papier-mâché models helped support African women's claims to civilization against established gendered and racial stereotypes, while at the same time the new European-style medical regime imposed restrictions on local women's access to abortions, thus enforcing a view of women's role in society as primarily reproductive.

Modelers' move to papier-mâché mattered in many ways. This move allowed modelers to counter critics' objections to the fragile, expensive, and problematically luxurious anatomical waxes. Models' relationships to other paper technologies were especially important for articulating the new artificial anatomies' superiority and utopian potential, and both Ameline and Auzoux discussed their products' relationship to textbooks, flap-anatomies, and other paper tools, while Auzoux in particular employed a wide range of paper technologies for his successful marketing campaign. Overall, the makers and promoters of papier-mâché models carefully balanced their comparisons: On the one hand, they highlighted similarities to well-established paper tools such as textbooks in order to appropriate for their own products the air of respectability that was associated with older forms of didactic material. Like textbooks, models could be carriers of text (in the form of labels) and incorporate the most up-to-date knowledge of human and animal bodies. Like other paper tools, models could be taken apart and juxtaposed with other objects such as synoptic tables or other models. On the other hand, when it was opportune to highlight the innovative nature of the new model makers stressed the differences between their products and existing teaching materials. The present case study highlights that paper was more than a carrier of text and images. The materiality of paper—its three-dimensionality, robustness, and detachability—was made visible and became central to models' success. Here, however, articulations of the models' utility had to be chosen with care. While producers themselves and users such as Hollick's female supporters were keen to single out models' machine-like ability to be handled, taken apart, and reassembled, and asserted that these features accounted for the models' success as teaching tools, they remained silent over the fact that this haptic appeal was potentially sensual and playful. Rather than dangerous dolls, tainted with effeminacy and delicacy, the new paper bodies were celebrated as wonderful machines devoid of moral or bodily corruption, ready to usher in a new age of public health and physical perfection.

TWELVE

Women Who Worked with Documents to Rationalize Reproduction

Dan Bouk

In the spring of 2016 three students joined me in a project building paper population models for 1930s America. My "history lab" holed up in an art studio, plundering a razor to cut our large slabs of paper board, as well as washers, bolts, and colored pencils—the various stores of each pointed out by a fellow faculty member. Abby Balfour, Erin Burke, and Christy Mills—my students—worked out a method for drawing a reliable circle and took turns hacking out the discs.[1] Then they set about drawing and shading in with rainbow colors the hundreds of sectors required by the wheels—each representing a white male's chance of living to a given age in 1930 (see plate 16). To guide their colored pencils they employed another paper tool, a printout of a pie graph showing the right size for each sector.

I had wanted to better understand how these peculiar paper population models worked, but the originals can no longer be discovered in the archives. Their absence inspired our recreational re-creation. The wheel we were building constituted one of five wheels devised by the biomathematician Alfred Lotka in the 1930s to help him test his theories of how and why the American population changed over time. With each spin of our rough-hewn wheel we would bring a fictional man into being. Another wheel would then be spun to determine the age at which that man married, which created a statistical woman—but we could not determine the source data necessary to reconstruct that disc. Nor could we build the wheel that set the woman's age at marriage. So next we made one that indicated how old the woman was when she gave birth to her first child, another act of statistical creation that required an auxiliary wheel to set the child's gender—either male or female. Finally, a wheel determined the age at which the statistical woman would die.

What escaped my attention at the time was the degree to which our hacking and shading depended fundamentally on yet another set of paper tools:[2] Elbertie Foudray's official national mortality tables prepared for the US census. These bound, bright-white pages listed long columns of printed black numbers arranged to make possible the study and prediction of rates of death by age, sex, and race. Paper cut to form Lotka's wheels had fascinated me and made paper into an interesting *material* to investigate. But Foudray's paper tables, stripped (to my eyes) of their materiality by their documentness, hid in plain sight.

Such was my blindness before I joined the project to write this book. I assumed the documentary manifestation to be paper's natural state and as with most things deemed natural or normal, I gave little further thought to paper meant to be typed or written on. The scales fell from my eyes, finally, when Elaine Leong explained what should have been obvious: that every morning I relied on paper to brew my daily cup of coffee. With that realization, it became possible for me to think about paper as a robust, flexible material and then to raise new questions about the ways the field of science and technology studies (STS) thinks about the uses of plain, old, boring documents. One discovery set in motion another: I also discovered (or finally paid attention to) Foudray and a host of other scientific women left out of our histories of population research.

In the 1930s Foudray became established as one of only a dozen or so permanent members of the Census's Division of Statistical Research—and one of only two permanent women professionals—but even while independently leading crucial work she always reported to a man and could not shake the "assistant" in her title, assistant economic analyst.[3] Her work has been forgotten. By contrast, Lotka is now one of the most celebrated figures in the early history of a twentieth-century scientific "interdiscipline" (to use a term from the historian Emily Merchant) called variously demography or population research, which played a prominent role in the history of the twentieth century. Some population researchers like Lotka served major life insurance corporations invested in predicting the course of lives while others conducted investigations funded by social elites and rich foundations into concerns that presumed racial inferiors might be on the verge of swamping native-born, white Protestants.[4] With time, population researchers gained bureaucratic power by offering their knowledge and methods to guide the building of the welfare state during the Great Depression and they won global attention by making population control (the decrease in fertility of the poorer, oppressed, or despised masses) a precondition for economic development. Without researchers like Foudray and Lotka there may have been no social security system, but also no mass programs of coerced sterilization around the world.[5]

Historians have not overlooked the power of paper in its documentary form. In his review of the literature on "paperwork," Ben Kafka traces a long history of the document stretching at least back to medieval England, and Jacob Eyferth's afterword to this volume makes clear that the history of paper documents (as well as other uses of paper) in China extends still further.[6] Paper documents afforded modern empires and corporations means to exert authority and influence across wide distances, defining "centers" and "peripheries" with paper, as JoAnne Yates and Theodore Porter have explained.[7] Another leading scholar of the document, Lisa Gitelman, asserts the primacy of the document as a genre worthy of the same attention as more charismatic forms, like the novel.[8] She builds on a foundation laid by Bruno Latour, whose work on the "immutable and combinable mobile" is a kind of obligatory passage concept in STS.[9] Both Latour and Gitelman endeavor to help us see past the document's representational work, revealing the papery materiality of documents in circulation (for Latour) or being reproduced (for Gitelman). Yet their conceptions of documents still, to a degree, overlook possibilities that paper affords the document.

This chapter argues that paper allowed documents to do much more than our scholarship usually appreciates, that paper documents were seldom immutable and were useful precisely because of the multiple purposes they could serve as they moved not only through space but through time. Many of those who best understood and took advantage of those multiple purposes were lower-status women knowledge workers. So in attending to the materiality of documents, we see also the gendering of knowledge.

It is well understood that the production, care, and reproduction of documents fell to women as bureaucracies grew in the late nineteenth century. The scholarly efforts of an earlier generation of labor and women's historians demonstrate how women in offices, aided and also burdened by new technologies—typewriters and then calculating and computing devices—became ensconced in low-status (but highly skilled) clerical work.[10] Today, exciting new work by historians like Marie Hicks, Jennifer Light, Lisa Nakamura, and Margot Lee Shetterly has brought to light forgotten contributions of women to the histories of calculation and computing—both histories suffused with paper.[11] Shetterly's tale of the heroic accomplishment of NASA's black female "computers" (people who computed before electronic machines took over) and mathematicians became a best seller and subsequently reached even larger audiences in the form of Theodore Melfi's feature film.[12] Yet many contributions by women remain obscured along with all that the study of such women's work can teach us about knowledge making.

This chapter turns now to consider in detail the gendered practices of population research taking place on and with paper at a key site for that field's development: in the offices of the Scripps Foundation for Research of Population Problems. Among the papers stored in the Scripps archives, we encounter not only the essential efforts of women researching population and rationalizing reproduction through social science but evidence of the extent to which those efforts involved the skilled use of that carefully camouflaged genre of paper tools, the document.

THE SCRIPPS FOUNDATION STUDIES

The Scripps Foundation located its central office in the small midwestern American town of Oxford, Ohio, on the campus of Miami University. There, in archival boxes, resides the vast store of paper tools generated in the development of population research. For this chapter, tools derived from two landmark studies—the Indianapolis Study and the Growth in American Families (GAF) study—will occupy most of our attention. Leafing through these archives, we can get a sense of Scripps's operations before exploring in the following two sections the possibilities of the class of paper tools that dominate the Scripps archive (a class of unparalleled modern significance: the document) and the overlooked women who turned those possibilities into knowledge.

Among the paper tools produced in the process of carrying out the Indianapolis Study one discovers a variety of printed forms, typed letters and memos, and an IBM tabulating card, all created sometime between 1941 and 1944. A four-line note typed on wrinkled paper introduces two key characters. A subordinate addresses her boss with "Dear Whelpton," walking the line between the informality of a first name and the more formal use of "Dr. Whelpton," while the letter closes with "Emily," signed by hand, nodding to her inferior academic status or his male privilege, or both.[13]

Emily Marks, whose story has yet to be included in histories of population research, had only graduated from Wellesley College in 1937 and yet four years later could already boast having won a research assistantship at Princeton and having supervised an "everyday life" study for George Gallup's public opinion institute.[14] In 1941 she coordinated a team of forty-three college-educated canvassers who scoured Indianapolis block by block (though skipping neighborhoods marked as housing almost entirely African Americans) asking questions about the couples in each house: if they had children, if they had been married before, how much they paid in rent or how much their housing was worth if rented, how long each had attended school, and what religion (Protestantism, Catholicism, or Judaism) they belonged to.[15] Marks accompanied her note to Whelpton with a copy of "Form 2" for instructing canvassers, which included guidelines on work hours ("at least 35"[16]) and tips for winning cooperation even when asking personal questions about age, children, religion, and finances. ("A matter-of-fact, pleasant approach will do much to dispel any suspicion or resentment."[17]) From that initial overview, a small group of around 1,500 native-born, white, Protestant, married women aged 27–44 who had completed at least the eighth grade were interviewed three times, and their husbands interviewed once in greater depth.[18]

Pascal K. Whelpton, the recipient of Marks's letter and a figure better represented in existing histories, was an agricultural economist who reinvented himself at Scripps, becoming the nation's foremost forecaster of the US population.[19] He joined other prominent population researchers, and eventually some psychologists too, in devising the Indianapolis Study. Whelpton oversaw its progress and published its results with Clyde V. Kiser, thanks to money from the Carnegie Corporation channeled through the Milbank Memorial Fund, another champion and organizer of population research. The study's architects believed that biological explanations for falling fertility had been discredited: instead, contraception explained why there were fewer babies—and a world filled with planned births demanded new studies of the "motivations" of parents.[20] They saw good omens for their potential to shape policy in their European peers' efforts to address falling birthrates and also in what they perceived as growing alarm among ordinary Americans about falling birthrates. Their study might lay the groundwork for a "national population commission" directing "certain changes in the number and quality of our people."[21] The word "quality" signaled the eugenic origins of the study, as the authors explained later that they needed to understand "the hereditary background and environmental surroundings that make for optimum quality in the population."[22] Even as the war against the Nazis raged, these researchers thought themselves well suited to decide the best kind of people to reproduce in America.

Turning to another major study of reproduction and another genre of paper tool/document, the Scripps archives preserve copies of paper questionnaires from the 1955 GAF study of a 2,713-woman-strong representative sample of all white American wives.[23] As early as 1938, the US government had granted population research a foothold in President Roosevelt's "New Deal" state through a "Committee on Populations Problems," which included Whelpton's Scripps colleague Warren S. Thompson and featured Whelpton's population projections.[24]

Whelpton subsequently won a post in the new United Nations as director of the Population Division from 1950 to 1953.[25] The Rockefeller Foundation, both directly and through its funding of the Population Council (founded in 1952), further organized population researchers into an informal national population committee.[26] Rockefeller funds provided over $100,000 to print the questionnaire surviving in this folder and to support Scripps in a joint project with the University of Michigan's Survey Research Center. The project was the 1955 GAF study, a more ambitious, national follow-up to the Indianapolis Study that promised to plug the leaks in demographic knowledge that had allowed the baby boom to so surprise prognosticators like Whelpton.[27] The GAF built the case that *Homo economicus* slept in every bedroom, that fertility curves were bound up with business cycles, and that a child should be seen as a "consumption good."[28] In years to come, the GAF inspired the so-called KAP (Knowledge, Attitudes, Practice) studies at the heart of programs to control populations—whether via coerced sterilization, mass IUD distribution, or less-aggressive contraceptive education—implemented by modernizing, economic-growth-hungry postcolonial states.[29]

THE POSSIBILITIES OF DOCUMENTS

In *Science in Action*, Bruno Latour tells the story of the French explorer Lapérouse's late eighteenth-century nautical expedition to the island of Sakhalin. According to Latour, Lapérouse copied islanders' drawings in the sand onto paper, transforming these maps of Sakhalin into "immutable and combinable mobiles" that could be carried back to a "center of calculation" to be processed and creatively combined. In Latour's model, which has been enormously influential in science and technology studies, paper documents stand at the center of the story because they afford writing, rescaling, and easy shipment (and so facilitate making knowledge as well as colonial conquest).

Paper, however, affords more possibilities than were dreamt of in Latour's anthropology. Scripps documents, for instance, could change their types: some began as a chart, table, or questionnaire and then mutated, unexpectedly, into quite different paper tools. Consider the way a survey questionnaire could become a site for calculation. Whelpton and his colleagues printed Form G-1 to explain to interviewers and analysts the primary hypotheses of the study, numbering twenty-three altogether, and to show them how the survey questions mapped onto those hypotheses. The form worked as a reference text. But as paper afforded the possibility for more writing on top of or around prior typing, the form could become a new kind of tool. In this case, Whelpton repurposed the form as a kind of index and spreadsheet. Next to each hypothesis he—or, perhaps someone else—handwrote in pencil an *H* (for husband) and *W* (for wife) with a number next to each, totaling the number of gendered questions related to each hypothesis. This exercise revealed the most-tested hypothesis to be "12. Family and childhood situations and attitudes affect the proportion of couples practicing contraception effectively and the size of planned families." Twenty-seven questions for both husbands and wives tested that hypothesis. To take an example with a large gender difference, the hypothesis that contraception correlated with a "feeling that children interfere with personal freedom" drew from eleven questions for husbands and fifteen for wives, an imbalance that implies an assumption that wives' freedom would be either more imperiled by children or would more directly drive contraception. A note in green pencil on the first page alerted the reader to new indexical qualities granted the form: "Basic CS and Col. shown in green." Each question on the form was then accompanied by handwritten numbers indicating the IBM card series (CS) and column number (Col.) where that question could be found. In this example, a supposedly *immutable mobile* became the *center of calculation*, a place where pretabulating activities—something akin to programming or algorithm development—could take place.[30]

Documents could be nearly silent and readily hidden. Therefore they acted in some cases as technologies of privacy. During the Indianapolis Study, husbands and wives were interviewed together at one point, but "each of them independently checked categories on a questionnaire (Forms B or B2 for the wife and C or C2 for the husband) dealing primarily with attitudes," and then spoke aloud when answering questions about their backgrounds and the family's economic circumstances. As a footnote explained: "To encourage frankness, the interviewer always emphasized at the outset that the husband would not see the wife's schedule nor the wife's the husband's."[31] Paper questionnaires made the promise of privacy plausible.

Paper documents were even capable of emotional labor. For the 1955 GAF study, interviewers brought two paper cards along with them to mediate discomfort and difference.[32] A little more than halfway through the interview, the conversation turned to specific means of contraception. Early tests of the survey in Detroit revealed that respondents *and interviewers* did not care to say aloud the names of contraceptive techniques.[33] Card 1 named a series of birth control techniques. Card 2 listed two sexual practices that respondents might not consider birth control (the "rhythm method" and douching after intercourse), but that "public health workers" did.[34] The cards were exceptional elements of the survey in the way they limited possible responses. Many other questions were open-ended; responses would have to be coded later for statistical analysis. Indeed, some questions invited very broad responses, such as "How do you feel about the way your life is turning out?"[35] When interviewers drew out these cards, they created a special paper interface, one that allowed for a frank but limited discussion of contraception for two women who may have lacked a shared vocabulary for talking about controlling sex or who felt visceral discomfort at the prospect of saying words like "rubber condom" or "withdrawal."

By design, documents frequently exerted power, including gendered forms of power, by virtue of being fundamentally mutable. The so-called blank form offers the clearest example. Paper blanks existed to be changed, to be filled. The numbers on computer cards fulfilled their purpose only when punched. Blank forms, as a genre, probably began with only a few blank spaces: for a name and a date, perhaps a place. Peter Stallybrass notes that our first evidence of Gutenberg's press comes from a printed indulgence, essentially a blank form.[36] Andrew Mendelsohn points to early printed blanks entering medical practice through leprosy inspection certificates, which were similarly text heavy and not all that blank.[37] By the nineteenth century, mass production of more complicated blank forms—with more blanks than text—accounted for a tenth of the printing industry, according to Gitelman.[38]

The blanks that Marks distributed to her canvassers had sixty-two empty fields. Some, like those for address or state of birth, allowed quite a bit of space for writing. Many, though, were only big enough for a check or an *X*. These smaller blanks exerted categorical power: they existed to force diverse individuals into preset groups and defined the acceptable categories of existence.[39] For the Indianapolis survey those groups broke down along religious lines or level of educational attainment, but the most important categorical binary blended gender and sexuality: H/W for husband or wife, a binary that organized the rest of the blank form. To be recorded by the survey, a person had first to be defined as a wife or a husband through an act of marking on paper.

By filling other blanks or affixing documents to one another, canvassers were guided in the project of making certain people disappear so far as the study was concerned. Marks's instructions explained to canvassers that "Negro, Japanese, Phillipine [*sic*], and all other people except whites are treated as colored."[40] On a pink paper, canvassers could indicate that they had skipped entire blocks because they were "partly or wholely colored."[41] A Works Progress Administration inventory of real estate guided these omissions: it specified those areas occupied by 91 percent or more "colored." Cards filled out for those people of color whose doorsteps canvassers accidentally visited were to be left almost entirely blank. A similar fate awaited houses with women deemed too old (over forty-five) or apartments housing people who did not cohabit according to heterosexual family norms.[42] "Husband or wife foreign born" warranted skipping much of the card for whites too, while other papers reveal concerns about "non-native" whites "passing" as American born.[43] Later interviews drew further limits, excluding all save Protestants, refusing those who did not graduate from elementary school, only speaking to women married before the age of thirty or men before forty, and excluding people who had remarried.[44]

Paper workers understood that documents made of paper could adapt to unforeseen circumstances. The Indianapolis researchers counted on that robustness. They expected human life to press and ultimately overflow against the preprinted bounds of their cards and questionnaires. Marks instructed her charges on how to channel that overflow: "Your own comments regarding any of the items should go only on the back of the card. Use foot-note form to indicate to which question they relate."[45] The back of the card could hold canvassers' estimates of an apartment's value, notes about peculiar rental arrangements, warnings about impending vacations that could interfere with follow-up interviews, and any number of other details.[46] Instructions to later interviewers authorized filling every nook and cranny of the provided forms: "Use margins and blank spaces provided on most forms." But Whelpton recognized that even that might not be enough. "You should always have with you paper on which to record significant comments for which there is not enough room on the schedule," he instructed.[47]

For the 1955 GAF study, the paper questionnaire drove the encounter and, much more than with the Indianapolis forms, it disciplined the women interviewers sent out into the field. It scripted every question, from the first ("Do you feel that it is easier or harder to raise a family these days than when you were a child?") to the last (the names of all children living in the house).[48] It instructed the interviewer on how to proceed based on possible responses, even supplying hints for getting better answers. After asking about the ideal number of children for the "*average* American family," for instance, the questionnaire anticipated an answer of "it depends," from a woman resisting the invitation to smooth out peculiarities among her neighbors. It instructed interviewers to reply: "As things are now for the *average* American family how many children would you say is the ideal number?"[49] Such a complicated questionnaire would still have required significant training to be used effectively, but as Merchant has argued, the 150 female interviewers operated with little room for judgment, especially when compared to their contemporaries working on the Kinsey sex study.[50]

Another folder of documents—these from the 1960 iteration of the study—reveals that even with such discipline and control, it remained difficult to fit complicated lives into fixed boxes. The Scripps office employed a series of small paper cards used to record and then resolve "coding" problems. How, for instance, in the case of a question asking, "What were your reasons for not having a larger family?" should one punch into a computer card the answer "They're a lot of trouble"?[51] In other cases, respondents named religions that the survey didn't encompass ("Holiness," "Mormon") and explained why they left jobs, quit school, or decided to stop having kids with hard-to-classify reasons: mothers who fell ill, cross-country moves, or quite often with claims not to know the reason. The office did its best to label each response with an appropriate code. Responses had to fit the classification system, just as individuals had to fit within an established gender and role—Scripps's paper work endeavored to hold the mess and variety of life within neat and rational lines, yet the very difficulty of their project suggests the instability of gender and gendered norms.

THE PAPER WORKERS

Men's names graced the bylines of studies and grants detailing the work at Scripps, but women working with paper contributed much more intellectual and physical labor to the project. For the Indianapolis initial household survey, Marks shepherded her team of four married women (identified by their husbands' names publicly as Mrs. Horace R. McClure, Mrs. John Paul Ragsdale, Mrs. T. M. Druliner, Mrs. Harvey N. Brown), two unmarried women (Miss Miriam Bintz, Miss Nontrew D. Goetz), and one man (Schafer Williams, who we assume is a man because he is allowed a name without a title).[52] As the canvass ramped up in the summer, Marks took on thirty-six more field staff, most of whom were college seniors or recent graduates "with a real interest in scientific study . . . highly recommended by competent persons"—and, I think, mostly women too.[53] Marks served as the public face of the operation. The *Indianapolis Star* profiled her in March 1941 before the survey began in earnest, printing a large photo of a smiling Marks pointing at a city map, while accompanying text emphasized her elite ties to Wellesley, Princeton, and Gallup.[54]

The Indianapolis survey preferred women surveyors and interviewers. Marks joined with another researcher, Martha Sampson Herrick, to train eleven women interviewers—each possessing an extensive education in the social sciences already—in the techniques necessary to conduct in-depth interviews between April 15, 1941, and January 31, 1942, with wives and their husbands about their pregnancies and children, the ones they had planned and the ones they had not. Whelpton and Kiser later credited the success of the study in gathering so much data to the interview skill of Mary M. Aikin, Miriam Bintz (a holdover from the survey), Frances N. Butts, Margaret Creviston, Gertrude D. Davis, Vida Davison, Dorothy McMillin Gross, Helen Jennings, Margaret A. McConnell, Ruth G. Moss, and Virginia Kahn White.[55] They singled out "Mrs. Gross and Miss Jennings" in particular for accumulating only 9 refusals against 206 completed interviews.[56]

Concern that women would balk at talking about their fertility with strange men, even if those men were scientifically trained, may have played a role in these gender preferences. Indeed, the GAF study made this preference and reasoning explicit. As one grant application explained: "In view of the nature of the subject matter, only women will be used as interviewers."[57] Yet this concern was neither natural nor universal. Alfred Kinsey's much more probing sex studies assumed that white, Protestant, heterosexual men were best suited to interview anyone, male or female.[58] Availability and cost may have been factors looming in the background. Highly educated young women would be found with apparent ease in large numbers and hired to do temporary work. Canvassers were paid sixty-five cents an hour, which would have made for an income of just over $1,000 a year if they worked the minimum suggested thirty-five hours a week.[59] Such an income would have placed these canvassers somewhere around the top quartile in terms of income levels for women nationally as reported in the 1940 census or approaching the median income for all men.[60] That is to say, these women were hardly paid badly in the context of the entire US population and yet were probably available at a discount for their class, race, education, and region because of their gender. As the Indianapolis survey model spread abroad to postcolonial spaces, the initial preference for women interviewers (whether its origin was ideological or financial) developed into a stricter gendering of interviews. In 1954 Dr. Radha Kamal Mukerjee of the J. K. Institute of Sociology and Human Relations led a survey of around one thousand couples in Uttar Pradesh, India, that employed a "lady health visitor" to speak with "wives" and a presumably male "economic investigator" to speak with "the men."[61] As fertility research began to assert the fundamental interconnection of economics and fertility, it also reinscribed a gendering of the productive sphere as male and the reproductive sphere as female.

Women crowded the Scripps offices and were crucial to its successes. A 1952 photograph published in the Rockefeller Foundation's annual report displays this fact and yet also obfuscates it (fig. 12.1).[62] Four clerks, all women, sit at desks. One woman stands to the side, apparently over a drafting table—names scrawled on the rear of the original image in the Rockefeller Archives identify her as either Jane Hartley or Elizabeth Bogue. A man in a suit and tie—the young researcher Donald Bogue, married to Elizabeth—leans over one of the desks toward the center of the picture. The photo suggests his authority—he appears to be the one overseeing the clerks.

What the photo gets right is the essential role women played in making calculations. Before them are spread broad sheets of paper to be used for calculating. Similar sheets preserved in the archive bear a fine grid mainly in blue, probably designed for ledger entries, but also well suited to graphing or hand calculation.[63] Mechanical calculators rest beside the papers and the clerks all hold pencils in hand.

A brief story sheds light on just how crucial women's intellectual labor

Figure 12.1. "Statistical Group preparing data for analysis." Jane Hartley, Elizabeth Bogue, Gracie van Winkle, Donald J. Bogue, Dorothy Harris, and Nettie Ulreich. Rockefeller Foundation, *Annual Report* (1952), 237. Original in box 42, folder 1023, Rockefeller Foundation Photograph Collection.

was. Whelpton had money—a $6,500 grant from the Rockefeller Foundation approved on June 26, 1944[64]—that he planned to use to complete a new "prolificacy study." The study aimed to help researchers better understand how a mother's age at marriage and upon the birth of her first child could be used to predict the total number of children she would have in her life. To do that, Whelpton's office would begin with historical data and construct a table presenting fertility prospects as a function of age. The tables amounted to a historical picture of women's reproductive choices employed to predict (or intervene in) future reproduction. They also, like Lotka's wheels, imagined the *end* of a woman's life to be marriage and childbearing. That is: the end as in the goal, but also the end as in the point at which a woman's life stops being interesting to the investigator.

But Scripps lacked computational capacity—which is to say that it lacked educated women. As Scripps's director, Warren Thompson, explained: "The most important factor in the situation was our inability to find competent statistical clerks under war conditions. Our community had been pretty well drained of such people by the attraction of war jobs in Cincinnati, Dayton and other centers of war production, more so than we had thought when we submitted the project for your consideration."[65] A slightly earlier project's records make clear that statistical clerks at Scripps were all women. Those clerks included Isabelle Brittain, Rosalind Ellis, Jane Frederikson, Evangelyn Minnis, Mary L. Smalley, and Ruth Smith.[66] Scripps could find only five clerks total in 1945: Brittain, joined by Doris Schaffer, H. B. Abbot, N. W. Edington, and H. S. Pollard.[67] They were not enough. Without more clerks, Whelpton's team could not do the necessary but laborious work of decomposing and recalculating its figures. Scripps's director scrambled to beg the Rockefeller Foundation to extend their grant. Crisis was averted.[68]

But the photo in figure 12.2 obscures the importance of women in *organizing* Scripps's computational labor. To oversee all computation work, Scripps employed one woman as a statistical supervisor. In 1943–1945 Schaffer filled that role and earned around $2,000 in grant money per year for her efforts, while the women she oversaw each earned a monthly wage around $100. Schaffer and her clerks earned their wages doing anything but rote work.

Men designed computing procedures at Scripps, but women required skill to translate those instructions into calculations. In 1955–1956 the man who appears to have worked most directly with the computers was a recent PhD named Norman Ryder, who would become one of the most influential demographers of his time. In the Scripps files, a typescript signed by Ryder narrates the procedures that could produce demographic tables.[69] In another file we find a similar handwritten procedure signed by Ryder.[70] The procedures explain a sequence of calculations to be carried forth and provide necessary equations and methods, but they assumed a skilled computing force. The procedures seldom specified where the data should come from. And they expected familiarity with fairly sophisticated techniques, as when Ryder wrote: "To obtain July 1 cohort sizes for intercensal years, interpolate (straight line) between adjacent census values for the cohort concerned."[71] Ryder could safely presume that the computers individually, or with the help of a supervisor like Schaffer or Brittain, could follow such instructions.

A series of tables and calculations from 1956 shows what it would have looked like to follow such a procedure, and demonstrates the ways the computers organized themselves and kept track of their work. The fourteen tables together lay out the stages necessary to take a group of five-year cohorts from census data and transform them into single-year cohorts. It begins with a table of contents written in blue ink on rough brown paper. This may be Ryder's writing, or perhaps more likely, the writing of one of the supervisors, organizing the tables after the fact. The table of contents details the tables that follow and the calculations used to complete them. Most of the rest of the sheets are ledger pages filled with handwritten figures. Check marks throughout the calculations suggest that computer supervisors had a system to check calculations, just as the ordered nature of the computer work shines through in regular notes next to each table showing where data from one table has been used to create a new table.[72]

Despite an apparently well-organized female computing operation, Ryder's authority still resurfaces throughout, in notes like "Mr. Ryder said to ignore corrections 9-13-56 RWS," where RWS may be the initials of Ruth Smith, one of the computers. Brown sheets and one typescript interrupt the ledger pages, each introducing new calculations into the series, suggesting that Ryder has injected new steps into the procedure. The typescript began as an instruction sheet, only to be filled by a fresh table of corresponding calculations.[73] Ryder, finally, emphasized his authority as he effaced the computational work of his staff within the passive voice. The typescript reads: "These computations were carried out for native white life tables."[74] The gendered division of labor, worked out on paper, allowed Ryder to claim authorship for this calculating work. In this sense, the 1952 photo's elision of women supervisors speaks to a truth of the gendered hierarchies of Scripps.

The Scripps archives offer the barest glimpses into the work and lives of that institution's female social surveyors and computers. Yet I felt a thrill, and I think my student researchers in our history lab did too, whenever we unearthed even a tiny new detail. When I found the letter from Marks to Whelpton, pleasure coursed through my veins to see "Emily" written in her own hand. The first moment I could, I shared the letter with Christy Mills, who had done remarkable work gathering every possible detail of Marks's life that could be found. She shared with me the sense that we were doing important work paying attention to these, as she put it, "forgotten female bureaucrats."[75]

WE, TOO, WERE WORKING WITH PAPER

We *were* doing important work as we endeavored to better understand the documents that allowed Scripps to create population knowledge and the skilled knowledge workers—most of whom were women—who knew that documents were seldom immutable and because of that fact were very useful. Yet our thrill of recovery, replete with a sense of some small justice done, must wrestle with the moral ambiguities of the case.

The movie adaptation of Shetterly's *Hidden Figures* chooses to tell the story of the race for space, seeing in it a goal that can be widely celebrated, even as the book upon which the film is based devotes most of its time and attention to the work these forgotten mathematicians, computers, and engineers did aiding in the development of military aircraft. Shetterly has much to celebrate in this book, and yet that celebration allows one to overlook the entire story's imbrication in the growth of the Cold War–era military-industrial complex. In the same way, our recovery of the Scripps interviewers' and computers' remarkable work must be understood with a certain tragic sensibility, with the understanding that their work did much to help us understand how ordinary people thought about their fertility, but also supported a global network of population control methods that often limited severely the fertility choices of poor women or women of color in the United States and throughout the Global South. As one of my students, Matthew Kato, noted upon reading a draft of this paper, the Scripps women are far from the most disadvantaged or oppressed in this story.

Future historians may happen upon this book and find themselves thinking too about the inscriptions of gender on and with paper. They will find some evidence that gendered power dynamics have shifted, that most of these chapters were written by scholars who were women. They may also note some continuities, that my little history lab included one white male (that's me) in a position of authority and three white, female students working just above the minimum wage as temporary labor. These details may not be on their own troubling, but given the history in this essay, they give me pause at least and impel me to think about what affirmative steps I might better take in the future to make reparation for the gendered power imbalances of the past. This essay, in historicizing partially the inscription of gendered distinctions between production and reproduction, is one step. But more should be taken. Our white-collar work offered us intellectual pleasure and satisfaction, as well as the opportunity to play with old documents and also razors, colored pencils, and spinning wheels—access to such opportunity must be made more widely available and equitably allocated.

AFTERWORD

Making and Using Paper in Late Imperial China

Comparative Reflections on Working and Knowing beyond the Page

Jacob Eyferth

In 1839 the papermakers of Macun and Zhongxing (Jiajiang County, Sichuan) sued the papermakers' association of neighboring Yingjiang town. The details of the case are murky: a few years earlier, the Yingjiang association had commissioned a statue of Cai Lun, the Han dynasty court eunuch who invented paper and is worshipped to this day as the protector of the papermaking trade. Every autumn, Yingjiang papermakers carried the Cai Lun statue through the villages of Jiajiang's paper districts and held a banquet in a temple, asking for Cai Lun's blessing. Yingjiang papermakers claimed that possession of the statue gave them ritual precedence over their neighbors in Zhongxing and Macun and demanded that those associations participate in their banquet and contribute to its costs. Papermakers in Zhongxing and Macun, on the other hand, insisted that they had always worshipped Cai Lun on their own and owed no allegiance to Yingjiang. The conflict escalated, leading to threats, scuffles, and raids on bamboo groves. After a particularly violent affray, the Yingjiang papermakers were brought to court.

I am less interested here in the details of the case than in the county magistrate's perception of the cultural significance of paper, recorded on a stele in the hills of Jiajiang. As Qing magistrates were wont to do, magistrate Ma Fenggao scolded the "ignorant" and "muddleheaded" litigants for failing to maintain harmony in their community. Did they not understand that all of them owed their livelihood to Lord Cai, who had taught them the art of papermaking? Were they not all disciples of Lord Cai, who wanted them to share the benefits of the trade? After a lengthy diatribe, the magistrate embarked on a general reflection on the importance of paper: "Under the Three Dynasties [China's mythical Golden Age] paper did not yet exist. Big matters were recorded on wooden boards, small matters on bamboo slips. In affairs of state, simplicity was cherished; therefore, wood and bamboo were sufficient. After the Qin and Han dynasties, changes and altercations became numerous. . . . Wooden boards and bamboo slips, when they accumulate, are less convenient than paper. If Heaven had not sent Lord Cai, who with divine intelligence invented paper, for all their weight wooden boards and bamboo slips would have been insufficient."[1] In magistrate Ma's view, paper existed to ease the work of bureaucrats struggling to keep up with a growing flood of documents. Cai Lun, himself a courtier and administrator, created paper not as a medium for writing letters or poetry, even less as a material from which windows, umbrellas, toys, playing cards, and a myriad other things were made, but solely to enable imperial administrators to better do their work. On reflection, magistrate Ma might have allowed that paper also has a role to play in the moral education of the populace and in the scholarly pursuits of men like himself: paper was needed to print the Confucian classics, and together with brushes, ink, and inkstones, it was essential to the complex of literate activities known as *wen*, a term whose meanings include writing, the written word, literature, and civilization. In the view of elite men like Ma, such pursuits were related to self-governance and self-perfection: paper work enabled literate men (and a few literate women) to rule themselves, and self-rule in turn qualified them to rule the unlettered masses.

In this view—and in commonplace modern understandings—paper is almost devoid of materiality. It is a surface without depth, a blank medium that exists only to be inscribed with signs. Flat, light, featureless, durable, and erasable, paper allows us to reproduce and simplify the complex social and material world that we inhabit. By arranging symbols on a page, we create a frictionless paper world in which paper facts and paper people can be manipulated provisionally, without fear of permanent real-world consequences. Complexity emerges not from the material itself, its physicality and depth, but from the inscription of words on a page; the ordering of pages into books, folders, card collections, etc.; and the linking of books and folders into a complex architecture of archives, libraries, reference systems, etc. Paper so understood is a tool of detached governance: by functioning as a distancing interface between our minds and the sensory impressions that threaten to overwhelm us, paper helps us to govern the material world, others, and ourselves.

Contrast this vision of detachment and control with the experience of the papermaker. People in the Jiajiang hills made paper from bamboo stalks, a stiff and resilient material that had to be coaxed through multiple transformations before it became clean, soft, white paper. After the bamboo harvest in early summer, the stalks were rotted in ponds until they were reduced to a black, smelly, slimy mass. Then they were steamed twice, first with quicklime, then a second time with potash, and pulped in foot-operated hammer mills. The resulting pulp was then fermented again, rinsed in water, bleached, and finally formed into sheets in the papermaker's vat. The materials papermakers worked with were the opposite of flat, featureless, and yielding: they were bulky and resistant, transforming (and indeed deforming) papermakers' bodies just as papermakers transformed these materials into paper. Papermakers' workshops were cold and drafty in the humid winter and mosquito-infested in the sweltering summers. The smell of rotting bamboo stung workers' nostrils, while lye and caustic soda caused their hands and feet to swell and their skin to crack. Bodies bent under heavy loads of coal, soda, and bamboo, carried on shoulder poles from the valleys to workshops in the hills. An average workshop needed fifteen to twenty tons of raw materials—bamboo, coal, soda, and quicklime—for a yearly output of two to three tons of paper, and papermakers and their hired porters carried loads that weighed more than their body weight up steep mountain paths. While papermakers handled (and understood, and cherished) paper, most of their work was not with paper sheets but with pulp—a bulky, shapeless, heavy substance. Paper in the workshop was a work in process, endlessly malleable (as pulp was dyed, bleached, and molded into different shapes), reversible (as reject paper was returned to the vat for pulping), and pregnant with spontaneous transformations (as changes in the weather could cause paper to rot or pulp to curdle in the vat).

I bring up the opposing worlds of papermakers and administrators to think more broadly about the "affordances" (a term frequently used in this volume) of paper. The affordances that first come to mind when thinking about paper have to do with its lack of fleshy physicality: its blankness, dryness, flatness; its light weight and small volume; its "open-endedness" and "incompleteness."[2] These are absences rather than presences: paper is valued because it offers no resistance to thought and action. Paper so understood is a data carrier, a precursor of the microchip, and as with the microchip, the materiality of the medium disappears behind the data that it carries. However, inscribability is not an intrinsic feature of paper: paper becomes suitable for writing only through the additional processes of sizing and glazing, and it seems that most early paper was intended for clothing and wrapping—a cheap substitute for hemp and silk—rather than writing.[3] As the chapters in this volume show, paper is a tangible material with mass and weight, and bulky materiality is essential to its functions. The Prussian census of 1871, Christine von Oertzen tells us, weighed 375 tons, and the distribution to census tabulators involved complex logistical operations—which makes one wonder how far modern Prussia had advanced over the China of the Three Dynasties, when bureaucrats tottered under the accumulated weight of bamboo slips and wooden tablets. Paper can also be worked into three-dimensional shapes (such as the anatomical models discussed in Maerker's chapter) that serve purposes other than writing. Paper, in short, may be a precursor of the microchip, but it is also an early cousin of celluloid, Bakelite, and other modern plastics.

In their introduction to this volume, Bittel, Leong, and von Oertzen show how our tendency to see paper as surface without depth leads us to overlook the sociomaterial processes that underlie knowledge production. As long as we see paper as data carrier, we do not see the hands that shape, mold, fold, cut, paste, and inscribe the paper and thus produce the preconditions for the production of new knowledge. Since the hands that perform such work are often female, a failure to acknowledge the materiality of paper results in an erasure of women's work—in particular, of the epistemic work done by women. My aim in the following pages is to reflect on the connections between materiality, epistemic work, and gender from a Chinese perspective. Such a perspective is helpful not only because China, as the world's oldest paper culture, offers plenty of scope for comparison but also because a generation of feminist historians of China has thought hard about gender, knowledge, and work and has produced insights that may be useful to historians of Europe and America. I draw three propositions from the work of these scholars. First, the different affordances of paper—its blankness, smoothness, and light weight, but also its bulk and plasticity—correspond to different skills. Paper is different things to different people, depending on the learned and embodied skills with which they approach it. Second, such skills, while individually embodied, are best understood as communal property: it is not the individual papermaker, bureaucrat, census tabulator, or phrenologist who is the ultimate bearer of knowledge but the wider community of skilled practitioners. Third, most societies have a tendency to map skills onto preexisting social divisions, such as gender, caste, or class. Francesca Bray speaks of *androtechnics* and *gynotechnics*: technologies that are so deeply gendered that they come to define what it is to be a man or woman. This gendering of technical and epistemic work may be more pervasive in traditional China than in modern and early modern Europe, but it is evident in most of the chapters of this book.

PAPER USAGES IN CHINA

Paper was used in China for a thousand years before it spread to Europe, and for much of recent history, paper was cheaper and more widespread in China than in Europe.[4] According to Chinese records, Cai Lun made the first paper from old rags and nets, but Chinese papermakers soon perfected methods of making paper directly from plant materials—in contrast to their European counterparts, who did not learn to produce pulp from plants until the late eighteenth century.[5] A second breakthrough came with the invention of bamboo paper in the Song dynasty. Early Chinese paper was made from the inner bark of the paper mulberry and other shrubs, and northern China, Korea, and Japan continued to rely on bark paper until the nineteenth century. Such paper is strong and durable but difficult to produce in quantity. In contrast, bamboo is ubiquitous in Asia, renewable, and fast-growing. Papermakers in Song-period Fujian discovered that its tough fibers could be broken down by double steaming with quicklime and potash—a process that did away with the manual beating and scraping that is necessary for making tree bark paper. By the late Ming dynasty, these processes had been perfected to a degree that made it possible to produce paper at an industrial scale. Bamboo paper technology spread across southern and central China in the Ming and Qing dynasty, leading to a great reduction in the price of paper and to an explosion of vernacular printing.[6]

Originally invented as a cheap substitute for silk and hempen fabrics, paper soon became the preferred writing material in China. Since the eighth century, if not earlier, the Chinese used marbled, embossed, perfumed, gold-sprinkled, and watermarked paper for letters and calling cards. However, paper was never the only medium of writing. Widespread graphomania led Chinese literati to inscribe virtually every material that could be written on, from silk fans and umbrellas to wooden furniture, porcelain vases, walls, and rock faces. While brush and paper played important roles in elite consciousness, practitioners of *wen* (writing, belles lettres, civilization) were just as likely to demonstrate their sophistication by carving characters on stone or wooden surfaces.[7] Writing and paper work, in other words, never fused into a single image, at least not to the extent that they did in Europe: the educated person (*wenren*) was of necessity a writer, but he might write on banana leaves, carve poems into stone, or embroider calligraphy on silk: all of these activities were equally manifestations of *wen*. The link between paper work and bureaucracy, in contrast, was quite firm: state functionaries from the emperor down to the lowly magistrate spent much of their time writing, and—in their capacity as bureaucrats, at least—they wrote on paper.

Other usages for paper included toys (kites, pinwheels, puppets, etc.), clothing, domestic furnishings, and toilet paper, widely used since the sixth century.

Paper reinforced cuffs and collars and gave garments shape; oiled or lacquered paper hats and coats protected their wearers from wind and rain. At home, paper covered cracks in walls, paper screens partitioned rooms, and paper windows let in light and kept out drafts. Chinese sources mention paper blankets and impenetrable paper armor, as strong as chain mail and more comfortable to wear.[8] Paper sealed jars and bottles and protected perishable food. Folded into envelopes or containers, paper was used to transport and store anything from silkworm eggs to clothing. In a usage apparently limited to Korea and Japan, paper strips were twisted into yarn and woven into shoes and belts, said to be as strong and flexible as leather; chamber pots woven from paper strips and made waterproof with lacquer were an indispensable part of a Korean dowry.[9]

One use that was widespread in China but virtually unknown in the West was paper as a means of communication with the heavens and the afterworld.[10] It is safe to assume that well into the twentieth century, the Chinese burned more paper in religious ceremonies than they used for all other purposes combined. Sacrificial paper came in hundreds of varieties—dyed in vibrant colors, cut to resemble strings of cash, printed in imitation of banknotes and official edicts, coated with tinfoil, and folded in the shape of gold or silver ingots. Paper was also stretched over bamboo frames to replicate houses, horses, and palanquins—or, more recently, luxury cars, smartphones, computers, and US or EU passports—so as to provide the deceased with comfort in the afterlife. In early twentieth-century Jiajiang, the customary standard for a funeral was to burn 144,000 sheets of paper money, as well as large quantities of paper replicas.[11] These offerings were repeated, at a reduced scale, on death anniversaries and holidays. Underlying this practice was the notion that each newborn child contracts a debt for its new life, which is reduced by virtuous conduct but never entirely repaid. The deceased, therefore, returns to the underworld indebted and cannot be released back into life until the debt is paid off. This is done by descendants who burn spirit money (the precise quantity depends on one's years of birth and death) together with supporting paperwork to ensure that the money goes to the right account in the underworld treasury.[12]

Obviously, such practices would not have developed without widespread familiarity with banknotes, drafts, and other instruments of payment, as well as familiarity with a paper-based bureaucracy. People sent money to the underworld in much the same way as they sent it to business partners in a distant province, with the difference that the "spirit money" was incinerated in the process. People assumed that the underworld was run by ghostly bureaucrats who, like their earthly counterparts, demanded documentation for claims to property and status. Excavated tombs from the Han dynasty contain land contracts (written, at this early stage, on wooden tablets rather than on paper) to prove that the deceased owned the land in which he was interred, that he had paid his earthly taxes and must not be taxed again or called up for corvée labor in the afterworld.[13] In late imperial times, people transmitted not only money but also contracts, petitions, and bureaucratic documents to their deceased relatives by burning them. Paper, in short, was understood as a universal medium of exchange and communication, capable of linking this world to the underworld, just as it tied different parts of the human world together.

Paper's association with the nonhuman realms was due in part to its physical features—or rather to its lack of physicality. Paper sheets are the product of a mysterious transmutation of rotting bamboo, dripping with black slime, into light, clean, disembodied paper. When paper is burned, a second transformation takes place: good spirit paper burns almost without residue, producing odorless white smoke that rises straight to heaven and leaving only very small amounts of white, powdery ash. Paper thus suggests purity and sublimation, a reduction of fleshy physicality into pure otherworldly essence.

AFFORDANCES, SKILL, AND COMMUNITIES OF SKILL

As Bittel, Bouk, Leong, and Werrett show in their chapters, paper "affords" different material features to its users. Brought into widespread use by ecological psychologist J. J. Gibson, "affordance" refers to the possibilities for action provided to a human or nonhuman animal by its environment. The same environment—for example, an urban park—exists as lifeworld for various organisms, each of which—microbes, mushrooms, ticks, squirrels, dogs, joggers, etc.—engages with it differently. Moving from the natural to the manmade world and from an ecological to a societal perspective, different individuals or groups find different affordances in the same environment because they use it differently and have developed skills that correspond to these uses. A papermaker in Jiajiang will engage with paper in ways that are quite different from a paper trader's, a calligrapher's, or a ritual expert's. The papermaker may examine a freshly molded paper sheet to determine if he should add more fiber, water, or mucilage to the pulp. The paper trader may touch the paper with a wet finger to test its absorbance, or fold and tear it to determine its tensile strength. The calligrapher may write on it to see if it will curl or tear under the impact of the brush. The ritual expert may burn the paper, inhale the smoke, and crumble the ashes between his fingers. These are all expert users, highly knowledgeable about paper, but to each of them paper appears as a different thing. Affordances, in short, are not simple attributes of objects but features that matter to actors in the context of their practical, purposive activity. They exist only for those skilled enough to see them; in fact, becoming skilled consists largely in learning to detect affordances—to interpret sensory signals (a sulfurous whiff in the workshop, a slight change in the tinge of the pulp, the gurgling sound that water makes as it is squeezed out of a pile of paper in the press) and to respond in appropriate ways.[14]

Skill in this sense is individually embodied, and the chapters of this book show numerous examples of individual skills, from the artistry of French papier-mâché model makers (Maerker) to the data-processing prowess of Berlin housewives (von Oertzen). What I want to highlight here is another, less obvious, aspect: skill, while individually embodied, is socially embedded in communities of specialized practitioners. In order to be reliably reproduced across time and space, skill needs to be distributed; a skill that exists in a single person disappears when that person dies. Skill, then, is best understood as pertaining to communities. This was pointed out by the anthropologist François Sigaut, who argued that "the skill-producing group is a basic social unit present in all societies because a society without techniques is inconceivable. This unit can take a wide variety of forms and enter into extremely diverse combinations with other units such as the family, the residence group, the age-group, etc. . . . Ideally, the morphology of all societies ought to be reconstructed from scratch, taking into account this necessary but, until now, unnoticed unit."[15]

The community of papermakers that I studied may be an extreme case of such a "skill-producing group." Tied together by kinship and coresidence, people in the Jiajiang hills have been making paper since the seventeenth century. Over time, almost all aspects of their social life were inflected by the need to reproduce and transmit their skills. To give just one example, Chinese kinship ideology can be harshly hierarchical in its insistence on the power of generational seniors over juniors. Papermakers in Jiajiang systematically deemphasized these hierarchical aspects; instead, they emphasized the equality of all descendants in the face of the ancestors and their shared responsibility for passing on the craft to future generations. One effect of this "horizontal" interpretation of kinship was that papermakers could not easily deny lineage cousins access to their workshops. Workshops exchanged labor and swapped raw materials and tools, and people visited each other's sheds without restraint. Skills were shared widely and innovations spread with great speed, in spite of a rhetorical insistence by workshop owners that their product was unique and that they possessed secret knowledge not shared by anybody else.

Returning to affordances, what makes paper work so interesting is that paper—like other complex and malleable materials such as cloth, metal, and clay—affords almost unlimited opportunities for the development of specialized skills, and that these skills form a basis for the development of intricately articulated networks of practitioners. The communities of skill discussed in the chapters of this book tend to be less long-lived and complex than the one I studied in Jiajiang. Nonetheless, it may be worthwhile in each of these cases to identify the skills needed to complete a certain task: to ask what array of skills is needed to compile a Prussian census, run an early modern household lab, or execute a paper trial; to delineate the various groups of skilled practitioners who are involved in these tasks; and to ask which formal and informal mechanisms ensure that all these actors are properly socialized into these tasks.

PAPER WORK, SKILL, AND GENDER

Not all skilled work is gendered, but most societies tend to ascribe specific skills to men and women and to divide work tasks accordingly. I take it for granted that such ascriptions are largely arbitrary—evidence not of different aptitudes of men and women but of a desire to sort individuals into categories, assign them identities that are more binary than human nature warrants, and socialize them into male and female genders that perform separate yet complementary tasks. Divisions of labor that are held to be natural and immutable in one place can be shown not to exist in other places, or to be abandoned quickly when conditions change. Gender, it is often said, is performance. More specifically, gender in late imperial China was *work*: as historian Dorothy Ko has argued, "one was not born a woman (nor man), but became one through work and learning, and had to work at maintaining the identity through one's life."[16] In a culture that did not subscribe to a model of fixed male and female essences (Chinese cosmology saw both sexes as composed of *yin* and *yang*, mixed in varying proportions), gender was understood not as innate difference but as social practice, as work, perhaps even as skill.[17]

In her work on gender and technology in imperial China, Bray proposes the terms *androtechnics* and *gynotechnics*, which she defines as "sets of technological attributes and practices that combine to shape and define female and male roles and hierarchies."[18] Her examples of gynotechnics include textile work, the organization and use of domestic space, fertility control, and the raising of offspring. These tasks are gynotechnics not because women monopolized them (they did not) but because they defined what it meant to be a woman. Similarly, literacy and agriculture were androtechnics not because writing and farming were exclusive male activities but because writing was central to the way elite men thought of themselves as men, just as farmwork defined masculinity for nonelite males. Andro- and gynotechnics should be understood not as rigid divisions of labor but as "material matrixes within which virtuous femininities [and masculinities] were developed, expressed, and judged."[19]

Defined in this way, these terms should be as applicable to modern and early modern Europe and America as they are to China. We see the same bifurcation in China and Euro-America between bureaucratic and epistemic paper work, gendered male; and work that uses paper not as a writing substrate but as a versatile material with myriad uses, gendered female. It is true that the chapters of this volume show women participating in epistemic paper work, but I would not hesitate to describe such work as androtechnics: even if women participated in it, it remained normatively male. Once we leave the laboratory and the study and enter the stillroom and the kitchen, paper work becomes gynocentric. The domestic technologies discussed by Leong, Werrett, and Wolfe—the use of paper to filter medicines and seal containers, to wrap and store objects, to light fires and cook meals—have close equivalents in China. Although we do not know for sure who papered walls in China, pasted oil paper over window lattices, made shoe soles from wastepaper, lined chests and boxes with paper, wrapped food and spices in paper envelopes, or raised silkworm eggs on paper sheets, we can assume that most such work was done by women, and that even if men participated in it, their work formed part of a feminine and domestic realm.

It bears repeating that andro- and gynotechnics do not describe an actual gendered division of labor but rather a systematic tendency to sort people into male and female categories, to train them in gender-specific skills and socialize them into gender-appropriate behavior. Normatively, "men farmed and women wove" (*nangeng nüzhi*) in late imperial China, but actual divisions of labor did not fit this simple formula. While actual labor divisions differed, the principle of complementarity in separation remained constant: as much as possible, men and women performed separate tasks, and women's work complemented and supported that of men. However, neat gender separation was not easy to accomplish: men and women often worked in close proximity, and their skills and activities seeped into one another's. It is precisely because of the blurriness of male and female roles that gender lines were often emphasized quite harshly. Let me illustrate this with three examples, one from my own research, two from the wider literature.

Ko's recent work on the "social life of inkstones" describes male literati's culture of skilled work and connoisseurship. Inkstones—palm-sized slabs of stone used to prepare ink for writing—were avidly collected, and throughout late imperial China, knowledge of inkstone types and pedigrees was an emblem of elite male identity and culture. Yet inkstone use and making—and with it the broader culture of *wen*, of refined literary pursuits—depended at all stages on the presence of women. As Ko explains, the persistent identification of literate pursuits as masculine "had to be reiterated all the more in an age when the first teachers of most educated men were their mothers, when women artists sold their paintings to support their men, and when women wrote, edited, and published books in substantial numbers."[20]

The same process of erasure can be observed in the practice of "cherishing lettered paper" (*xi zizhi*). Chinese elites always treated the written word with great reverence. Beginning in the early Ming (1368–1633), devout Confucians collected written paper, burned it in special furnaces, and buried the ash on sacred ground to prevent it from defilement. In the eighteenth and nineteenth century, "cherishing paper associations" were established in urban centers. The cult of written paper was an elite male concern, closely associated with the imperial examination system: by insisting that paper should be treated with respect, literate elites demanded respect for themselves, at a time when their power and prestige were declining.[21] Originally, the emphasis of "paper cherishers" was on collecting scraps of discarded paper, but Confucian literati soon realized that if they wanted to protect *all* lettered paper from being used as shoe soles and bottle stoppers (or put to even less mentionable uses), they had to change the attitudes of the uneducated masses. The medium they employed was Buddhist-inspired storybooks that laid out a calculus of rewards for those who cherished lettered paper, and threatened divine retribution for those who abused such paper. According to one such pamphlet, persons who spend their entire life "buying lettered paper and bringing it home, washing it with fragrant water and burning it" would earn ten thousand merits, have twelve years added to their life, enjoy prosperity and honor, and pass on their good fortune to sons and grandsons. Lesser efforts reaped lesser rewards, while the abuse of lettered paper led to impoverishment, disfiguring disease, death by lightning, reincarnation as a dog, or the end of the family line.[22]

Attempts to safeguard paper from defilement ultimately depended on the people who were most likely to use and abuse paper; that is, women. The mid-nineteenth century—a time of wrenching social change—saw the publication of "precious scrolls" (*baojuan*: morality tales aimed at mostly female audiences that were read or chanted in private gatherings) that exhorted women to cherish grain and paper. Cherishing grain, like cherishing paper, was less concerned with the practicalities of household management than with the extreme elevation of an emblematic object, and the rewards that came from saving this object from defilement. The women in these stories earned karmic merit and long life by collecting soiled paper from outhouses or picking rice kernels from dog feces, and then washing, drying, and ritually disposing them.[23] It is no coincidence, I think, that grain and paper correspond to the two pursuits that defined respectable masculinity in China—writing and agriculture—and that women's best hope for salvation came from suffering pollution while glorifying the products of male work.

The above examples show the intermingling of male and female activities in the household, and an attempt to erase women's work or to cherish it only to the extent that it enhances and adds value to the work of men. Let me cite one more example from my own research, which I chose because it lends itself

Figure A.1. Working at the steamer, Jiajiang, ca. 1990. Photo by Jacob Eyferth.

for comparison with Szalay's chapter. Most paper workshops in Jiajiang were household-based, employing male and female members of the family and a few hired workers (see plate 17). The gendering of tasks was similar to that described by Szalay for early modern German paper workshops: men made the pulp and molded paper, while women dried, sorted, and finished the sheets. Male labor in the workshops outweighed female by a ratio of about three to two, but women's work was no less skilled than men's, and since men and women shared the same workshop space, women were knowledgeable about all aspects of men's work. It was because women were so centrally involved in the technical procedures that men used symbolic means to claim exclusive ownership of the craft. This was most evident in the context of steaming, the crucial operation that transformed rotted bamboo into clean, white fiber. During the ten days or so when the steamers were in operation, men abstained from sexual intercourse for fear that women's polluting influence might affect the pulp and result in blackened paper. Women participated in steamer work in an auxiliary function: men stood atop the huge wooden steamers, pounding the steaming fibers with long pestles, while women stood at the steamers' feet and rinsed the fibers in running water. This was arduous work, and men were given license to tease the women on the ground in improvised work songs (fig. A.1).[24] We do not know how women experienced these taunts (the practice had stopped by the time of my interviews), but we do know that they were not supposed to answer back. I see this as a way to signal to women that steaming and pulp preparation were male work, that men were in control, and women were at best helpmates in the process. The aggressive marking of papermaking skills as male appears to have been effective: none of the women I interviewed expressed the sense of confidence in their skills—of being "owners" of a craft—that men routinely demonstrated.[25]

CONCLUSIONS AND FURTHER QUESTIONS

How similar or different, then, were the worlds of paper work in China and the West? Let me begin with commonalities that are so obvious that they are easily overlooked. In both China and the West, women performed a large part of the paper work that underpinned commerce, administration, cultural production, and everyday life; in both parts of the world, their paper work generated knowledge that could be both mundane and practical and highly specialized. With some exceptions, women's work tended to fall on the manual/applied side of the hand–mind, practice–theory spectrum; it tended to take place under male supervision and tutelage, and the symbolic and material value produced by it tended to be appropriated by men. Next, let me list two obvious differences. First, it is difficult to find evidence in pre-twentieth-century China for women doing paper work that can be clearly labeled epistemic. Until the rise of missionary science in the late nineteenth century, I see no Chinese parallels to Linker's posture scientists, Maerker's midwife-anatomists, or von Oertzen's housewife-tabulators. Vernacular science of the type discussed by Leong, Werrett, and Wolfe—the knowledge of material transformations in the context of the domestic economy—was common in China, and while I haven't looked for evidence, I suspect that a study of Chinese recipe books would reveal a similar range of knowledges and practices. However, if we were looking for women's participation in the production of formal knowledge, we may have to look not to science (in the sense of the systematic study of the natural world) but to those fields of knowledge that enjoyed most prestige in early modern China: philology, antiquarianism, connoisseurship, etc. Ko's recent work has shown that women were heavily involved not only in the making and use of inkstones but also in the reproduction of a wider culture of knowledge about texts and artifacts. I am sure similar examples exist in other fields. If they haven't been found yet, it is because few scholars have looked for them.

A second obvious difference lies in the greater development of public institutions in Euro-America, and in the greater likelihood that Western women participated in paid, public work outside the home. Both normatively and in actual practice, most Chinese women before the mid-twentieth century worked at home. This volume shows that this is true also for women in the early modern West, who participated in the world of science, moral discourse, and administration without leaving their homes (von Oertzen, Leong, Eddy, Yale). Nonetheless, it seems evident that Western women had access to roles outside the family earlier and to a greater extent than their Chinese counterparts. Lacking expertise in the subject, I take Karin Hausen's 1976 article on the polarization of sexual stereotypes as my guide to trends in Europe.[26] Until circa 1780, Hausen argues, men's and women's roles were defined by their position in the household *oeconomy*—as they were in pre-twentieth-century China. Women were subordinate to men, but gender did not operate in isolation: men's and women's roles were defined by their position in the large, multigenerational household composed of family, remote kin, and servants. From the 1780s on, status-based identities were replaced by notions of innate sexual natures: men were active, rational, and goal-oriented; women passive, emotional, and dependent. Early modern notions of personal autonomy applied to men alone, who participated as citizens in the public sphere and as contracting agents in the marketplace. The family was deinstitutionalized, depoliticized, and redefined as the realm where the individualizing, maximizing logic of public life did not apply. This new gender dichotomy was practiced in full only by professionals and public servants of bourgeois or aristocratic background; it was infeasible in practice but nonetheless powerful as an ideal for proletarians and farmers.

In China, gender polarization was just as strong, but it was contained within the household. As I mentioned above, the ruling paradigm was one of complementarity in separation: men and women worked side by side, and while divisions were quite blurred in practice, the ideological emphasis was always on the distinctiveness of male and female work. This was not a discourse of innate, physiological difference: the Western notion of strong natural differences between men and women had no equivalent in premodern China. Yet there was a strong sense that the separation of tasks in the home mirrored a broader moral and cosmic order: if men and women pursued their proper tasks in separation, prosperity and good governance would automatically follow. Europe, too, experienced moral panics when traditional gender divisions of labor were challenged, but few European authors can have matched the incomprehension that Confucian moralists showed when faced with women who, perversely, chose to work in the fields rather than to spin or weave.[27]

However, the main reason why Chinese women did not often work outside the home was not Confucian moralism but the simple fact that wage work as a permanent condition (as opposed to seasonal employment) was rare for men *and* women: almost everybody worked at home. State officials worked for a salary and urban shops employed clerks and accountants, but the vast majority of the population worked on farms or in family firms and received a share of the household's income stream rather than a wage. As late as 1950, even in the great metropolis of Shanghai, many more people were self-employed or worked in small handicraft and commerce than in factories;[28] in a medium-size city like Chengdu—more typical for China as a whole—90 percent of the workforce worked in small shops and workshops.[29] For China's vast rural population, almost all work took place in the context of the home.

I conclude with three questions. The first is about long-term historical trajectories. If the contrast I have sketched is real, what did it mean for women? Was the growth of a new sphere of public work a net loss to women, who were—initially at least—excluded from it and relegated to a degraded domesticity? The chapters of this volume are too rich and complex to provide an unambiguous answer. It is certainly possible to read a story of decline out of our chapters, from an early modern world in which the mistress of the household wields her knowledge in relative autonomy, to the world of big science and big bureaucracy in which women are reduced to supporting roles in projects led and designed by men. Yet there is also evidence to the contrary: women quickly find new roles for themselves as scientists, medical practitioners, and managers of public institutions. The comparison with China suggests that a world in which the household remains the uncontested nexus of social organization is not necessarily a better one for women. But perhaps it is best to set aside the household/wage work, private/public binaries, and instead to focus on how modern institutions were rooted in the household and its culture. This mutual dependence comes out most clearly in von Oertzen's chapter: the Prussian census seems to epitomize the modern, masculine logic of centralized power and control, yet it relied on the punctuality and meticulousness of women working at home. The question, perhaps, is not if, when, and how an older household-based mode of life gave way to modern institutions but how the household and its habits continue to underpin modernity.

My second question is about masculinity, a topic explicitly addressed in Szalay's chapter. Here I am struck by the contrast between early modern German papermakers and the artisans I studied in Jiajiang. Work routines for the two groups were similar: in both cases, men and women worked together, and men controlled the crucial tasks of pulping and molding while women were relegated to "lesser" (though by no means unskilled) tasks. Yet the ways workers experienced the gendered nature of their work appear quite different. Masculinity among early modern German papermakers is multiple: men defined themselves not against women but against other men. Owners competed with journeymen, journeymen with interloping jobbers, apprentices with unskilled workers, masters with lettered oeconomists in a contest for "honor," measured in terms of skill and status. Nothing comparable existed in Jiajiang: men were as defensive about their masculinity as in the German case, but their masculinity is less fractured. There were bosses and workers in the paper workshops, but the difference was not institutionalized: most workers expected to become owners at some point in time, and owners and workers thought of themselves as members of what Hill Gates calls a "patricorporation"—a community of related men who, despite internal rivalry, would ultimately stand up for one another.[30] Status in this community depended to some extent on skill and business acumen, but it was never divorced from men's position in the family and the kinship network. What I find striking from a Chinese perspective is how much men's roles in Szalay's chapter are governed by professional codes that owe nothing to the dyadic logic of the household, in which male and female roles define and stabilize one another. Male paper work, with its wide range from sweat-drenched labor in the paper mill to the cerebral tasks of scribes, scientists, and bureaucrats, appears to be a promising site for exploring how male identities changed when men became wage and salary earners.

A final question relates to the role of paper as a means of governance and control. Paper "disciplines" female labor (Bouk); the subdivided page serves to monitor the thoughts and actions of young women (Eddy); tracing the outline of female bodies on paper corrects their posture and comportment (Linker). Much recent literature, from Michel Foucault to James Scott, has ascribed great power to bureaucratic overlords who control and structure information flows, simplify complex worlds into charts, and dominate millions by skillfully manipulating paper tools. If this volume has shown us anything, it is that the eye that dominates the millions depends on thousands of supporting hands, and that paper power, while immense, is capillary, dispersed, and contingent.

NOTES

Introduction. Paper, Gender, and the History of Knowledge

This volume originated in a working group on "Working with Paper," a project at the Max Planck Institute for the History of Science in Berlin. We thank the authors of this collaborative project, members of Department II (Daston), and the anonymous reader for their enormously helpful and constructive comments.

1. See Heather Wolfe, this volume; as well as Wolfe, "'Neatly Sealed, with Silk, and Spanish Wax or Otherwise': The Practice of Letter-Locking with Silk Floss in Early Modern England," in *In the Prayse of Writing: Early Modern Manuscript Studies*, ed. S. P. Cerasano and Steven W. May (London: British Library, 2012), 169–89.

2. Wanda Orlikowski, "Sociomaterial Practices: Exploring Technology at Work," *Organization Studies* 28, no. 9 (2007): 1437; Lissa L. Roberts and Simon Werrett, "Introduction: 'A More Intimate Acquaintance,'" in *Compound Histories: Materials, Governance and Production, 1760–1840*, ed. Simon Werrett and Lissa L. Roberts (Leiden: Brill, 2017), 1–33; Simon Werrett, "Household Oeconomy and Chemical Inquiry" in Werrett and Roberts, *Compound Histories*, 35–56; Werrett, this volume.

3. Joan W. Scott, "Gender: A Useful Category of Historical Analysis," *American Historical Review* 91, no. 5 (1986): 1053–75; Rebecca M. Herzig, "Does Google™ Have Gender? Technologies of Everyday Life in Affluent Industrial Societies," *Icon* 16 (2010): 92–97.

4. Nina E. Lerman, "Categories of Difference, Categories of Power: Bringing Gender and Race to the History of Technology," *Technology and Culture* 51, no. 4 (2010): 893–918; Nina E. Lerman, Ruth Oldenziel, and Arwen P. Mohun, eds., *Gender and Technology: A Reader* (Baltimore: Johns Hopkins University Press, 2003), 2.

5. Ruth Schwartz Cowan, *More Work for Mother: The Ironies of Household Technology from the Open Hearth to the Microwave* (New York: Basic Books, 1983); and Cowan, "The Consumption Junction: A Proposal for Research Strategies in the Sociology of Technology," in *The Social Construction of Technological Systems: New Directions in the Sociology and History of Technology*, ed. Wiebe E. Bijker, Thomas P. Hughes, and Trevor Pinch (Cambridge, MA: MIT Press, 1989), 261–80. Francesca Bray, *Technology, Gender and History in Imperial China: Great Transformations Reconsidered* (London: Routledge, 2013), 16.

6. Our intervention is an explicit contribution to the relatively young and dizzying field of the history of knowledge. As to how this field may change the history of science, see Lorraine Daston, "The History of Science and the History of Knowledge," *KNOW: A Journal on*

the Formation of Knowledge 1, no. 1 (2017): 131–54. See also Peter Burke, *What is the History of Knowledge?* (Cambridge, UK: Polity Press, 2015).

7. Don Norman, *The Design of Everyday Things* (New York: Basic Books, 2013).

8. Jacob Eyferth, *Eating Rice from Bamboo Roots: The Social History of a Community of Handicraft Papermakers in Rural Sichuan, 1920–2000* (Cambridge, MA: Harvard Asia Center, 2009). See also Eyferth, this volume. On the Japanese context, see, for example, Peter Kornicki, *The Book in Japan: A Cultural History from the Beginnings to the Nineteenth Century* (Leiden: Brill, 1998), chapter 2; and Sukey Hughes, *Washi, the World of Japanese Paper* (Tokyo: Kodansha International, 1978); Cynthia J. Brokaw, *Commerce in Culture: The Sibao Book Trade in Qing and Republican Periods* (Cambridge, MA: Harvard University Press, 2007).

9. Timothy Barrett, Mark Ormsby, and Joseph B. Lang, "Non-Destructive Analysis of 14th–19th Century European Handmade Papers," *Restaurator* 37, no. 2 (2016): 93–135; Dard Hunter, *Papermaking: The History and Technique of an Ancient Craft*, 2nd ed. (London: Alfred A. Knopf, 1947).

10. Scholars such as D. C. Coleman, Richard Leslie Hills, and Alfred Shorter, for example, have crafted together detailed histories of the paper industry in Britain: Coleman, *The British Paper Industry, 1495–1860: A Study in Industrial Growth* (Oxford: Clarendon Press, 1958); Hills, *Papermaking in Britain, 1488–1988: A Short History* (London: Athlone Press, 1988); Shorter, *Paper Mills and Paper Makers in England, 1495–1800* (Hilversum, Netherlands: Paper Publications Society, 1957).

11. See, for example, D. F. McKenzie, *Bibliography and the Sociology of Texts: The Panizzi Lectures* (London: British Library, 1986).

12. James Raven, *The Business of Books: Booksellers and the English Book Trade, 1450–1850* (New Haven, CT: Yale University Press, 2007), 49–51. Donald Francis McKenzie, Peter D. McDonald, and Michael Felix Suarez, eds., *Making Meaning: "Printers of the Mind" and Other Essays* (Amherst: University of Massachusetts Press, 2002); Adrian Johns, *The Nature of the Book: Print and Knowledge in the Making* (Chicago: University of Chicago Press, 1998); Rima D. Apple, Gregory J. Downey, and Stephen L. Vaughn, *Science in Print: Essays on the History of Science and the Culture of Print* (Madison: University of Wisconsin Press, 2012).

13. Bernard V. Lightman, *Victorian Popularizers of Science: Designing Nature for New Audiences* (Chicago: University of Chicago Press, 2007); James A. Secord, *Victorian Sensation: The Extraordinary Publication, Reception, and Secret Authorship of* Vestiges of the Natural History of Creation (Chicago: University of Chicago Press, 2003); Secord, *Visions of Science: Books and Readers at the Dawn of the Victorian Age* (Chicago: University of Chicago Press, 2015). On how Victorians engaged with books more broadly, see Leah Price, *How to Do Things with Books in Victorian Britain* (Princeton, NJ: Princeton University Press, 2012).

14. Within this volume, Gabriella Szalay offers an in-depth description of the papermaking process. For a detailed and up-to-date description of European papermaking methods, see Timothy Barrett, "European Papermaking Techniques 1300–1800," *Paper through Time: Nondestructive Analysis of 14th- through 19th-Century Papers*, University of Iowa Libraries, July 14, 2014, accessed January 8, 2018, http://paper.lib.uiowa.edu/european.php.

15. Szalay, this volume.

16. For an overview on the Anglo-American development see Mark Kurlansky, *Paper: Paging through History* (New York: Norton, 2017); Judith McGaw, *Most Wonderful Machine: Mechanization and Social Change in Berkshire Paper Making, 1801–1885* (Princeton, NJ:

Princeton University Press, 1987); for the German context, see Heinz Schmidt-Bachem, *Aus Papier: Eine Kultur- und Wirtschaftsgeschichte der Papier verarbeitenden Industrie in Deutschland* (Berlin: de Gruyter, 2011); Lothar Müller, *Weiße Magie: Die Epoche des Papiers* (Munich: Hanser, 2012), 251–71; Jürgen Blechschmidt and Alf-Mathias Strunz, "Der Beginn eines neuen Zeitalters der Papierstoff-Erzeugung: Die Erfindung des Holzschliff-Verfahrens durch Friedrich Gottlob Keller," *Papiergeschichte* (1996): 136–50.

17. See, for example, Heidi Brayman, Jesse M. Lander, and Zachary Lesser, eds., *The Book in History, the Book as History: New Intersections of the Material Text. Essays in Honor of David Scott Kastan* (New Haven, CT: Yale University Press, 2016); James Daybell, *The Material Letter in Early Modern England: Manuscript Letters and the Culture and Practices of Letter-Writing, 1512–1635* (London: Palgrave Macmillan, 2012); Heidi Brayman Hackel, *Reading Material in Early Modern England: Print, Gender, and Literacy* (Cambridge: Cambridge University Press, 2005); Alison Wiggins, *Bess of Hardwick's Letters: Language, Materiality, and Early Modern Epistolary Culture* (New York: Routledge, 2016).

18. Nancy Pollard Brown, *Paperchase: The Dissemination of Catholic Texts in Elizabethan England*, vol. 1 of *English Manuscript Studies 1100–1700*, ed. Peter Beal and Jeremy Griffiths (Oxford: B. Blackwell, 1989), 120–43. For classic studies on early modern English manuscript culture, see Harold Love, *The Culture and Commerce of Texts: Scribal Publication in Seventeenth-Century England* (Amherst: University of Massachusetts Press, 1998); Arthur Marotti, *Manuscript, Print, and the English Renaissance Lyric* (Ithaca, NY: Cornell University Press, 1995); David McKitterick, *Print, Manuscript and the Search for Order, 1450–1830* (Cambridge, UK: Cambridge University Press, 2003); H. R. Woudhuysen, *Sir Philip Sidney and the Circulation of Manuscripts, 1558–1640* (Oxford: Clarendon Press, 1996).

19. Joshua Calhoun, "The Word Made Flax: Cheap Bibles, Textual Corruption, and the Poetics of Paper," *PMLA* 126, no. 2 (2011): 327–44; Peter Stallybrass, Roger Chartier, J. Franklin Mowery, and Heather Wolfe, "Hamlet's Tables and the Technologies of Writing in Renaissance England," *Shakespeare Quarterly* 55, no. 4 (2005): 379–419; Juliet Fleming, "Damask Papers," in *The Elizabethan Top Ten: Defining Print Popularity in Early Modern England*, ed. Andy Kesson and Emma Smith (Aldershot: Ashgate, 2013), 179–92.

20. Helen Smith, "'A Unique Instance of Art': The Proliferating Surfaces of Early Modern Paper," *Journal of the Northern Renaissance* 8 (2017), http://www.northernrenaissance.org/a-unique-instance-of-art-the-proliferating-surfaces-of-early-modern-paper/.

21. Tim Ingold, "Materials against Materiality," *Archaeological Dialogues* 14, no. 1 (June 2007): 1–16.

22. See Werrett, this volume.

23. Wolfe, this volume. We might do well to consider letter paper alongside the various gifts and "little presents" exchanged and proffered as part of patronage networks, alliance formation, and family strategy. See Gadi Algazi, Valentin Groebner, and Bernhard Jussen, eds., *Negotiating the Gift: Pre-modern Figurations of Exchange* (Göttingen: Vandenhoeck & Ruprecht, 2003); Ilana Krausman Ben-Amos, *The Culture of Giving: Informal Support and Gift-Exchange in Early Modern England* (Cambridge, UK: Cambridge University Press, 2008); Natalie Zemon Davis, *The Gift in Sixteenth-Century France* (Madison: University of Wisconsin Press, 2000); Felicity Heal, *The Power of Gifts: Gift Exchange in Early Modern England* (Oxford: Oxford University Press, 2014).

24. Paper historians speak of an expansion and differentiation of the paper-producing industry during the nineteenth century, stating that the availability and variability of paper products from postcards and cardboard packaging to papier-mâché, to stained paper, to dis-

posables such as envelopes or paper bags to newspaper, book print, or luxury and official archive paper grew exponentially. See Schmidt-Bachem, *Aus Papier*, 27–40.

25. Peter Stallybrass, "Printing and the Manuscript Revolution," in *Explorations in Communication and History*, ed. Barbie Zelizer (London: Routledge, 2008), 111–18; Lisa Gitelman, *Paper Knowledge: Toward a Media History of Documents* (Durham, NC: Duke University Press, 2014); H. J. Jackson, *Marginalia: Readers Writing in Books* (New Haven, CT: Yale University Press, 2001).

26. For an overview of recent work in the field, see Elizabeth Yale, "The History of Archives: The State of the Discipline," *Book History* 18, no. 1 (2015): 332–59; and Alexandra Walsham, "The Social History of the Archive: Record-Keeping in Early Modern Europe," *Past and Present* 230, suppl. 11 (2016): 9–48. See also Kathryn Burns, *Into the Archive: Writing and Power in Colonial Peru* (Durham, NC: Duke University Press, 2010); Markus Friedrich, *Die Geburt des Archivs: Eine Wissengeschichte* (Munich: de Gruyter, 2013); Lorraine Daston, ed., *Science in the Archives: Pasts, Presents, Futures* (Chicago: University of Chicago Press, 2017). Significantly, paper is also used for saving, storing, and archiving a variety of materials beyond the written text, such as medicines and specimens. See, for example, Leong and Yale, both in this volume.

27. Norman, *Design of Everyday Things.*

28. Boris Jardine characterizes a "flickering materiality" as a central condition of paper tools, as they function as "bearers of meaning and of paper as a manipulable substance," and thus fade out of and come back into view. Jardine, "State of the Field: Paper Tools," *Studies in History and Philosophy of Science* 64 (2017): 54. While we do not object to his observation, we maintain that in order to analyze the sociomateriality of paper historically, it is essential to not let it escape from view.

29. Daston, "History of Science"; Burke, *What Is the History of Knowledge?*

30. Robert E. Kohler and Kathryn M. Olesko, eds., *Clio Meets Science: The Challenges of History*, Osiris 27 (Chicago: Chicago University Press, 2012).

31. See, for example, Ursula Klein and Emma C. Spary, eds., *Materials and Expertise in Early Modern Europe: Between Market and Laboratory* (Chicago: University of Chicago Press, 2010); Pamela H. Smith, Amy R. W. Meyers, and Harold J. Cook, eds., *Ways of Making and Knowing: The Material Culture of Empirical Knowledge* (Ann Arbor: University of Michigan Press, 2014); Simon Werrett, "Matter and Facts: Material Culture and the History of Science," in *Material Evidence: Learning from Archaeological Practice*, ed. Robert Chapman and Alison Wylie (New York: Routledge, 2014), 339–52; Roberts and Werrett, *Compound Histories.*

32. For example, see Seth Rockman's introduction to "Forum: The Paper Technologies of Capitalism": Rockman, "Introduction," *Technology and Culture* 58, no. 2 (2017): 487–505.

33. Bruno Latour, "Drawing Things Together," in *Representation in Scientific Practice*, ed. Michael Lynch and Steve Woolgar (Cambridge, MA: MIT Press, 1990), 19–68. See also Latour, *Science in Action: How to Follow Scientists and Engineers through Society* (Cambridge, MA: Harvard University Press, 1988).

34. Ursula Klein, "Paper Tools in Experimental Cultures," *Studies in History and Philosophy of Science* 32, no. 2 (2001): 265–302. Klein, *Experiments, Models, Paper Tools: Cultures of Organic Chemistry in the Nineteenth Century* (Stanford, CA: Stanford University Press, 2002); on Klein, see also Mary Joy Nye and Stephen J. Weininger, "Paper Tools from the 1780s to the 1960s: Nomenclature, Classification, and Representations," *Ambix* 65, no. 1 (2018): 1–8.

35. Klein, *Experiments, Models, Paper Tools*, 245. By way of contrast, Christoph Hoffmann and others used the notion of paper tools as scribal procedures and explicitly "non-material devices of concretion," which should be seen as "an instrument of research itself, structuring practical tinkering, organizing the outcomes, and intervening in the more abstract work of reasoning and reflecting." See Hoffmann, "Processes on Paper: Writing Procedures as Non-Material Research Devices," *Science in Context* 26, no. 2 (2013): 280.

36. Anke te Heesen, "The Notebook: A Paper-Technology," in *Making Things Public: Athmospheres of Democracy*, ed. Bruno Latour and Peter Weibel (Cambridge, MA: MIT Press, 2003), 584. See also te Heesen, *The Newspaper Clipping: A Modern Paper Object* (Manchester: Manchester University Press, 2014).

37. Ann Blair, *Too Much to Know: Managing Scholarly Information before the Modern Age* (New Haven, CT: Yale University Press, 2010); Fabian Kraemer and Helmut Zedelmaier, "Instruments of Invention in Renaissance Europe: The Cases of Conrad Gesner and Ulisse Aldrovandi," *Intellectual History Review* 24, no. 3 (2014): 321–41; Elaine Leong, "Collecting Knowledge for the Family: Recipes, Gender and Practical Knowledge in the Early Modern Household," *Centaurus* 55, no. 2 (2013): 81–103; Lorraine Daston, "Taking Note(s)," *Isis* 95, no. 3 (2004): 443–48; Richard Yeo, *Notebooks, English Virtuosi, and Early Modern Science* (Chicago: University of Chicago Press, 2014); Isabelle Charmantier and Staffan Müller-Wille, "Carl Linnaeus's Botanical Paper Slips (1767–1773)," *Intellectual History Review* 24, no. 2 (2014): 215–38.

38. See, for example, Daniel Rosenberg, "Early Modern Information Overload," *Journal of the History of Ideas* 64, no. 1 (2003): 1–9.

39. Charmantier and Müller-Wille, "Carl Linnaeus's Botanical Paper Slips." For lists, see James Delbourgo and Staffan Müller-Wille, "Introduction," *Isis* 103, no. 4 (2012): 710–15. However, paper instruments find mention by Jardine, who refers to paper cuttings used by sixteenth-century mathematicians to measure surface areas by way of weighing paper. See Jardine, "State of the Field," 54. A brilliant study on how paper engineering of movable flaps and dials in books redefined the relationship of early modern viewers to art, humanism, and science is provided by Suzanne Karr Schmidt, *Interactive and Sculptural Printmaking in the Renaissance* (Leiden: Brill, 2018). Other examples of material paper use include Judith Walsh, "The Use of Paper Splitting in Old Master Print Restorations," *Print Quarterly* 17, no. 4 (2000): 383–90.

40. Volker Hess and J. Andrew Mendelsohn, "Case and Series: Medical Knowledge and Paper Technology, 1600–1900," *History of Science* 48, no. 3–4 (2010): 287.

41. Markus Krajewski and Peter Krapp, *Paper Machines: About Cards & Catalogs, 1548–1929* (Cambridge, MA: MIT Press, 2011); Gitelman, *Paper Knowledge*; Matthew S. Hull, "The File: Agency, Authority, and Autography in an Islamabad Bureaucracy," *Language and Communication* 23, no. 3–4 (2003): 287–314; Peter Becker and William Clark, eds., *Little Tools of Knowledge: Historical Essays on Academic and Bureaucratic Practices* (Ann Arbor: University of Michigan Press, 2001); Miles Ogborn, *Indian Ink: Script and Print in the Making of the English East India Company* (Chicago: University of Chicago Press, 2007).

42. John Brewer, *The Sinews of Power: War, Money, and the English State, 1688–1783* (Cambridge, MA: Harvard University Press, 1990). See also Paul M. Dover, ed., *Secretaries and Statecraft in the Early Modern World*, 1st ed. (Edinburgh: Edinburgh University Press, 2016); Megan Williams, "Unfolding Diplomatic Paper and Paper Practices in Early Modern Chancellery Archives," in *Praktiken der Frühen Neuzeit: Akteure—Handlungen—Artefakte*, ed. Arndt Brendecke (Munich: Böhlau, 2015), 496–508.

43. Ben Kafka, "Paperwork: The State of the Discipline," *Book History* 12, no. 1 (2009): 341.

44. Cornelia Vismann, *Files: Law and Media Technology* (Stanford, CA: Stanford University Press, 2008); Peter Becker, ed., *Sprachvollzug im Amt. Kommunikation und Verwaltung im Europa des 19. und 20. Jahrhunderts* (Bielefeld: Transkript, 2011).

45. Jacob Soll, *The Information Master: Jean-Baptiste Colbert's Secret State Intelligence System* (Ann Arbor: University of Michigan Press, 2009); Ben Kafka, *The Demon of Writing: Powers and Failures of Paperwork* (New York: Zone Books, 2012); for the twentieth century, JoAnne Yates, *Structuring the Information Age: Life Insurance and Technology in the Twentieth Century* (Baltimore: Johns Hopkins University Press, 2005); Jon Agar, *The Government Machine: A Revolutionary History of the Computer* (Cambridge, MA: MIT Press, 2003).

46. With this terminology, Latour adopted the language that the clerks themselves used to describe their work. See Latour, *The Making of Law: An Ethnography of the Conseil d'Etat* (Cambridge, UK: Polity Press, 2009), 82.

47. See first and foremost Bhavani Raman, *Document Raj: Writing and Scribes in Early Colonial South India* (Chicago: University of Chicago Press, 2012); for the nineteenth century, Ann Laura Stoler, *Along the Archival Grain: Epistemic Anxieties and Colonial Common Sense* (Princeton, NJ: Princeton University Press, 2010).

48. In this, we join arguments made by Jardine in his recent essay, largely written at the Max Planck Institute for the History of Science, Berlin, in the same year this working group was formed: Jardine, "State of the Field," 53–63.

49. te Heesen, "Notebook," 589.

50. We are grateful to Lucia Dacome for her commentary on a set of our essays presented at the History of Science Society annual meeting in Toronto in 2017 and for her suggestions that focusing on the materiality of paper and gender has this impact on our analysis of inscriptions.

51. There are some notable exceptions. In the history of science, see, for example, Victoria E. Burke, "'The art of Numbering well': Late Seventeenth-Century Arithmetic Manuscripts Compiled by Quaker Girls," in *Material Readings of Early Modern Culture: Texts and Social Practices, 1580–1730*, ed. James Daybell and Peter Hinds (Basingstoke: Palgrave Macmillan, 2010), 246–65; and Elaine Leong, "Collecting Knowledge for the Family"; and Leong, "'Herbals she peruseth': Reading Medicine in Early Modern England," *Renaissance Studies* 28, no. 4 (2014): 556–78. For a recent overview of research in literary studies on commonplace book cultures, including women's participation, see Burke, "Recent Studies in Commonplace Books," *English Literary Renaissance* 43, no. 1 (2013): 153–77. On women activists, paper, and "self-creation," see Ellen Gruber Garvey, *Writing with Scissors: American Scrapbooks from the Civil War to the Harlem Renaissance* (New York: Oxford University Press, 2013), chapter 5.

52. See the special issue of *Icon* on "Technology in Everyday Life," and the introduction by Timo Myllyntaus, "Prologue: Constructing Technology for Everyday Life," *Icon* 16 (2010): 3–21.

53. For an overview, see Sally Gregory Kohlstedt and Helen E. Longino, eds., *Women, Gender, and Science: New Directions*, Osiris 12 (Chicago: University of Chicago Press, 1997); Kohlstedt, ed., *History of Women in the Sciences: Readings from Isis* (Chicago: University of Chicago Press, 1999); on masculinities, see Erika Lorraine Milam and Robert A. Nye, eds., *Scientific Masculinities*, Osiris 30 (Chicago: University of Chicago Press, 2015).

54. Donald L. Opitz, Staffan Bergwik, and Brigitte Van Tiggelen, eds., *Domesticity in the*

Making of Modern Science (Basingstoke, Hampshire: Palgrave Macmillan, 2016). See also Melanie Keene, "Familiar Science in Nineteenth-Century Britain," *History of Science* 52, no. 1 (2014): 53–71.

55. Roger Cooter and Stephen Pumfrey, "Separate Spheres and Public Places: Reflections on the History of Science Popularization and Science in Popular Culture," *History of Science* 32, no. 3 (1994): 237–67; Christine von Oertzen, Maria Rentetzi, and Elizabeth Siegel Watkins, "Finding Science in Surprising Places: Gender and the Geography of Scientific Knowledge. Introduction to 'Beyond the Academy: Histories of Gender and Knowledge,'" *Centaurus* 55, no. 2 (2013): 73–80.

56. Francesca Bray, *Technology, Gender and History in Imperial China: Great Transformations Reconsidered* (London: Routledge, 2013); Eyferth, *Eating Rice from Bamboo Roots*. See also Eyferth, this volume.

57. Bray, *Technology, Gender and History*, 5, 17, 6.

58. Lerman, Oldenziel, and Mohun, *Gender and Technology*; Judy Wajcman, *Feminism Confronts Technology* (University Park: Pennsylvania State University Press, 1991); Roger Horowitz and Arwen Mohun, eds., *His and Hers: Gender, Consumption, and Technology* (Charlottesville: University of Virginia Press, 1998).

59. Judith A. McGaw, "Why Feminine Technologies Matter," in Lerman, Oldenziel, and Mohun, *Gender and Technology*, 13–36; and Ruth Oldenziel, "Why Masculine Technologies Matter," in Lerman, Oldenziel, and Mohun, *Gender and Technology*, 37–71.

60. For studies in gender and dress/textile history, see Barbara Burman and Carole Turbin, eds., "Material Strategies," special issue, *Gender & History* 14, no. 3 (2002). See also Burman and Turbin, "Material Strategies Engendered," *Gender & History* 14, no. 3 (2002): 371–81. There is also a vast scholarship on gender, embodiment, and materiality influenced by Judith Butler, *Bodies that Matter: On the Discursive Limits of "Sex"* (New York: Routledge, 1993). On gender and material culture, see Hannah Greig, Jane Hamlett, and Leonie Hannan, eds., *Gender and Material Culture in Britain since 1600* (London: Palgrave, 2016).

61. Lerman, "Categories of Difference," 893–918.

62. As Carolyn de la Peña reminds historians, "white people have race," and "constructing and protecting whiteness" is part of our technological histories, whether explicitly stated by the historical actors under study or not. Carolyn de la Peña, "The History of Technology, the Resistance of Archives, and the Whiteness of Race," *Technology and Culture* 51, no. 4 (2010): 936.

Chapter 1. Letter Writing and Paper Connoisseurship in Elite Households in Early Modern England

I am extremely grateful to the editors of this volume and the members of the Working with Paper working group at the Max Planck Institute for the History of Science for their astute comments and feedback as I developed this essay. All mistakes and shortcomings are my own.

1. Robert Cecil to Thomas Windebank, February 26, 1600, National Archives, SP 12/274, fol. 69. I am grateful to Anna Bertholet for this reference. The "House of Florence" refers to the Medici family. "Buchanan's scholar" refers to James VI, who was tutored by the great Latin scholar George Buchanan. The "son of Catherine de Medici" is an oblique reference connecting the House of Florence to Francis, Duke of Anjou, Elizabeth's suitor in the late 1570s. Sir John Smith was sent on an embassy to Spain in 1576. He passed through France and met the French king, the Queen Mother, the queen, and Duke of Anjou on the way. In

a letter describing this visit to Elizabeth, he describes how much more beautiful she is than the women of the French court (National Archives, SP 70/141, fol. 50).

2. Cecil quotes Psalm 101, verse 8, "Passer solitarius in tecto" (a lone sparrow on the rooftop).

3. Utensil, n. (2a): "any article useful or necessary in a household; a domestic implement, vessel, or article of furniture," *Oxford English Dictionary Online*, accessed December 2018, Oxford University Press.

4. Preshaped, machine-made envelopes were a mid-nineteenth-century invention.

5. In most libraries and archives, letters are housed as unfolded documents (often after having been flattened by caretakers), either in individual folders or hinged into large volumes. Therefore, it is surprisingly difficult to reimagine the original three-dimensional packet as sent, to imagine seals and paper strips where none may remain, to interpret layers of fold lines, and to see slits and holes, unless a simulacrum is created. In the process of binding or conservation, important evidence is often removed or altered in irrecoverable ways.

6. For bearers, see Daniel Starza Smith, "Postage and Bearers," in *Bess of Hardwick's Letters: The Complete Correspondence, c.1550–1608*, ed. Alison Wiggins, Alan Bryson, Daniel Starza Smith, Anke Timmermann, and Graham Williams, University of Glasgow, web development by Katherine Rogers, University of Sheffield Humanities Research Institute, April 2013, accessed September 6, 2017, https://www.bessofhardwick.org/background.jsp?id=160; and Mark Brayshay, "Conveying Correspondence: Early Modern Letter Bearers, Carriers, and Posts," in *Cultures of Correspondence in Early Modern Britain*, ed. James Daybell and Andrew Gordon (Philadelphia: University of Pennsylvania Press, 2016), 48–65.

7. For descriptions, technical terms, and instructional videos of different categories of letterlocking, see Jana Dambrogio, Daniel Starza Smith, et al., *Dictionary of Letterlocking* (DoLL), last updated December 30, 2017, http://www.letterlocking.org/dictionary.

8. See Heather Wolfe, "'Neatly Sealed, with Silk, and Spanish Wax or Otherwise': The Practice of Letter-Locking with Silk Floss in Early Modern England," in *In the Prayse of Writing*, ed. S. P. Cerasano and Steven W. May (London: British Library, 2012), 169–89; and Wolfe, "A Letter from Queen Anne to Buckingham Locked with Silk Embroidery Floss," *The Collation*, Folger Shakespeare Library, last modified January 2013, https://collation.folger.edu/2013/01/a-letter-from-queen-anne-to-buckingham-locked-with-silk-embroidery-floss.

9. Alison Wiggins uses the layout of the rooms in Bess of Hardwick's household as a case study for how letters would have been produced in *Bess of Hardwick's Letters: Language, Materiality, and Early Modern Epistolary Culture* (New York: Routledge, 2016), 89–105. For the interconnections between sewing and literacy, see Susan Frye, *Pens and Needles: Women's Textualities in Early Modern England* (Philadelphia: University of Pennsylvania Press, 2010); and Bianca F.-C. Calabresi, "'You sow, Ile read': Letters and Literacies in Early Modern Samplers," in *Reading Women: Literacy, Authorship, and Culture in the Atlantic World, 1500–1800*, ed. Heidi Brayman Hackel and Catherine E. Kelly (Philadelphia: University of Pennsylvania Press, 2008), 79–104.

10. Sir Thomas Lake to Robert Cecil, January 28, 1595/6, Hatfield House, Cecil Papers 30/29.

11. Elizabeth to Princess Catherine of Navarre (Catherine of Bourbon), January 21, 1595/96. A scribal copy of this letter is Hatfield House, Cecil Papers 133/133.

12. Wiggins, *Bess of Hardwick's Letters*, 187.

13. Rayne Allinson, *A Monarchy of Letters: Royal Correspondence and English Diplomacy in the Reign of Elizabeth I* (New York: Palgrave Macmillan, 2012), 91.

14. See Andrew Zurcher and Christopher Burlinson, "'Secretary to the Lord Grey Lord Deputie here': Edmund Spenser's Irish Papers," *Library* 6, no. 1 (2005): 30–75, for the complicated, multi-peopled routine of writing, addressing, sealing, and sending letters in Spenser's secretariat in Ireland.

15. Oldenziel, "Why Masculine Technologies Matter," in *Gender and Technology: A Reader*, ed. Nina E. Lerman, Ruth Oldenziel, and Arwen P. Mohun (Baltimore: Johns Hopkins University Press, 2003), 50.

16. Wiggins, *Bess of Hardwick's Letters*, 188.

17. She suggests that with an analysis of the watermarks in the Portland literary collection at the University of Nottingham, "it should be possible to suggest which watermark forms were most commonly used in the paper available in any given period in England, in London or in the provinces. Other issues to be answered might concern patterns of paper importation from Europe, and the preference of particular people or professionals for one type of paper." Thompson, "Watermarks and Other Physical Evidence from the Portland Literary Manuscripts," in *Puzzles in Paper: Concepts in Historical Watermarks*, ed. Daniel W. Mosser, Michael Saffle, and Ernest W. Sullivan II (New Castle, DE: Oak Knoll, 2000), 197.

18. E. J. Labarre observes that there were many names for paper sizes in English, and a single name could represent "anything from 2 to 15 slight variations in size." Labarre, "The Sizes of Paper, Their Names, Origin and History," in *Buch und Papier*, ed. H. Kunze (Leipzig: Harrassowitz, 1949), 247.

19. Helen Smith, "'A Unique Instance of Art': The Proliferating Surfaces of Early Modern Paper," *Journal of the Northern Renaissance* 8 (2017), par. 11, http://www.northernrenaissance.org/a-unique-instance-of-art-the-proliferating-surfaces-of-early-modern-paper/.

20. Nehemiah Rogers, *A sermon preached at the second trienniall visitation* (1632), sig. C2v–C3r, quoted in Smith, "Unique Instance of Art," par. 11.

21. Antoine de Courtin, *The Rules of Civility* (1675). Bess of Hardwick used extra-large paper for three of her most critical and deferential letters—to the queen, to her husband, and to the Earl of Leicester—complementing "the letters' linguistic content." See Alison Wiggins, "Material Features," *Bess of Hardwick's Letters: The Complete Correspondence, c.1550–1608*, ed. Alison Wiggins, Alan Bryson, Daniel Starza Smith, Anke Timmermann, and Graham Williams, University of Glasgow, web development by Katherine Rogers, University of Sheffield Humanities Research Institute (April 2013), https://www.bessofhardwick.org/background.jsp?id=189.

22. Thomas Docksie to Walter Bagot, March 10, 1611/12, Folger Shakespeare Library, MS L.a.414.

23. Sir Philip Sidney to Queen Elizabeth, November 10, 1585, Hatfield House, Cecil Papers 14/19.

24. The Duke of Anjou to Queen Elizabeth, [1581?], Hatfield House, Cecil Papers 148/193. The original French begins, "Je vous supplie tres humblemant me pardonner si sete letre n'est escripte en si beau papier ni avec que telle desliquatese et dignite que je le dessirerois . . ."

25. Dr. Pierre Beutterich to D. Rogers, November 10, 1578, National Archives, SP 83/10, fol. 22.

26. William Herle to Bishop of Ross, 1571, British Library, Cotton Calig., C. III., fol. 69b; George Hackert's receipt to the Bishop of Ross, 1571, British Library, Cotton Calig., C. III, fol. 2; Sir William Galley to Richard Harvey, June 2, 1640, National Archives, SP 16/456, fol. 19.

27. Chatsworth inventory for July 25, 1592: Chatsworth House, Hardwick MS 7, fol. 30, but taken from a typescript transcription: Nottingham University Library, Special Collections, Durant Papers MS 663/2/1/13/1-8. I am grateful to Wiggins for sharing her notes.

28. For the price of everyday paper in early modern England, see Heather Wolfe, "Was Early Modern Writing Paper Expensive?," *The Collation*, Folger Shakespeare Library, February 13, 2018, https://collation.folger.edu/2018/02/writing-paper-expensive.

29. Letters from John Donne to various recipients: Folger Shakespeare Library, MSS L.b.526–30, 532–34.

30. See Angela Andreani, *The Elizabethan Secretariat and the Signet Office: The Production of State Papers, 1590–1596* (New York: Routledge, 2017), 90–98 and 130–32, for the types of paper used by Elizabeth's secretaries, including Venetian paper with the flag watermark or with a crossbow or lion watermark.

31. Thomas Birch, *The History of the Royal Society of London* (London: Strand, 1744), 2:279. I am grateful to Gabriella Szalay for this reference.

32. See Thomas Fulton, "Gilded Monuments: Shakespeare's Sonnets, Donne's Letters, and the Mediated Text," in *Comparative Textual Media: Transforming the Humanities in the Postprint Era*, ed. N. Katherine Hayles and Jessica Pressman (Minneapolis: University of Minnesota Press, 2013), 221–53. In a gilt-edged letter written from France on February 7, 1612, to his brother-in-law, Sir Robert More, Donne jokes about the poor quality of French velvet/Catholicism compared to Italian and Spanish velvet/Catholicism (Folger Shakespeare Library, MS L.b.535; see Folger MS L.b.539 for another letter from Donne to More on gilt-edged Venice paper, written August 10, 1614). This letter originally enclosed the only surviving example of his poetry in his own handwriting—the verse epistle "A Letter to the Lady Carey, and Mrs Essex Riche, from Amyens" (Bodleian Library, MS Eng. poet. d. 197). See Laetitia Yeandle, "Watermarks as Evidence for Dating and Authenticity in John Donne and Ben Franklin," in Mosser, Saffle, and Sullivan, *Puzzles in Paper*, 85–87. In a letter the following year to Sir Robert Harley, April 7, 1613, he wrote on gilt-edged Venetian paper and closed the letter with silk floss (British Library, Add MS 70001, fol. 124).

33. Jane Skipwith to Lewis Bagot, Folger Shakespeare Library, MSS L.a.851–L.a.853; Jane Skipwith to Walter Bagot, Folger MS L.a.854.

34. Frances Howard to Simon Forman, Bodleian Library, MS Ashmole 208, fols. 143–44. I am grateful to Henry Woudhuysen for this reference.

35. Laetitia Yeandle, editor, *Sir Edward Dering, lst bart., of Surrenden Dering and his 'Booke of Expences' 1617–1628*, Kent Archaeological Society: Online PDF of transcription of Dering's account book, Centre for Kentish Studies, U350 M4, fols. 6r, 68r; "An accompt of money received and disbursed by me Edward Hamond for Mr. Arthur Capell, esquire," May 26, 1629, Hatfield House, Cecil Papers 200/108a; Stationery requisition signed by Francis Windebank, 1638, among papers relating to the Exchequer, British Library, Add MS 31,043, fol. 21.

36. Nadine Akkerman, ed., *The Correspondence of Elizabeth Stuart, Queen of Bohemia*, vol. 1, *1603–1631* (Oxford: Oxford University Press, 2015).

37. John Donne to Sir Robert More, July 14, 1614, Folger Shakespeare Library, MS L.b.537. Perhaps not coincidentally, in it he makes a joke about the blue clothing of Danish courtiers. On the connection between the blue edges and the blue clothing, as well as a description of green-edged paper used by George Garrard, see Daniel Starza Smith, "Paper," *Bess of Hardwick's Letters*, last modified April 2013, https://www.bessofhardwick.org/background.jsp?id=164.

38. Described in *Bess of Hardwick's Letters*, letter 237, https://www.bessofhardwick.org/letter.jsp?letter=237, from Sotheby's sale catalog, June 26, 1974, Lot 2840.

39. For the fermesse as a symbol in letters, see Claude Dulong-Sainteny, "Les signes cryptiques dans la correspondance d'Anne d'Autriche avec Mazarin: Contribution à l'emblématique du XVIIe siècle," *Bibliothèque de l'école des chartes* 140, no. 1 (1982): 61–83, esp. 70. Orest Ranum describes it as "'S,' an enigmatic symbol of fidelity and love known since the fourteenth century," popularized in France in the mid-sixteenth century, in Ranum, "The Refuges of Intimacy," in *Passions of the Renaissance*, ed. Roger Chartier, trans. Arthur Goldhammer, vol. 3 of *A History of Private Life*, ed. Philippe Ariès and Georges Duby (Cambridge, MA: Belknap Press of Harvard University Press, 1993), 246.

40. Wiggins, *Bess of Hardwick's Letters*, 234; Jonathan Gibson, "Significant Space in Manuscript Letters," *Seventeenth Century* 12, no. 1 (1997): 1–10.

41. Conceited, adj. and n. (3c., now obsolete): "Of a material object: elaborately or ingeniously wrought; fancifully elaborated," in *Oxford English Dictionary Online*, December 2018, Oxford University Press.

42. Francis Gainsford to the Earl of Essex, 1598, Hatfield House, Cecil Papers 178/74. Image included in the Cecil Papers [Database], Hatfield House Archives, 2010.

43. For the interplay between prints, the decorative arts, and manuscript culture, see Anthony Wells-Cole, *Art and Decoration in Elizabethan and Jacobean England: The Influence of Continental Prints, 1558–1625* (New Haven, CT: Yale University Press, 1997); and the introduction to Heather Wolfe, ed., *The Trevelyon Miscellany of 1608: A Facsimile of Folger Shakespeare Library MS V.b.232* (Washington, DC: Folger Shakespeare Library, 2007).

44. Amanda Flather, *Gender and Space in Early Modern England* (Woodbridge, UK: Boydell, 2007), 14.

Chapter 2. Papering the Household

My thanks to members of the "Working with Paper" working group at the Max Planck Institute for the History of Science and to seminar participants at the Folger Shakespeare Library, the Humboldt Universität, the University of York, and the Wellcome Library for their constructive comments and feedback. My research here is indebted to the work of citizen transcribers on the Early Modern Manuscripts Online Collective (https://emroc.hypotheses.org) and Shakespeare's World (https://www.shakespearesworld.org) projects. Thank you—for your transcriptions and for #paper!

1. Margaret Pelling, "Thoroughly Resented? Older Women and the Medical Role in Early Modern London," in *Women, Science and Medicine 1500–1700: Mothers and Sisters of the Royal Society*, ed. Lynette Hunter and Sarah Hutton (Stroud: Sutton, 1997), 70.

2. Mary Fissell, "Women, Health and Healing in Early Modern Europe," *Bulletin of the History of Medicine* 82, no. 1 (2008): 1–17; Alisha Rankin, *Panaceia's Daughters: Noblewomen as Healers in Early Modern Germany* (Chicago: University of Chicago Press, 2013); Lucinda McCray Beier, "In Sickness and in Health: A Seventeenth Century Family's Experience," in *Patients and Practitioners: Lay Perceptions of Medicine in Pre-industrial Society*, ed. Roy Porter, Cambridge Studies in the History of Medicine (Cambridge: Cambridge University Press, 1986), 101–28; Alun Withey, *Physick and the Family: Health, Medicine and Care in Wales, 1600–1750* (Manchester: Manchester University Press, 2011); Olivia Weisser, *Ill Composed: Sickness, Gender, and Belief in Early Modern England* (New Haven, CT: Yale University Press, 2015); and Anne Stobart, *Household Medicine in Seventeenth-Century England* (London: Bloomsbury Publishing, 2016).

3. Rankin, *Panaceia's Daughters*; Meredith K. Ray, *Daughters of Alchemy: Women and Scientific Culture in Early Modern Italy* (Cambridge, MA: Harvard University Press, 2015); Stobart, *Household Medicine*; Elaine Leong, *Recipes and Everyday Knowledge: Medicine, Science, and the Household in Early Modern England* (Chicago: University of Chicago Press, 2018); and Michelle DiMeo and Sara Pennell, eds., *Reading and Writing Recipe Books, 1550–1800* (Manchester: Manchester University Press, 2013).

4. Here, I also take inspiration from David Arnold, *Everyday Technology: Machines and the Making of India's Modernity* (Chicago: University of Chicago Press, 2013); and Projit Bihari Mukharji, *Doctoring Traditions: Ayurveda, Small Technologies, and Braided Sciences* (Chicago: University of Chicago Press, 2016).

5. There is a large body of literature on the topic. For recent overviews, see Carsten Timmermann and Julie Anderson, eds., *Devices and Designs: Medical Technologies in Historical Perspective* (Basingstoke: Palgrave Macmillan, 2006); Stefan Timmermans and Marc Berg, "The Practice of Medical Technology," *Sociology of Health and Illness* 25, no. 3 (2003): 97–114; Harry M. Marks, "Medical Technologies: Social Contexts and Consequences," in *Companion Encyclopedia of the History of Medicine*, ed. William Bynum and Roy Porter (New York: Routledge, 1993), 2:1592–618; and Jennifer Stanton, "Making Sense of Technologies in Medicine," *Social History of Medicine* 12, no. 3 (1999): 437–48.

6. Liliane Hilaire-Pérez and Christelle Rabier, "Self-Machinery? Steel Trusses and the Management of Ruptures in Eighteenth-Century Europe," *Technology and Culture* 54, no. 3 (2013): 460–502; Anna Maerker, "Anatomizing the Trade: Designing and Marketing Anatomical Models as Medical Technologies, ca. 1700–1900," *Technology and Culture* 54, no. 3 (2013): 531–62; Claire L. Jones, "Instruments of Medical Information: The Rise of the Medical Trade Catalog in Britain, 1750–1914," *Technology and Culture* 54, no. 3 (2013): 563–99; and François Zanetti, "Curing with Machines: Medical Electricity in Eighteenth-Century Paris," *Technology and Culture* 54, no. 3 (2013): 503–30.

7. See Christelle Rabier, "Introduction: The Crafting of Medicine in the Early Industrial Age," *Technology and Culture* 54, no. 3 (2013): 445.

8. In fact, the term *technology* did not take its current meaning until the 1930s. See Eric Schatzberg, "*Technik* Comes to America: Changing Meanings of *Technology* before 1930," *Technology and Culture* 47, no. 3 (2006): 486–512; and Schatzberg, "From Art to Applied Science," *Isis* 103, no. 3 (2012): 555–63.

9. Francesca Bray, *Technology, Gender and History in Imperial China: Great Transformations Reconsidered* (London: Routledge, 2013), 5.

10. David Edgerton, *The Shock of the Old: Technology and Global History since 1900* (London: Profile Books, 2006), xi–xii; Judith A. McGaw, "Why Feminine Technologies Matter," in *Gender and Technology: A Reader*, ed. Nina E. Lerman, Ruth Oldenziel, and Arwen P. Mohun (Baltimore: Johns Hopkins University Press, 2003), 13–36; and Bray, *Technology, Gender and History*. See also Ruth Schwartz Cowan, "The Consumption Junction: A Proposal for Research Strategies in the Sociology of Technology," in *The Social Construction of Technological Systems: New Directions in the Sociology and History of Technology*, ed. Wiebe E. Bijker, Thomas P. Hughes, and Trevor Pinch (Cambridge, MA: MIT Press, 1989), 261–80.

11. Fissell, "Women;" Lisa Smith, "The Relative Duties of a Man: Domestic Medicine in England and France, ca. 1685–1740," *Journal of Family History* 31, no. 3 (2006): 237–56; and Leong, "Collecting Knowledge for the Family: Recipes, Gender and Practical Knowledge in the Early Modern English Household," *Centaurus* 55, no. 2 (2013): 81–103.

12. This is particularly evident in literary studies. See, for example, Harold Love's discussion in his essay "Oral Culture and the Manuscript Economy," in *Thomas Middleton and Early Modern Textual Culture*, ed. Gary Taylor and John Lavagnino (Oxford: Oxford University Press, 2007), 101. I thank Heather Wolfe for sharing this reference in earlier versions of her essay in this volume.

13. See, for example, John Bidwell, "French Paper in English Books," in *The Cambridge History of the Book in Britain*, vol. 4, *1557–1695*, ed. John Barnard and D. F. McKenzie (Cambridge: Cambridge University Press, 2002), 587.

14. For the term "paper technology," see Anke te Heesen, "The Notebook: A Paper-Technology," in *Making Things Public: Atmospheres of Democracy*, ed. Bruno Latour and Peter Weibel (Cambridge, MA: MIT Press, 2005), 582–89. Recent work on the topic includes: Ann Blair, *Too Much to Know: Managing Scholarly Information before the Modern Age* (Philadelphia: University of Pennsylvania Press, 2008); Isabelle Charmantier and Staffan Müller-Wille, "Worlds of Paper," special issue, *Early Science and Medicine* 19, no. 5 (2014); Ann Blair and Richard Yeo, eds., "Note-Taking in Early Modern Europe," special issue, *Intellectual History Review* 20, no. 3 (2010); and Volker Hess and J. Andrew Mendelsohn, eds., "Paper Technology in der Frühen Neuzeit," special issue, *NTM. Zeitschrift für Geschichte der Wissenschaften, Technik und Medizin* 21, no. 1 (2013).

15. See, for example, Richard Yeo, *Notebooks, English Virtuosi and Early Modern Science* (Chicago: University of Chicago Press, 2014); Isabelle Charmantier and Staffan Müller-Wille, "Carl Linnaeus's Botanical Paper Slips (1767–1773)," *Intellectual History Review* 24, no. 2 (2014): 215–38; Volker Hess and J. Andrew Mendelsohn, "Paper Technology und Wissensgeschichte," *NTM. Zeitschrift für Geschichte der Wissenschaften, Technik und Medizin* 21, no. 1 (2013): 1–10; Michael Stolberg, "Medizinische *Loci Communes*: Formen und Funktionen einer ärztlichen Aufzeichnungspraxis im 16. und 17. Jahrhundert," *NTM. Zeitschrift für Geschichte der Wissenschaften, Technik und Medizin* 21, no. 1 (2013): 37–60; and Matthew C. Hunter, *Wicked Intelligence: Visual Art and the Science of Experiment in Restoration London* (Chicago: University of Chicago Press, 2013).

16. See also Helen Smith, "'A Unique Instance of Art': The Proliferating Surfaces of Early Modern Paper," *Journal of the Northern Renaissance* 8 (2017), http://www.northernrenaissance.org/?p=4847; Heather Wolfe, "Was Early Modern Writing Paper Expensive?," *The Collation*, Folger Shakespeare Library, February 13, 2018, https://collation.folger.edu/2018/02/writing-paper-expensive/; and Wolfe, this volume.

17. Sandra Cavallo, *Artisans of the Body in Early Modern Italy: Identities, Families and Masculinities* (Manchester: Manchester University Press, 2007); Celeste Chamberland, "Partners and Practitioners: Women and the Management of Surgical Households in London, 1570–1640," *Social History of Medicine* 24, no. 3 (2011): 554–69; Seth LeJacq, "The Bounds of Domestic Healing: Medical Recipes, Storytelling and Surgery in Early Modern England," *Social History of Medicine* 26, no. 3 (2013): 451–68; Margaret Pelling, "Appearance and Reality: Barber-Surgeons, the Body and Disease," in *London 1500–1700: The Making of the Metropolis*, ed. A. L. Beier and Roger Finlay (London: Longman, 1986), 82–112; Margaret Pelling, "Occupational Diversity: Barber Surgeons and the Trades of Norwich, 1550–1640," *Bulletin of the History of Medicine* 56 (1982): 484–511; and Olivia Weisser, "Boils, Pushes and Wheals: Reading Bumps on the Body in Early Modern England," *Social History of Medicine* 22, no. 2 (2009): 321–39.

18. Folger Shakespeare Library (hereafter Folger), MS v.a. 388, 16.

19. Wellcome Library (hereafter Wellcome), Western MS 4338, fol. 18v.

20. D. C. Coleman, *The British Paper Industry, 1495–1860: A Study in Industrial Growth* (Oxford: Clarendon Press, 1958), chapter 1, esp. table 1, "Imports of Paper into London, 1560–1720," 13.

21. *At the Marine Coffee-house in Birchin-Lane, near Corn-Hill, will be sold by Aunction these following Goods on Friday, Octob. 25, 1695*, British Library Harley 5942, fol. 35. I thank Heather Wolfe for bringing this source to my attention.

22. Judith M. Spicksley, ed., *The Business and Household Accounts of Joyce Jeffreys, Spinster of Hereford, 1638–1648* (Oxford: Oxford University Press, 2012).

23. Spicksley, *Business*, 146, 147, 153, 155, and 156.

24. Spicksley, *Business*, 281, 278, 253, and 246.

25. Spicksley, *Business*, 153, 156, 172, 177, 180, 183, 187, 188, 191, 211, 214, 215, 220, 225, 229, 246, 255, 256, 258, 283, and 287.

26. Folger, MS v.b. 366. fol. 62v.

27. Wellcome, Western MS 4338, fol. 15v.

28. Rose Marie San Juan, "Contaminating Bodies: Print and the 1656 Plague in Naples," in *New Approaches to Naples, c.1500–c.1800: The Power of Place*, ed. Melissa Calaresu and Helen Hills (Aldershot, UK: Ashgate, 2012), 68; and San Juan, "Corruptible Bodies and Contaminating Technologies: Jesuit Devotional Print and the 1656 Plague in Naples," in *Imagining Contagion in Early Modern Europe*, ed. Claire L. Carlin (Basingstoke: Palgrave Macmillan, 2005), 107–23. I am grateful to Mechthild Fend for bringing San Juan's work to my attention.

29. Orietta Da Rold, "Paper as Commodity in Medieval Magical and Medical Practices," *The Recipes Project*, last edited August 16, 2016, https://recipes.hypotheses.org/8358.

30. John Woodall, *The Surgeons Mate or Military &Domestic Surgery* (London, 1617), sig. A1v and 37. It is worth noting that paper is listed as "of other necessaries" (36); that is, it was considered secondary in importance to knives, cupping glasses, lancets, and other such instruments.

31. The *unguentum tripharmacon* addresses shingles, burns, scalds, and many more ailments. *Mel saponis* is, unsurprisingly, a mix of honey and soap and also good for burns, Woodall, *Surgeons Mate*, 46–48. The advice to young surgeons appears in the addendum "Viaticum Being the Pathway to the Surgeon's Chest," dated to 1628 on the subtitle page and included in the 1639 and 1655 editions, 310.

32. The narrative of medical professionalization tends to start in the nineteenth century. While some structures of self-regulation and professional identity come into play in the early modern period, many historians of early modern medicine prefer the terms "occupation" or "trade" to describe the constructed identities of various practitioners. For an overview, see Margaret Pelling, "Trade or Profession? Medical Practice in Early Modern England," in *The Common Lot: Sickness, Medical Occupations and the Urban Poor in Early Modern England* (London: Longman, 1998), 230–58.

33. *The Royal Pharmacopœa, Galenical and Chymical according to the Practice of the most Eminent and Learned Physitians of France* (London, 1678), 48. The same quotation can be found in the French edition of the work: "On employe le papier à filtrer diverse liqueurs, à couvrir les pots & les bouteilles, & à envelopper les remedes; On se sert enfin de 'mille matieres, & de mille inventions.'" *Pharmacopée royale galénique et chimique* (Paris, 1676), 98.

34. Wellcome, Western MS 4338, fol. 180v; and Folger MS v.b. 366, p. 77.

35. See, for example, Wellcome, Western MS 4338, fol. 207r-v.

36. Folger, MS v.a. 619, fols. 120r–121r.

37. Charas, *Royal Pharmacopœea*, 164.

38. Folger, MS v.a. 215, p. 256.

39. Folger, MS v.b. 366, p. 203.

40. Folger, v.b. 366, p. 121.

41. See for example, Folger, MS v.a. 619, fol. 95v (recipe for jumballs) and Folger, MS v.a. 19, fols. 32, and 152 (recipes "to make very fine cakes" and "to make fine Cracknells").

42. Spicksley, *Business*, 146, 147, 153, 155, and 156.

43. Folger, MS v.a. 8, p. 127.

44. Folger, MS v.a. 8, p. 100.

45. On the notion of paper tools, see Ursula Klein, *Experiments, Models, Paper Tools: Cultures of Organic Chemistry in the Nineteenth Century* (Stanford, CA: Stanford University Press, 2002).

46. Wellcome, Western MS 4338, fol. 18r.

47. Yale University Library, Osborn fb255, p. 8. I am grateful to Vera Keller for bringing this recipe to my attention. Another recipe for "Puppy water" can be found in Mary Doggett's recipe book, British Library, Additional MS 27466, p. 14.

48. See, for example, Fissell, "Women"; and Elaine Leong, "Making Medicines in the Early Modern Household," *Bulletin of the History of Medicine* 82, no. 1 (2008): 145–68.

49. Blair, *Too Much to Know*; Yeo, *Notebooks*; Michael Stolberg, "Medizinische *Loci Communes*"; and Hess and Mendelsohn, "Paper Technology."

50. Te Heesen, "Notebook"; and Charmantier and Müller-Wille, "Carl Linnaeus's Botanical Paper Slips."

51. Mary Fissell, "The Marketplace of Print," in *Medicine and the Market in England and Its Colonies, c. 1450–c. 1850*, ed. Mark Jenner and Patrick Wallis (Basingstoke: Palgrave Macmillan, 2007), 108–32; Mary Fissell, "Popular Medical Writing," in *The Oxford History of Popular Print Culture*, vol. 1, *Cheap Print in Britain and Ireland to 1660*, ed. Joad Raymond (Oxford: Oxford University Press, 2011), 417–30; and Paul Slack, "Mirrors of Health and Treasures of Poor Men: The Uses of the Vernacular Medical Literature of Tudor England," in *Health, Medicine and Mortality in the Sixteenth Century*, ed. Charles Webster (Cambridge, UK: Cambridge University Press, 1979), 237–73.

52. For the Renaissance scholar's feeling of information overload, see Blair, *Too Much to Know*; and Daniel Rosenberg, ed., "Early Modern Information Overload," special section, *Journal of the History of Ideas* 64, no. 1 (2003).

53. Blair, *Too Much to Know* (for an overview of note-taking systems, see chapter 2). See also essays in Alberto Cevolini, ed., *Forgetting Machines: Knowledge Management Evolution in Early Modern Europe* (Leiden: Brill, 2016).

54. Unlike their modern-day counterparts, these indices or tables of contents were not necessarily organized alphabetically but were more commonly merely lists of the recipe titles with page numbers. See Ann Blair, "Annotating and Indexing Natural Philosophy," in *Books and the Sciences in History*, ed. Marina Frasca-Spada and Nicholas Jardine (Cambridge: Cambridge University Press, 2000), 69–89, for a discussion of the practice of indexing in the early modern period.

55. Early modern English recipe collections arranged alphabetically include Wellcome, Western MS 761–62. A few books emphasize this organizational strategy by using a cutaway alphabetical index, much like the ones used in modern address books. These include Cambridge University Library (CUL), MS Dd 8 36; CUL MS Dd 5 18; and Royal College of Physicians Library, MS 502.

56. See, for example, Fabian Kraemer, "Ulisse Aldrovandi's *Pandechion Epistemonicon* and the Use of Paper Technology in Renaissance Natural History," *Early Science and Medicine* 19, no. 5 (2014): 398–423; Fabian Kraemer and Helmut Zedelmaier, "Instruments of Invention in Renaissance Europe: The Cases of Conrad Gesner and Ulisse Aldrovandi," *Intellectual History Review* 24, no. 3 (2014): 321–41; and Volker Hess and J. Andrew Mendelsohn, "Case and Series: Medical Knowledge and Paper Technology, 1600–1900," *History of Science* 48, no. 3–4 (2010): 288–314.

57. Owen Wood, *An Alphabetical Book of Physicall Secrets* (London, 1631), sig. A3r.

58. Thomas Willis appears to have been Johanna St. John's regular physician, as various prescriptions and health-related information make frequent appearences within the recipe book. For biographical information see, Robert L. Martensen, "Willis, Thomas (1621–1675)," *Oxford Dictionary of National Biography*, last edited October 4, 2007, http://www.oxforddnb.com/view/10.1093/ref:odnb/9780198614128.001.0001/odnb-9780198614128-e-29587.

59. The term *medical marketplace* was first coined by Harold Cook in his *Decline of the Old Medical Regime.* For a critique see Mark S. R. Jenner and Patrick Wallis, "The Medical Marketplace," in Jenner and Wallis, *Medicine and the Market*, 1–23.

60. Wellcome, Western MS 4338, fols. 203r, 203v, 205v, and 209v.

61. On the notion of women's secrets in the early modern period, see, for example, Katharine Park, *Secrets of Women: Gender, Generation, and the Origins of Human Dissection* (New York: Zone Books, 2006); and Monica H. Green, *Making Women's Medicine Masculine: The Rise of Male Authority in Pre-modern Gynaecology* (Oxford: Oxford University Press, 2008).

62. Wellcome, Western MS 4338, fols. 205v and 212r.

63. Angus Vine, "Commercial Commonplacing: Francis Bacon, the Waste-Book, and the Ledger," in *Manuscript Miscellanies, c. 1450–1700*, ed. Richard Beadle and Colin Burrow (London: British Library, 2011), 197–218; Kraemer and Zedelmaier, "Instruments of Invention," 328; and Kraemer, "Ulisse Aldrovandi's *Pandechion Epistemonicon.*"

64. These papers evidently were passed on to Margaret Boscawen's daughter Bridget Boscawen Fortescue and are now part of the Fortescue family papers at the Devon Heritage Centre. The three folders are Devon Heritage Centre 1262M/0/FC/6–8. I am grateful to Anne Stobart for introducing me to the Fortescue archive. For more information on the Boscawen/Fortescue archive see Stobart, *Household Medicine.*

65. A number of the slips have the original addressee information. For example, one recipe is addressed to "ye Rt Hon[ora]bble ye Margreat Boscawen Tregotnan to be left [at the] post House in Trego[tnan] Cornwall" and endorsed as "Mr Hancokes Herbs for to draw downe paine from the head & swelling." Internal evidence suggests that the bundle was largely created during the 1680s. For example, one recipe is labeled as "The Receite to make ye plague water from my Neice Trefusis Aprill: 1683." Unless otherwise noted, all examples of loose slips are from Devon Heritage Centre, 1262/0/M/FC 8. The folder contains numerous bundles of paper slips, all unfoliated.

66. The bundle contains two recipes for plague water (one is the abovementioned recipe from "Neice Trefusis"); a recipe for the "never failing cordiall,' another for "Banniter's Cordial Powder" (presumably, this refers to the surgeon John Bannister [1533–1610]); "Receipt of the lady Allins Cordial Water"; lemon water; black cherry water; "Doctor Adertons directions for ye makeing of Cinamon water"; and a long slip of paper with numerous recipes for medicinal waters.

67. See, for example, Kraemer and Zedelmaier, "Instruments of Invention"; Charman-

tier and Müller-Wille, "Carl Linnaeus's Botanical Paper Slips"; and Yeo, *Notebooks*, chapter 6: "Robert Boyle's Loose Notes."

68. Like many others, the endorsement of this slip is written in two different hands. One labeled it as a 'Diet Drinke for the chilldren yt hath the Ricketts" and another hand in a darker ink wrote "this reset I find my mothers Large boke."

69. On the practice of cutting, clipping, and pinning of loose papers, see, in the early modern English context, Adam Smyth, *Autobiography in Early Modern England* (Oxford: Oxford University Press, 2010), 38–39; and Smyth, "Little Clippings: Cutting and Pasting Bibles in the 1630s," *Journal of Medieval and Early Modern Studies* 45, no. 3 (2015): 595–613. For examples of clippings in other contexts see, for example, Ellen Gruber Garvey, *Writing with Scissors: American Scrapbooks from the Civil War to the Harlem Renaissance* (New York: Oxford University Press, 2012); and Anke te Heesen, *The Newspaper Clipping: A Modern Paper Object* (Manchester: Manchester University Press, 2014).

Chapter 3. The Sociomateriality of Waste and Scrap Paper in Eighteenth-Century England

Epigraph: Joseph Priestley, *Lectures on History, and General Policy; To Which is Prefixed, an Essay on a Course of Liberal Education for Civil and Active Life*, 2 vols. (London, 1793), 2:218.

1. See, e.g., Adrian Johns, *The Nature of the Book: Print and Knowledge in the Making* (Chicago: University of Chicago Press, 1998); Marina Frasca-Spada and Nick Jardine, eds., *Books and the Sciences in History* (Cambridge: Cambridge University Press, 2000); Jonathan Topham, "Anthologizing the Book of Nature: The Circulation of Knowledge and the Origins of the Scientific Journal in Late Georgian Britain," in *The Circulation of Knowledge between Britain, India, and China: The Early-Modern World to the Twentieth Century*, ed. Bernard Lightman, Gordon McOuat, and Larry Stewart (Leiden: Brill, 2013), 119–52.

2. Wanda J. Orlikowski, "The Sociomateriality of Organisational Life: Considering Technology in Management Research," *Cambridge Journal of Economics* 34, no. 1 (2009): 125–41; Wanda J. Orlikowski and Susan V. Scott, "Sociomateriality: Challenging the Separation of Technology, Work and Organization," *Academy of Management Annals* 2, no. 1 (2008): 433–74.

3. On gender and sociomateriality, see, e.g., Johanna Sefyrin, "Entanglements of Participation, Gender, Power and Knowledge in IT Design," in *PDC 2010: Proceedings of the 11th Biennial Participatory Design Conference* (New York: Association for Computing Machinery, 2010), 111–20; Joan H. Fujimura, "Sex Genes: A Critical Sociomaterial Approach to the Politics and Molecular Genetics of Sex Determination," *Signs* 32, no. 1 (2006): 49–82.

4. John Beale concluded a letter to Boyle by saying, "But I have filled my paper." John Beale to Robert Boyle, April 28, 1666, in *The Correspondence of Robert Boyle*, 6 vols., ed. Michael Hunter, Antonio Clericuzio, and Lawrence M. Principe (London: Pickering & Chatto, 2001), 3:160.

5. Henry Peacham, *The Worth of a Penny: or, A Caution to Keep Money* (London, 1703), 11.

6. [Richard Burridge], "Oeconomie and Extravagance," *Gentleman's Magazine*, November 1731, 489.

7. *World (1787)* (London), June 4, 1789.

8. Simon Werrett, *Thrifty Science: Making the Most of Materials in the History of Experiment* (Chicago: University of Chicago Press, 2019), 45.

9. Knorr-Cetina, "Objectual Practice," in *The Practice Turn in Contemporary Theory*, ed. Theodore R. Schatzki, Karin Knorr-Cetina, and Eike von Savigny (London: Routledge,

2001), 190; Knorr-Cetina draws on the idea of "epistemic things" in Hans-Jörg Rheinberger, *Toward a History of Epistemic Things: Synthesizing Proteins in the Test Tube* (Stanford, CA: Stanford University Press, 1997).

10. John Aikin, *Evenings at Home; or, The Juvenile Budget Opened*, 3rd ed., vol. 2 (London, 1795), 140.

11. Gervase Markham, *The English Housewife*, ed. Michael R. Best (Montreal: McGill-Queen's University Press, 1994), 146.

12. Georg Andreas Agricola, *A Philosophical Treatise of Husbandry and Gardening*, ed. Richard Bradley (London, 1721), 133; William Amos, *The Theory and Practice of the Drill-Husbandry* (London, 1794), 30; John Beale Bordley, *Essays and Notes on Husbandry and Rural Affairs* (Philadelphia, 1799), 105.

13. Karen Harvey, *The Little Republic: Masculinity and Domestic Authority in Eighteenth-Century Britain* (Oxford: Oxford University Press, 2012), 27–28.

14. A folk verse compiled in Achim von Arnim and Clemens Brentano, *Des Knaben Wunderhorn* (1806–1881), quoted in Lothar Müller, *White Magic: The Age of Paper*, trans. Jessica Spengler (Cambridge: Polity Press, 2015), 49.

15. William Emerson, *Calculation, Libration, and Mensuration; or, The Arts of Reckoning, Weighing, and Measuring* (London, 1770), 37.

16. Alexander Craig, *The Amorose Songes, Sonets, and Elegies: of M. Alexander Craige, Scoto-Britane* (London, 1606), n.p.

17. Thomas Day, *The Children's Miscellany: In Which Is Included the History of Little Jack* (London, 1790), 107; William Wynne Ryland, *Authentic Memoirs of William Wynne Ryland, Containing a Succinct Account of the Life and Transactions of that Great but Unfortunate Artist* (London, 1784), 25; using books as ciphers for people was of course a common conceit in early modern satire, as in Jonathan Swift's "The Battle of the Books," for example.

18. Thomas Salusbury, Mathematical Collections and Translations (London, 1661), 329–30; Jan Swammerdam, *Ephemeri vita, or, The Natural History and Anatomy of the Ephemeron, a Fly That Lives but Five Hours* (London, 1681), 17.

19. Edward Fountaine, *Melancholys Bane: or, Choice, Pleasant, and Profitable Recreations* (London, 1654), n.p.

20. Isaac Newton, *Opticks, or A Treatise of the Reflections, Refractions, Inflections, & Colours of Light* (New York: Dover, 1979), 20.

21. Karen Harvey, *Reading Sex in the Eighteenth Century: Bodies and Gender in English Erotic Culture* (Cambridge: Cambridge University Press, 2004), 90–100; Amelia Rauser, "Hair, Authenticity, and the Self-Made Macaroni," *Eighteenth-Century Studies* 38, no. 1 (2004): 101–17; Gillian Perry, "Staging Gender and 'Hairy Signs': Representing Dorothy Jordan's Curls," *Eighteenth-Century Studies* 38, no. 1 (2004): 145–63; Edith Snook, "Beautiful Hair, Health, and Privilege in Early Modern England," *Journal for Early Modern Cultural Studies* 15, no. 4 (2015): 22–51; Susan Vincent, "Men's Hair: Managing Appearances in the Long Eighteenth Century," in *Gender and Material Culture in Britain since 1600*, ed. Hannah Greig, Jane Hamlett, and Leonie Hannan (London: Palgrave, 2016), 49–67.

22. Legros de Rumigny, *L'Art de la coëffure des dames françoises, avec des estampes* (Paris, 1768).

23. William Moore, *The Art of Hair-dressing, and Making it Grow Fast, together, with a Plain and Easy Method of Preserving It* (Bath, ca. 1780).

24. Alexander Stewart, *The Art of Hair Dressing, or, The Gentleman's Director; Being a Concise Set of Rules for Dressing Gentlemen's Hair* (London, 1788); Alexander Stewart, *The*

Art of Hair Dressing, or, The Ladies Director; Being a Concise Set of Rules for Dressing Ladies Hair (London, 1788).

25. James Stewart, *Plocacosmos: or, The Whole Art of Hair-Dressing, Wherein is Contained Ample Rules for the Young Artizan, More Particularly for Ladies Women Valets &c. &c.* (London, 1782).

26. Stewart, *Plocacosmos*, 245.

27. Deborah Simonton, *A History of European Women's Work: 1700 to the Present* (London: Routledge, 1998), 61.

28. James MacDonald et al., *The British Legacy; or, Fountain of Knowledge* (London, 1754), 122–23.

29. R. Turner Wilcox, *The Mode in Costume: A Historical Survey with 202 Plates* (Mineola, NY: Dover, 2008), 212.

30. Margaret K. Powell and Joseph Roach, "Big Hair," *Eighteenth-Century Studies* 38, no. 1 (2004): 80.

31. Powell and Roach, "Big Hair," 83.

32. Stewart, *Plocacosmos*, quoted in Powell and Roach, "Big Hair," 84.

33. K. D. M. Snell, *Annals of the Labouring Poor: Social Change and Agrarian England, 1660–1900* (Cambridge: Cambridge University Press, 1987), 273.

34. Stewart, *Plocacosmos*, 256.

35. Penny Jenkins, "India Paper Proofs," *Paper Conservator* 14 (1990): 46–57; John Krill, *English Artists' Paper: Renaissance to Regency* (New Castle, DE: Oak Knoll Press, 2002).

36. "Woman," *European Magazine* 85 (1824): 226.

37. M. Maty, *Miscellaneous Works of the Late Philip Dormer Stanhope, Earl of Chesterfield* (Dublin, 1777), 2:180–81.

38. Maty, *Miscellaneous Works*, 2:181.

39. Maty, *Miscellaneous Works*, 2:182–83.

40. Maty, *Miscellaneous Works*, 2:183.

41. Alexander Pope, *The Rape of the Lock*, canto 4:97–102, quoted in Tita Chico, *Designing Women: The Dressing Room in Eighteenth-Century English Literature and Culture* (Lewisburg, PA: Bucknell University Press, 2005), 119. "Fillets" were headbands.

42. Rauser, "Hair," 101–17; Jennifer Jones, "*Coquettes* and *Grisettes*: Women Buying and Selling in Ancien Régime Paris," in *The Sex of Things: Gender and Consumption in Historical Perspective*, ed. Victoria de Grazia (Berkeley: University of California Press, 1996), 40–41.

43. If it counts as wrapping, then Wolfe's work on envelopes is an important exception. See chapter 1 in this volume.

44. Tobias Smollett, *The Adventures of Sir Launcelot Greaves* (London, 1762), 2:24–25; Gilles Deleuze considered "the fold" a central element of the baroque. Gilles Deleuze, *The Fold: Leibniz and the Baroque*, trans. Tom Conley (London: Athlone Press, 1993).

45. Edward Watkinson, *Essay upon Oeconomy* (York, 1767), 22, 55.

46. John Adams, *Woman. Sketches of the History, Genius, Disposition, Accomplishments, Employments, Customs and Importance of the Fair Sex in All Parts of the World* (London, 1790), 52.

47. Charles Bryant, *Flora diaetetica; or, History of Esculent Plants* (London, 1783), 45.

48. Benjamin Franklin, "Paper, a Poem," *Monthly Visitor, and Entertaining Pocket Companion* vol. 10, 1800, 134–35.

49. "Inventory, James Watt's Workshop, Objects presented by Maj. J. M. Gibson Watt 1924, R.P.Sc.M.1814," Science Museum, London.

50. Jonathan Swift, *A Tale of a Tub and Other Works*, ed. Marcus Walsh (Cambridge: Cambridge University Press, 2010), 23. Lord Chesterfield read, then used pages from Horace as toilet paper. Kirstin Olsen, *Daily Life in 18th-Century England* (Westport, CT: Greenwood Press, 1999), 269.

51. Richard Briggs, *The English Art of Cookery, According to the Present Practice* (Dublin, 1798), 120, 310.

52. Elizabeth Raffald, *The Experienced English Housekeeper* (London, 1806), 222–23.

53. Patrick Lamb, *Royal Cookery; or the Complete Court-Cook* (London, 1710), 46–47.

54. "Caesar (Sir Julius)," *Encyclopaedia Britannica; or, a Dictionary of Arts, Sciences, and Miscellaneous Literature*, vol. 4 (Dublin, 1791), 15.

55. John Aubrey, *Scientific Lives* (London: Hesperus, 2011), 109.

56. Gabriel Plattes, *The Profitable Intelligencer, Communicating His Knowledge for the Generall Good of the Common-Wealth and All Posterity* (London, 1644), n.p.

57. Joseph Warton, *An Essay on the Genius and Writings of Pope* (London, 1782), 2:216. See also Capt. Lemuel Gulliver [Jonathan Swift], *Travels into Several Remote Nations of the World* (London, 1727), 15.

58. "Clerc (Sebastien Le)," *A New and General Biographical Dictionary* (London, 1798), 4:78–79.

59. "Schurman (Anna Maria)," *An Historical, Genealogical, and Classical Dictionary: Containing the Lives and Characters of All the Illustrious Personages in the Several Ages and Nations of the World* (London, 1743), 2:n.p.

60. John Locke, *Some Thoughts Concerning Education*, 5th ed. (London, 1705), 235.

61. Locke, *Some Thoughts Concerning Education*, 235.

62. Locke, *Some Thoughts Concerning Education*, 237.

Chapter 4. Paper Trials, Multiple Masculinities, and the *Oeconomy* of Honor

1. Jacob Christian Schäffer, *Versuche und Muster ohne alle Lumpen oder doch mit einem geringen Zusatze derselben Papier zu machen* (Regensburg, 1765).

2. Schäffer, *Versuche und Muster*, 3.

3. Schäffer, *Versuche und Muster*, 3.

4. See for example, Joachim Radkau, *Holz: Ein Naturstoff in der Technikgeschichte* (Reinbeck bei Hamburg: Rowohlt, 1987).

5. René Antoine Ferchault de Réaumur, *Mémoires pour servir à l'histoire des insectes* (Paris: L'imprimerie royale, 1742), 155–247.

6. Edgar Zilsel, "The Sociological Roots of Science," *American Journal of Sociology* 47, no. 4 (1942): 544–62. On the influence of Zilsel's thesis see Pamela O. Long, *Artisans/Practitioners and the Rise of the New Sciences, 1400–1600* (Corvallis: Oregon State University Press, 2011).

7. Henrique Leitão, Antonio Sánchez, "Zilsel's Thesis, Maritime Culture, and Iberian Science in Early Modern Europe," *Journal of the History of Ideas* 78, no. 2 (2017): 191–210.

8. While recent years have seen a number of important studies on "scientific masculinities," there has yet to be any discussion of the scholar-artisan relationship from a gendered perspective. For the former, see in particular the essays edited by Erika Lorraine Milam and Robert A. Nye, *Scientific Masculinities*, Osiris 30 (Chicago: University of Chicago Press, 2015).

9. Ute Frevert, *Mann und Weib, und Weib in Mann: Geschlechter-Differenzen in der Mod-*

erne (Munich: Beck, 1995); Karen Harvey and Alexandra Shepard, "What Have Historians Done with Masculinity? Reflections on Five Centuries of British History, circa 1500–1950," *Journal of British Studies* 44, no. 2 (2005): 274–80.

10. Scholarship coming out of queer studies, such as Judith Halberstam's *Female Masculinity* (Durham, NC: Duke University Press, 1998) has done much to open up this area of research.

11. Hanna Hodacs, "Linnaean Scholars Out of Doors: So Much to Name, Learn and Profit From," in *Naturalists in the Field: Collecting, Recording and Preserving the Natural World from the Fifteenth to the Twenty-First Century*, ed. Arthur MacGregor (Leiden: Brill, 2018), 240–57.

12. I borrow the concept of "gestural knowledge" from Heinz Otto Sibum, "Reworking the Mechanical Value of Heat: Instruments of Precision and Gestures of Accuracy in Early Victorian England," *Studies in the History and Philosophy of Science* 26, no. 1 (1995): 76.

13. While the importance of honor in "polite" society has been extensively treated by historians, its role among the "ruder" classes has received far less attention. See, for example, Ute Frevert, *Ehrenmänner: Das Duell in der bürgerlichen Gesellschaft* (Munich: C. H. Beck, 1991); and Kathy Stuart, *Defiled Trades and Social Outcasts: Honor and Ritual Pollution in Early Modern Germany* (Cambridge: Cambridge University Press, 1999).

14. Leonard N. Rosenband, *Papermaking in Eighteenth-Century France: Management, Labor, and Revolution at the Montgolfier Mill, 1761–1805* (Baltimore: Johns Hopkins University Press, 2000), 30–49.

15. Andre Wakefield, *The Disordered Police State: German Cameralism as Science and Practice* (Chicago: University of Chicago Press, 2009), 3–5.

16. An important exception is Marion Gray's *Productive Men, Reproductive Women: The Agrarian Household and the Emergence of Separate Spheres during the German Enlightenment* (New York: Berghahn Books, 2000).

17. Born in Erfurt to a Protestant archdeacon and his wife, Schäffer enrolled at the University of Halle in 1736 in order to study theology. Poor health and financial constraints forced him to abandon his studies in 1738 and accept a position as a tutor in Regensburg. In 1741 he was named pastor at a local church and from 1779 until his death served as superintendent of the Protestant community.

18. The earliest mention of such experiments I have been able to locate appears in a letter by Edward Lloyd (1660–1709) to the Royal Society that concerns paper made out of asbestos. "An Account of a Sort of Paper Made of *Linum Asbestinum* Found in Wales in a Letter to the Publisher, from Edward Lloyd of Jesus Coll. Oxon," *Philosophical Transactions* 14 (1684): 823–24.

19. Mary Terrall, "Gendered Spaces, Gendered Audiences: Inside and Outside the Paris Academy of Sciences," *Configurations* 3, no. 2 (1995): 207–32.

20. Among others, Schäffer belonged to the Leopoldina in Berlin, the Academy of Sciences in Paris, and the Royal Society in London. Angelika Reich, ed., *Jacob Christian Schaeffer: Superintendent, Naturforscher, Erfinder* (Regensburg: University Library, Regensburg, 1993), exhibition catalog.

21. Schäffer, Jacob Christian, Briefe 1766/3 nrs. 28–42, Bavarian Academy of Arts and Sciences, Munich.

22. "Jacob Christian Schäffers Nachricht von dem dritten und vierten Versuch aus Pappelwolle Papier zu machen" and "Anleitung zur Herstellung von Tapeten" 1764, Folio 2 Nr. 3137, Das Niedersächsische Landesarchiv Standort Bückeburg.

23. The popularity of Schäffer's books is attested to by the appearance of a Dutch translation in 1770, and a second, German edition in 1772.

24. Schäffer, *Versuche und Muster*, 1:17. "Und so bald ich sehen oder hören werde, daß auch nur ein oder der andere Papiermacher wirklich mir nachzuarbeiten anfangen sollte; so bald werde ich die weitern Versuche aufgeben, und die Feder niederlegen, weil ich meine Absicht als denn werde erreichet haben."

25. Georg Christoph Keferstein, *Unterricht eines Papiermachers an seine Söhne diese Kunst betreffend* (Leipzig: Breitkopf und Sohn, 1766), 64.

26. See for example, S. R. Epstein, "Craft Guilds in the Pre-modern Economy: A Discussion," *Economic History Review* 61, no. 1 (2008): 155–74.

27. Schäffer deliberately reserved copies of his books for papermakers. Since each experiment yielded around fifty sheets of paper, this was no small sacrifice.

28. The following description is based on Dard Hunter's *Papermaking: The History and Technique of an Ancient Craft*, 3rd ed. (New York: Alfred A. Knopf, 1974).

29. Women also assisted in drying the paper, after it had been formed with the aid of a mold, as well as after it was sized, and were responsible for polishing the individual sheets near the end of the papermaking process.

30. Johann Georg Krünitz, "Gautscher," *Oekonomische Encyclopädie oder allgemeines System der Staats-, Stadt-, Haus- und Landwirthschaft*, accessed August 15, 2017, http://www.kruenitz1.uni-trier.de/.

31. Here I follow Richard Sennett's definition of skill as "a source of the legitimacy of command of the dignity of obedience." Sennett, *The Craftsman* (New Haven, CT: Yale University Press, 2008), 54.

32. Johann Georg Krünitz, "Papiermacherordnung," *Oekonomische Encyclopädie*, accessed August 15, 2017. The same held true for sons of journeymen.

33. Keferstein, *Unterricht eines Papiermachers*, 58.

34. Rosenband, *Papermaking in Eighteenth-Century France*, 57–60.

35. Krünitz, "Gautscher."

36. Sibum, "Reworking the Mechanical Value of Heat," 76.

37. Sibum, "Reworking the Mechanical Value of Heat," 76; Krünitz, "Büttgesell," *Oekonomische Encyclopädie*, accessed August 15, 2017. "Er kann einer verstümmelten Bogen Papier machen, wenn er seine Materie nicht genug ausbreitet . . . wenn er zu steife Arme hat, wenn er einen übeln Stoß gibt."

38. Sibum, "Reworking the Mechanical Value of Heat," 76; Krünitz, "Büttgesell"; Krünitz, "Gautscher." "Legt er den Papierbogen zu stark auf, dergestalt, daß das Wasser des Filzes in den darauf gelegten Bogen dringt, so macht er das Papier wasserschadhaft."

39. For the issue of control over one's body see Lissa Roberts, Simon Schaffer, and Peter Dear, eds., *The Mindful Hand: Inquiry and Invention from the Late Renaissance to Early Industrialization* (Amsterdam: Royal Netherlands Academy of Arts and Sciences, 2007).

40. There is growing interest in early modern workshops as spaces of identity formation. See for example: Sandro Cavallo, *Artisans of the Body in Early Modern Italy: Identities, Families and Masculinities* (Manchester: Manchester University Press, 2007); Rosa Salzberg, "Masculine Republics: Establishing Authority in the Early Modern Venetian Printshop," in *Governing Masculinities in the Early Modern Period: Regulating Selves and Others*, ed. Susan Broomhall and Jacqueline van Gent (Farnham, UK: Ashgate, 2011), 47–65; and Merry E. Wiesner, "*Wandervögels* and Women: Journeymen's Concepts of Masculinity in Early Modern Germany," *Journal of Social History* 24, no. 4 (1991): 767–82.

41. The following account is based on "Entwurf einer Papiermüllerordnung für die Churmark Brandenburg 1745," in Daniel Gottfried Schreber, *Neue Sammlung Verschiedener in die Cameralwissenschaften einschlagender Abhandlungen und Urkunden*, vol. 8 (Bützow: Wismar, 1765), 115–16.

42. On shame as it related to skinners and executioners, see Stuart, *Defiled Trades*, 69–94.

43. "Gebräuche und Mißbrauche der Papiermacher," *Journal von und für Deutschland* 1 (1786): 37.

44. Lissa Roberts, "Practicing *Oeconomy* during the Second Half of the Long Eighteenth Century: An Introduction," *History and Technology* 30, no. 3 (2014): 133–48.

45. The classic statement on rulers in the German-speaking territories as "Hausväter" remains Otto Brunner, "Das 'Ganze Haus' und die alteuropäische 'Ökonomik,'" in *Neue Wege der Verfassungs- und Sozialgeschichte*, 2nd ed. (Göttingen: Vandenhoeck & Ruprecht, 1968), 103–27.

46. Wilhelm Wegener, "Der Reichsschluß von 1731," *Annales Universitatis Saraviensis, Rechts- und Wirtschaftswissenschaften* 5 (1956/57): 36–47.

47. Friedrich Harbich, "Von Besserer Einrichtung der Papiermühlen in Deutschland," *Hannöversche Nützliche Samlungen* 49, Stück (1756): 762.

48. Hunter, *Papermaking*, 162–65.

49. Rosenband, *Papermaking in Eighteenth-Century France*, 39–40.

50. Jacob Christian Schäffer, "Erfolg der Versuche die Saamenwolle der Schwarzpappel und des Wollengrasses," *Abhandlungen der Churfürstlich-Baierischen Akademie die Wissenschaften* 2 (1764): 263. "Nicht nur etwann in den Gegenden hiesigen Ortes, sondern fast in ganz Deutschland, werden über den allgemeinen Abgang des Papierzeuges, oder der sogenannten Lumpen . . . seit einiger Zeit ehr, als jemals, die größten Klagen geführt."

51. Matt Schumann, "The End of the Seven Years' War in Germany," in *The Seven Years' War: Global Views*, ed. Mark H. Danley and Patrick J. Speelman (Boston: Brill, 2012), 487–518.

52. Günter Bayerl, *Die Papiermühle: Vorindustrielle Papiermacherei auf dem Gebiet des alten deutschen Reiches* 1 (Frankfurt: Lang, 1987), 380.

53. Viktor Thiel, "Papiererzeugung und Papierhandel vornehmlich in den deutschen Landen von den ältesten Zeiten bis zum Beginn des 19. Jahrhunderts," *Archivalische Zeitschrift* 41 (1932): 147–50.

54. Compare the prices in *Handlungszeitung oder wöchentliche Nachrichten von Handel, Manufacturen, Künste und neuen Erfindungen* (1785), 200; with Krünitz "Papier," *Oekonomische Encyclopädie*, accessed August 15, 2017.

55. Keferstein, *Unterricht eines Papiermachers*, 64–65. "Denn so lange Menschen auf den deutschen Boden sind, so gebrauchen dieselben Kleider; man rechnet, daß jeder Mensch jährlich zwey Hemden zerreißt; nach gerade kommen dieselben doch in die Papiermühle, ohne daß ich mich ängstlich darum zu bemühen nöthig habe."

56. Wolfang Ilg, ed., *Geschichte der Botanik in Regensburg: 200 Jahre Regensburgische Botanische Gesellschaft* (Regensburg: Museen der Stadt Regensburg, 1990), exhibition catalog.

57. Anne Secord, "Corresponding Interests: Artists and Gentlemen in Nineteenth-Century Natural History," *British Journal for the History of Science* 27, no. 4 (1994): 383–408.

58. Jacob Christian Schäffer, *Erläuterte Vorschläge zur Ausbesserung und Förderung der Naturwissenschaft* (Regensburg: Zunkel, 1763), 16. "Der Ungelehrte muss aufsuchen, dem Gelehrten zutragen, und in die Hände liefern" so that the learned man can "beobachten, nachsehen, prüfen, beschreiben, bestimmen, auseinandersetzen, in Ordnung bringen."

59. Schäffer, *Erläuterte Vorschläge*, 27.

60. Schäffer, *Erläuterte Vorschläge*, 5. "Ich ließ von jeder Saamenwolle zusammenlesen, das Gesamelte nach Hause tragen, und besprach mich darüber mit dem hiesigen Papiermacher Meckenhäuser."

61. Steven Shapin, "The Invisible Technician," *American Scientist* 77, no. 6 (1989): 554–63.

62. Schäffer, *Versuche und Muster*, 6.

63. Schäffer, *Versuche und Muster*, 6. "Hatte keine rechte Steife und Festigkeit, und war überdem noch voller braunen knötgen, als den Ueberbleibseln der zerstrossenen Saamenkörnergen."

64. Schäffer, *Versuche und Muster*, 9–10. "Vom Anfange bis zum Ende, in meinem eigenen Hause, und unter meiner beständigen Aufsicht und Anweisung." On experiments in the household, see Steven Shapin, "The House of Experiment in Seventeenth-Century England," *Isis* 79, no. 3 (1988): 373–404; Alix Cooper, "Homes and Households," in *The Cambridge History of Science*, vol. 3, *Early Modern Science*, ed. Katharine Park and Lorraine Daston (Cambridge: Cambridge University Press, 2008), 224–37; Mary Terrall, *Catching Nature in the Act: Réaumur and the Practice of Natural History in the Eighteenth Century* (Chicago: University of Chicago Press, 2014).

65. Jacob Christian Schäffer, *Versuche mit dem beständigen Electricitätträger* (Regensburg: Montag, 1780); Jacob Christian Schäffer, "Erdäpfelgrütze zu machen," *Hannoverisches Magazin* vol. 10, 1772, 415–16.

66. Jacob Christian Schäffer, *Empfehlung, Beschreibung und erweiterte Gebrauch des sogennanten und zur Erspahrung des Holzes höchstvortheilhasten Backofens* (Regensburg: Johann Christoph Keyser, 1770); Jacob Christian Schäffer, *Nachtrag zu dem Gebrauche und Nutzen der verbesserten Waschmaschine* (Regensburg: Johann Leopold Montag, 1767).

67. The stamper, which measures 142.5 × 122.5 × 118 cm, is now in the collection of the Museen der Stadt, Regensburg. I am grateful to Karin Geiger for information pertaining to its operation and current condition.

68. Jacob Christian Schäffer, *Neue Versuche und Muster des Pflanzenreich zum Papiermachen und andern Sachen wirtschaftsnützlich zu gebrauchen* (Regensburg, 1767), preface, 17–22.

69. Schäffer, *Versuche und Muster*, 10. "Meinen Bedienten . . . die ganze Verfarhungsart und Handgriffe des Papiermachens."

70. Schäffer, *Versuche und Muster*, 10.

71. In the second edition of his books on paper trials, Schäffer mentions the journeyman by name: Heinrich Barth from Saxony. Jacob Christian Schäffer, *Sämtliche Papierversuche* (Regensburg: Zunkel, 1772), "Vorrede zur zwoten Auflage."

72. Schäffer, *Sämtliche Papierversuche*. This accusation became the basis for a later investigation by Wilhelm Herzberg, which showed that in some cases the paper samples produced by Schäffer contained as much as 95 percent linen! Wilhelm Herzberg, "Die Schäfferschen Papierversuche," *Mittheilungen aus den königlichen technischen Versuchsanstalten zu Berlin* 3, Heft (1898): 143–60.

73. Anne Goldgar, *Impolite Learning: Conduct and Community in the Republic of Letters, 1680–1750* (New Haven, CT: Yale University Press, 1995); Steven Shapin, *A Social History of Truth: Civility and Science in Seventeenth-Century England* (Chicago: University of Chicago Press, 1994).

74. Shapin, *Social History of Truth*, 107–14.

75. Schäffer, *Versuche und Muster*, "Vorbericht" [foreword].

76. See for example the short article in the *Leipziger Intelligenzblatt*, 1771, 98.

77. Among his most recent accolades was a gold chain from the Holy Roman emperor.

78. Keferstein, *Unterricht eines Papiermachers*, 63.

79. Unsigned review of *Unterricht eines Papiermachers an seine Söhne diese Kunst betreffend, Neue critische Nachrichten* 2 (1765): 328.

80. Keferstein, *Unterricht eines Papiermachers*, 51. "Macht mit einem Wort den letzten Bogen im Post mit eben dem Fleiß als den ersten."

81. Keferstein, *Unterricht eines Papiermachers*, 20.

82. Harbich, "Von Besserer Einrichtung der Papiermühlen in Deutschland," 762.

83. Keferstein, *Unterricht eines Papiermachers*, 4. For a similar use of the term *republic* among papermakers in France, see Rosenband, *Papermaking in Eighteenth-Century France*, 57–59.

Chapter 5. Bookkeeping for Caring

I am most grateful to the volume editors, Carla Bittel, Elaine Leong, and Christine von Oertzen, for their zest and intelligence in driving this project and their insightful suggestions, and to the wonderful working group in the MPIWG. My heartfelt thanks to Sebastian Felten, Antoni Malet, and Senthil Babu for their careful reading and commentaries, and very especially to Oriana Walker and Department I of the MPIWG-Berlin.

1. 9054/1, Archivo Regional de la Comunidad de Madrid, Fondo Diputación de Madrid, Inclusa de Madrid (hereafter ARM), p. 326. The Madrid Foundling House was near La Montera, on Preciado Street. On September 2, 1801, it moved to the El Soldado 3–4 (abutting today's La Libertad street). In Ignacio Maria Ruiz de Luzuriaga, *Estadística Politico Médica* (Mss, 1817–) V.1, 39–62, *Signatura: 21–3ª Biblioteca / 15*, Archivo de la Real Academia de Medicina, Madrid (hereafter ARAM). I would like to thank the staff of the Archivo Regional de la Comunidad de Madrid for their generous help, as well as the librarian of Archivo de la Real Academia de Medicina.

2. "Reglamento de las Hermanas de la Caridad, 1799," leg. 164/11, Sociedad Economica Matritense (hereafter SEM); and "Borrador a los artículos para formar la contrata de las hermanas de la caridad," 8881/14, ARM.

3. See the complaints of the clerk Juan González, 5109/2, ARM, p. 17. See also "Report of the treasurer on the tasks of the bookkeeper clerk [*oficial de libros*]," 8416/14, ARM. "Memorias instructivas de los negocios de la Junta de Honor y Mérito" (years 1800 and 1801), 8880/17, ARM. On baptism, see the Countess of Montijo's translation of Nicolas Letourneux, *Instrucciones cristianas sobre el sacramento del matrimonio* (Barcelona: Bernardo Pla, 1774), 116–215.

4. Mortality rates were high in contemporary European Foundling Houses, especially in the bigger institutions in Paris, Dublin, and Moscow. Rachel Ginnis Fuchs, *Abandoned Children: Foundlings and Child Welfare in Nineteenth-Century France* (Albany: State University of New York, 1984), 19–27. For the mortality rates in other Spanish Foundling Houses, see Isabel Ilzarbe, *Brocar*, n° 41 (2017): 89-115.

5. Sometimes there were falsifications, so careful checking was necessary. See 7721/7, ARM.

6. *Kalendario Manual y Guía de Forasteros para 1801* (Madrid, 1801), 208. According to this source, 4,443 children were baptized in the Madrid parishes that year.

7. On sociotechnical systems, see von Oertzen, this volume.

8. Leg. 105/6, SEM. See also Paula Demerson, *María Francisca de Sales Portocarrro, Condesa del Montijo: Una figura de la Ilustración* (Madrid: Editora Nacional, 1975), 387.

9. Joan Sherwood, *Poverty in Eighteenth-Century Spain: The Women and Children of the Inclusa* (Toronto: University of Toronto Press, 1988), 9–11.

10. The children of the Foundling Houses came mostly from poor families. Especially in epochs of famine and bad harvests, families abandoned newborns in the hope of preserving the rest of their children. Poor mothers also abandoned their children to seek employment, and these abandonments increased in the busy season of summertime. Finally, imprisoned pregnant women (mostly in jails accused of vagrancy) were forced to leave their newborns in the care of the House. Beginning in 1796, Spanish families who could prove they were "in extreme necessity" when leaving their children could later recover them: Santos Sánchez, *Colección de las Pragmáticas, Cédulas, Provisiones [. . .] del Señor Don Carlos IV,* vol. 2 (Madrid: Imprenta de la Viuda e Hijo de Marín, 1798), 278.

11. Alan Forrest, "Poverty," in *The Oxford Handbook of the Ancien Régime,* ed. William Doyle, 167–82 (Oxford: Oxford University Press, 2012), 179; Sandra Cavallo, "The Motivations of Benefactors: An Overview of Approaches to the Study of Charity," in *Medicine and Charity before the Welfare State,* Studies in the Social History of Medicine, ed. Jonathan Barry and Colin Jones (London: Routledge, 1991), 46–62; Barry and Jones, "Introduction," in Barry and Jones, *Medicine and Charity,* 1–14; Antonio Carreras Panchon, *El problema del niño expósito en la España Ilustrada* (Salamanca: Universidad de Salamanca, 1977).

12. Andrea Rusnock, "Quantifying Infant Mortality in England and France, 1750–1800," in *Body Counts: Medical Quantification in Historical and Sociological Perspective,* ed. Gérard Jorland, Annick Opinel, and Geroge Weisz, 65–86 (London: McGill-Queen's University Press, 2005), 66.

13. Francesca Bray, *Technology, Gender and History in Imperial China: Great Transformations Reconsidered* (London: Routledge, 2013), 17.

14. Peter Becker and William Clark, "Introduction," in *Little Tools of Knowledge: Historical Essays on Academic and Bureaucratic Practices,* ed. Peter Becker and William Clark (Ann Arbor: University of Michigan Press, 2001), 1–35. See also Ben Kafka, "Paperwork: The State of the Discipline," *Book History* 12, no. 1 (2009): 340–53; and von Oertzen, this volume.

15. For instance, in the Coram Foundling Hospital in London, children were routinely renamed as a way to erase their past. In the Hospital degli Innocenti, in Florence, illegitimate children were obscured in the pages of the registry books, as the indexes were organized by family names. In the French Hôpital des Enfants Trouvés, in the hands of the revolutionaries, the families had the right to retrieve their children if they paid a fee. Frances Miley and Andrew Read, "Go gentle babe: Accounting and the London Foundling Hospital, 1757–97," *Accounting History* 21, no. 2–3 (2016): 167–84. On relation to the *Hôpital des Enfants Trouvés,* see Rachel Fuchs, *Abandoned Children,* 17, 19–27. From 1801, an administrative commission and the Conseil Général des Hospices (set up as part of the Ministry of Interior) ran the hospices.

16. About paper affordances, see Eyferth, this volume.

17. Demerson, *María Francisca de Sales,* 141; Theresa Anne Smith, *The Emerging Female Citizen: Gender and Enlightenment in Spain* (Berkeley: University of California Press, 2003), 138 (table 5). Some of the more active members were María Josefa Alonso Pimentel y Borja, Condesa-Duquesa de Benavente (1752–1834); María Francisca de Sales Portocarrero, Countess of Montijo (1754–1808); the learned Rosario Cepeda (1756–1816); Francisca María

Dávila, Countess of Truillas (1753–1820); María Lorenza de los Ríos y Loyo Marchioness of Fuerte Híjar (1768?–1817?), and Josefa Lezo Pacheco, Marchioness of Espeja (fl. 1805).

18. Koen Stapelbroek and Jani Marjanen, *The Rise of Economic Societies in the Eighteenth Century: Patriotic Reform in Europe and North America* (New York: Palgrave Macmillan, 2012).

19. Josefina Méndez Vázquez, *Formación professional de las mujeres en las escuelas de la Matritense: Un Proyecto politico-económico en la España Ilustrada* (Madrid: Ediciones Trabe, 2016).

20. Elena Serrano, "Chemistry in the City: The Scientific Role of Female Societies in Eighteenth-Century Spain," *Ambix* 60, no. 2 (2013): 146–52.

21. Joaquín Fernández Pérez, "La ciencia ilustrada y las sociedades económicas de amigos del país," in *Carlos III y la ciencia de la Ilustración*, ed. Manuel Selles, José Luis Peset, and Antonio Lafuente (Madrid: Alianza Universidad, 1988), 217–45; Luis Miguel Recio Enciso, *Las sociedades económicas en el siglo de las luces* (Madrid: Real Academia de la Historia, 2010); Vicente Llombart Rosa and Jesús Astigarraga Goenaga, "Las primeras 'antorchas de la economía': Las sociedades económicas de amigos del país en el siglo XVIII," in *Economía y Economistas Españoles: La Ilustración*, ed. Enrique Fuentes Quintana (Barcelona: Galaxia Gutenberg, 2000), 677–707.

22. Smith, *Emerging Female Citizen.*

23. Discourses in Olegario Negrín, *Ilustración y educación: La Sociedad Económica Matritense* (Madrid: Editora Nacional, 1984).

24. Josefa Amar y Borbón, "Discurso en defensa del talento de las mugeres, y de su aptitud para el gobierno, y otros cargos en que se emplean los hombres", "In Defence of Women's Intelligence and Their Aptitudes for Government and Other Employments for Men," *Memorial literario*, August 1786, 400–430.

25. *Memorial Literario*, October 1787, 204–9; and *Gazeta de Madrid*, October 12, 1787, 688–89.

26. See *Estatutos de la Junta de Socias de Honor y Mérito de la Real Sociedad Económica de Madrid* (Madrid: Sancha, 1794). Also in Negrín, *Ilustración y educación*, 60–72.

27. "Proposición de la condesa de Montijo, julio 1789," leg. 92–93, SEM.

28. "Proposición," leg. 92–93, SEM. See also Demerson, *María Francisca de Sales*, 201–13.

29. Leg. 105/6, SEM. Gérard Jorland and George Weisz, "Introduction: Who Counts?," in Jorland, Opinel, and Weisz, *Body Counts*, 1–15.

30. The report gives no information on the source of these "tables of general mortality."

31. Leg. 105/6, SEM.

32. However, looking at the children's record books, one can deduce them. Luzuriaga, *Estadística político médica.*

33. "Libro de Juntas que mandó celebrar el actual protector de la Casa de la Inclusa Gonzalo Josef Vilches," 5109/2, ARM.

34. 8342/9, February 18, 1802, ARM.

35. "Memoria instructiva de 1798," 8880/17, ARM.

36. "Memoria instructiva de 1799," 8880/17, ARM.

37. Josefa Díez de la Cortina y Morales, *Elogio de la Reyna Nuestra Señora [. . .] leído en la junta publica de distribución de premios en 18 de diciembre de 1799* (Madrid: Imprenta Real, 1800), 15.

38. Lissa Roberts, "Practicing *Oeconomy* during the Second Half of the Eighteenth Century: An Introduction," *History and Technology* 30, no. 3 (2014): 133–48.

39. Josefa Amar y Borbón, *Discurso sobre la educación física y moral de las mujeres* (Madrid: Benito Cano, 1790), 158.

40. I quote the Spanish version of François de Salignac de La Mothe Fénelon's *Traité de l'éducation des filles. Education of Girls: Tratado de la educación de las hijas* (Madrid: Viuda de Eliseo Sánchez, 1769), 4.

41. The classical reference was Xenophon's *Oeconomy*, which was reread and very much quoted in eighteenth-century Spain. It was again translated in 1786: *La economía y los medios de aumentar las rentas públicas de Atenas*, trans. Griego al Castellano (Madrid: Benito Cano, 1786), 36–37.

42. This did not mean that accounting was considered a gendered practice. In fact, balancing incomes and outcomes, keeping the books, and "reflecting on figures" was advised for both men and women and thought equally important for the material and spiritual health of everyone. See Amanda Vickery, *The Gentleman's Daughter: Women's Lives in Georgian England* (New Haven, CT: Yale University Press, 2003), esp. the chapter "Prudent Oeconomy."

43. Caroline-Stéphanie-Félicité du Crest, Comtesse de Genlis, *Adela y Teodoro, o Cartas sobre la educación* (Madrid: Imprenta Real, 1792). See also, Pedro de Montegón, *Eudoxia, hija de Belisario* (Madrid: Sancha, 1793).

44. The secretary of the Madrid Economic Society, Pedro Cevallos, made a report on each of the four memoirs. See Smith, *Emerging Female Citizen*, 152–53.

45. Rosario de Cepeda was accountable for the Monte Pío de Hilazas. The Countess of Sonora, who would later have a key role in Madrid Foundling House, was the treasurer of the Asociación de Señoras, a female association for taking care of female inmates in Madrid jails.

46. Leg. 164/10 (1799), SEM.

47. Sherwood, *Poverty*, 11, 12, 17, and 47.

48. "Annual Report of 1801," 8880/17, ARM; "Papel presentado por la Montijo a la Junta de Damas de 1800," 8342, ARM; and "Instrucciones de la Condesa de Montijo para habilitar una sala arriba para niños enfermos," 8355, ARM.

49. The surgeon Tomas Mas, for example, was fired because he treated wet nurses badly; see 8476/2, July 30, 1800, ARM.

50. 8881/14, ARM.

51. Amar y Borbon, *Discurso*, 178.

52. François de Salignac de La Mothe Fénelon, *The Education of Girls*, trans. Kate Lupton (Boston: Ginn, 1891), 105.

53. "Borrador de los artículos para formar la contrata a las hermanas," 8881/14, ARM.

54. "Razón de las criaturas (Marzo)," 8456/2, ARM.

55. Luzuriaga, *Estadística.*

56. Usually, wet nurses were supposed to care for only two babies at a time (ideally of different sexes to avoid confusion), but in actual fact Teresa Facho was busy looking after four children. Teresa Facho also signed the registers of three other babies that entered that week. From 1802 onward the books were no longer signed by wet nurses.

57. The actual parchment of María Salomé has not been preserved. The information is taken from parchments in 7721/1, ARM; and 7721/9, ARM.

58. Libro de Salidas de Criaturas 169, 9216/1, ARM, p. 149.

59. Luis Luque, *Arte de partida doble ilustrado* (Cádiz: Calle de la Compañía, 1783).

60. Luque, *Arte de partida doble*, 34.

61. Note also the connections between mercantile and scholarly bookkeeping in the practices of world travelers (especially botany). Anke te Heesen, "Accounting for the Natural World: Double-Entry Bookkeeping in the Field," in *Colonial Botany: Science, Commerce, and Politics in the Early Modern World*, ed. Londa Schiebinger and Claudia Swan (Philadelphia: University of Pennsylvania Press, 2007), 237–51.

62. Anke te Heesen, "The Notebook: A Paper-Technology," in *Making Things Public: Atmospheres of Democracy*, ed. Bruno Latour and Peter Weibel (Cambridge, MA: MIT Press, 2005), 582–89.

63. See for example *Constituciones, ordenanzas y reglamentos del Real Hospicio [. . .] de la Ciudad de Badajoz* (Madrid: Josef Collado, 1804), 61–64.

64. Thanks to Maria Rentetzi for her witty comment about this.

65. See the letter that the Junta sent to D. Josef de Santa María, 8416/14, ARM.

66. "Memoria Instructiva de 1801," 8880/17, ARM.

67. Crisanto de Bazo brought a letter of a clergyman with him testifying that the family's circumstances had forced them to abandon their child and that they wanted to have her back. 9054/1, ARM, p. 326.

68. For example, "Memoria Instructiva de 1803," 880/17, ARM; "Memoria de 1805," 196/19, SEM.

69. "[. . .] sobre las disposiciones dadas por la Junta a los oficiales de libros," 8342/9, ARM, especially "Carta al Rector de la Inclusa, Madrid 2 Marzo 1802."

70. "Memoria Instructiva de 1802," 8888/17, ARM.

71. 8881/14, ARM.

72. 10264/2, ARM; and 1064/3, ARM. The first one is from 1802 to 1808 and the second one from 1809 to 1812.

73. "Memoria Instructiva de 1806," 8880/17, ARM.

74. Theodore M. Porter, "Making Things Quantitative," *Science in Context* 7, no. 3 (1994): 399.

75. Porter, "Funny Numbers," *Culture Unbound* 4 (2012): 585–98.

76. Porter, "Making Things Quantitative," 389. Porter, *Trust in Numbers: The Pursuit of Objectivity in Science and Public Life* (Princeton, NJ: Princeton University Press, 1995).

77. Ulrich Tröhler, "Quantifying Experience and Beating Biases: A New Culture in Eighteenth-Century British Clinical Medicine," in Jorland, Opinel, and Weisz, *Body Counts*, 19–50.

Chapter 6. Unpacking the Phrenological Toolkit

Many thanks to the Max Planck Institute for the History of Science in Berlin and members of the Department II working group, "Working with Paper." I am especially grateful to my coeditors, Elaine Leong and Christine von Oertzen. I also wish to thank my writing group in Los Angeles, Emily K. Abel, Janet Farrell Brodie, Lisa Forman Cody, Sharla Fett, and Alice Wexler, for their comments on early versions of this piece. Melissa Lo and Courtney Thompson provided valuable feedback in the final stages. Research on this article was supported by the National Endowment for the Humanities (Summer Stipends), the New York State Library, and Loyola Marymount University. Special thanks also to Russell Johnson and Teresa Johnson at the UCLA Library, History and Special Collections for the Sciences.

1. On phrenology, see John Dunn Davies, *Phrenology, Fad and Science: A 19th-Century American Crusade* (New Haven, CT: Yale University Press, 1955); Madeleine B. Stern, *Heads*

and Headlines: The Phrenological Fowlers (Norman: University of Oklahoma Press, 1971); Steven Shapin, "Phrenological Knowledge and the Social Structure of Early Nineteenth-Century Edinburgh," *Annals of Science* 32, no. 3 (1975): 219–43; Roger Cooter, *The Cultural Meaning of Popular Science: Phrenology and the Organization of Consent in Nineteenth-Century Britain* (Cambridge: Cambridge University Press, 1984); Michael M. Sokal, "Practical Phrenology as Psychological Counseling in the 19th-Century United States," in *The Transformation of Psychology: Influences of 19th-Century Philosophy, Technology, and Natural Science*, ed. Christopher D. Green, Marlene Shore, and Thomas Teo (Washington, DC: American Psychological Association, 2001), 21–44; John Van Wyhe, *Phrenology and the Origins of Victorian Scientific Naturalism* (Aldershot: Ashgate, 2004); Daniel Patrick Thurs, *Science Talk: Changing Notions of Science in American Popular Culture* (New Brunswick, NJ: Rutgers University Press, 2007).

2. Wanda Orlikowski, "Sociomaterial Practices: Exploring Technology at Work," *Organization Studies* 28, no. 9 (2007): 1435–48; Werrett, this volume.

3. James Poskett, "Phrenology, Correspondence, and the Global Politics of Reform, 1815–1848," *Historical Journal* 60, no. 2 (2017): 409–42; John van Wyhe, "The Diffusion of Phrenology through Public Lecturing," in *Science in the Marketplace: Nineteenth-Century Sites and Experiences*, ed. Aileen Fyfe and Bernard V. Lightman (Chicago: University of Chicago Press, 2007), 60–96; Fenneke Sysling, "Science and Self-Assessment: Phrenological Charts 1840–1940," *British Journal for the History of Science* 51, no. 2 (2018): 261–80.

4. Scholarship that addresses certain gender elements of phrenology includes: Cynthia Eagle Russett, *Sexual Science: The Victorian Construction of Womanhood* (Cambridge, MA: Harvard University Press, 1989); Jan Todd, *Physical Culture and the Body Beautiful: Purposive Exercise in the Lives of American Women, 1800–1870* (Macon: Mercer University Press, 1998); Helen Lefkowitz Horowitz, *Rereading Sex: Battles over Sexual Knowledge and Suppression in Nineteenth-Century America* (New York: Knopf Doubleday, 2002); Carla Bittel, "Woman, Know Thyself: Producing and Using Phrenological Knowledge in 19th-Century America," *Centaurus* 55, no. 2 (2013): 104–30; Erica Lilleleht, "'Assuming the Privilege' of Bridging Divides: Abigail Fowler-Chumos, Practical Phrenology, and America's Gilded Age," *History of Psychology* 18, no. 4 (2015): 414–32.

5. On the history of the "interface" in physics, see Peter Schaefer, "Interface: History of a Concept, 1868–1888," in *The Long History of New Media: Technology, Historiography, and Contextualizing Newness*, ed. David W. Park, Nicholas W. Jankowski, and Steve Jones (New York: Peter Lang, 2011), 163–75.

6. "Interface," *Oxford English Dictionary*, 1882.

7. Rachel Plotnick, "At the Interface: The Case of the Electric Push Button, 1880–1923," *Technology and Culture* 53, no. 4 (2012): 815–45.

8. Fenneke Sysling also describes phrenological charts as mediators between science and the self. See Sysling, "Science and Self-Assessment."

9. Eyferth, this volume. On affordances, see Donald A. Norman, *The Design of Everyday Things*, rev. ed. (New York: Basic Books, 2013).

10. On paper as a technology, see Nina E. Lerman, Ruth Oldenziel, and Arwen Mohun, eds., *Gender and Technology: A Reader* (Baltimore: Johns Hopkins University Press, 2003); Anke te Heesen, "The Notebook: A Paper Technology," in *Making Things Public: Atmospheres of Democracy*, ed. Bruno Latour and Peter Weibel (Cambridge, MA: MIT Press, 2005), 582–89; Volker Hess and J. Andrew Mendelsohn, "Case and Series: Medical Knowledge and Paper Technology, 1600–1900," *History of Science* 48, no. 3–4 (2010): 287–314;

Lauren Kassell, "Casebooks in Early Modern England: Medicine, Astrology, and Written Records," *Bulletin of the History of Medicine* 88, no. 4 (2014): 595–625.

11. Te Heesen, "Notebook," 584; Anke te Heesen, *The World in a Box: The Story of an Eighteenth-Century Picture Encyclopedia* (Chicago: University of Chicago Press, 2002).

12. As historian Carolyn de la Peña reminds us, "White people have race. And they make it, sustain it, and protect it in part through technology." De la Peña, "The History of Technology, the Resistance of Archives, and the Whiteness of Race," *Technology and Culture* 51, no. 4 (2010): 923.

13. For some of the most recent work on race and phrenology, see Susan Branson, "Phrenology and the Science of Race in Antebellum America," *Early American Studies* 15, no. 1 (2017): 164–93; Ann Fabian, *The Skull Collectors: Race, Science, and America's Unburied Dead* (Chicago: University of Chicago Press, 2010); Cynthia S. Hamilton, "'Am I Not a Man and a Brother?' Phrenology and Anti-slavery," *Slavery & Abolition* 29, no. 2 (2008): 173–87; James Poskett, "National Types: The Transatlantic Publication and Reception of *Crania Americana* (1839)," *History of Science* 53, no. 3 (2015): 264–95.

14. Henry B. Gibbons, collection no. 404, Henry B. Gibbons Papers, Research Library, Fenimore Art Museum, Cooperstown, NY; L. M. Hammond, *Trials and Triumphs of an Orphan Girl: Or The Biography of Mrs. Deiadamia Chase, Physician and Phrenologist* (Cortland, NY: Van Slyck & Hitchcock, 1859).

15. Gibbons Diary, vol. 2, September 8, 1845, Henry B. Gibbons Papers.

16. Gibbons Diary, vol. 2, December 1845, Henry B. Gibbons Papers.

17. On using blanks, see Bouk, this volume; and von Oertzen, this volume.

18. John Harley Warner, *The Therapeutic Perspective: Medical Practice, Knowledge, and Identity in America, 1820–1885* (Cambridge, MA: Harvard University Press, 1986), 15.

19. John Cutting, testimonial for Henry B. Gibbons, Queens, December 19, 1843, Henry B. Gibbons Papers, #404/12.

20. Reuel Payne et al., from Broad Brook, January 9, 1845. In favor of both Gibbons and Buell, Henry B. Gibbons Papers, #404/12. Emphasis mine.

21. Sokal, "Practical Phrenology"; Sysling, "Science and Self-Assessment."

22. Lauren Kassell discusses the role of casebooks in the astrological medical encounter. See Kassell, "Casebooks in Early Modern England."

23. On phrenological charts and notions of the self, see Sysling, "Science and Self-Assessment."

24. For example, see O. S. Fowler and L. N. Fowler, *The Illustrated Self-Instructor in Phrenology and Physiology, with One Hundred Engravings, and a Chart of the Character of _____ as Given by _____* (New York: Fowlers and Wells, 1854).

25. Sharrona Pearl, *About Faces: Physiognomy in Nineteenth-Century Britain* (Cambridge, MA: Harvard University Press, 2010).

26. Henry B. Gibbons, diaries, vol. 1: February 10, 1844; vol. 2: June 30, August 9, September 22, 1845, Henry B. Gibbons Papers.

27. Henry B. Gibbons, diaries, vol. 4, January–July 1847, Henry B. Gibbons Papers.

28. Printed phrenological organ labels, ephemera, #404/2, Henry B. Gibbons Papers.

29. Henry B. Gibbons, diaries, vol. 10, September 4, 1856, Henry B. Gibbons Papers.

30. Hammond, *Trials and Triumphs.*

31. "Biographical Sketch of Mrs. Deiadamia Chase," *American Phrenological Journal (1838–1869)* 15, no. 5 (1852): 99–100.

32. Hammond, *Trials and Triumphs*, 182.

33. Hammond, *Trials and Triumphs*, 186.

34. Regina Markell Morantz-Sanchez, *Sympathy and Science: Women Physicians in American Medicine* (New York: Oxford University Press, 1985).

35. Hammond, *Trials and Triumphs*, 124.

36. Hammond, *Trials and Triumphs*, 159.

37. See copy of Hammond, *Trials and Triumphs* at WZ 100c386W, 1859, History and Special Collections for the Sciences, University of California, Los Angeles Library (UCLA hereafter), Special Collections.

38. Hammond, *Trials and Triumphs*, 118.

39. Hammond, *Trials and Triumphs*, 181–82, 198.

40. Hammond, *Trials and Triumphs*, 197–230.

41. Robert H. Collyer, *Manual of Phrenology, or the Physiology of the Human Brain* (Cincinnati: Alexander Flash, 1838). See Kirtland copy at Dittrick Medical History Center, Case Western Reserve University.

42. Collyer, *Manual of Phrenology* (Kirtland copy), 84, 71, 105.

43. Judith Walzer Leavitt, "'A Worrying Profession': The Domestic Environment of Medical Practice in Mid-Nineteenth-Century America," *Bulletin of the History of Medicine* 69, no. 1 (1995): 1–29.

44. T. Barlow, *A Phrenological Guide, Designed for Students of Their Own Character* (New York: T. Barlow and L. N. Fowler, 1838), 16–17.

45. See also Charles Colbert, *A Measure of Perfection: Phrenology and the Fine Arts in America* (Chapel Hill: University of North Carolina Press, 1997).

46. William Leach, *True Love and Perfect Union: The Feminist Reform of Sex and Society* (New York: Basic Books, 1980).

47. L. N. Fowler, Collection of Phrenological Cards, 1834–1835 (inclusive), 1834, Countway Library of Medicine. I have assigned a number to each card, in order of appearance.

48. On the prison as scientific research site and laboratory, see Courtney Thompson, "'Directly at War with the Gallows'? The Prison and Phrenological Criminal Science," presented at the History of Science Society, Chicago, 2014.

49. Fowler, card #43, Collection of Phrenological Cards, 1834–1835.

50. Fowler, card #197, Collection of Phrenological Cards, 1834–1835.

51. Fowler, card #203, Collection of Phrenological Cards, 1834–1835.

52. Thankful B. Paige, "Phrenological Chart by Reverend A. Paige, January 7, 1845," in *A Phrenological Guide, Designed for Students of Their Own Character* (New York: L. N. Fowler, 1844), personal collection of Carla Bittel.

53. O. S. Fowler and L. N. Fowler, *The New Illustrated Self-Instructor in Phrenology and Physiology; with over One Hundred Engravings; Together with the Chart and Character of* [Emily Alverson] *as Marked by* [L. N. Fowler, June 2, 1859] (New York: Fowler and Wells, 1859), History and Special Collections for the Sciences, UCLA Library Special Collections, copy no. 1: BF870.F787i 1859.

54. O. S. Fowler and L. N. Fowler, *The Illustrated Self-Instructor in Phrenology and Physiology, with One Hundred Engravings, and a Chart of the Character of* [L. M. Ninde and B. C. Ninde] *as Given by* [O.S. Fowler, March 6, 1851] (New York: Fowlers and Wells, 1850), History and Special Collections for the Sciences, UCLA Library Special Collections, BF870. F787i 1850.

55. On the Nindes, see Peggy Seigel, "Winning the Vote in Fort Wayne, Indiana: The

Long, Cautious Journey in a German American City," *Indiana Magazine of History* 103, no. 3 (2006): 220–57.

56. Fowler and Fowler, *The Illustrated Self-Instructor, 1850*.

57. Deborah Jean Warner, *Graceanna Lewis: Scientist and Humanitarian* (Washington, DC: Smithsonian Institution Press, 1979).

58. Graceanna Lewis to Esther Lewis, October 7, 1842, Lewis-Fussell Family Papers, RG5/087, Friends Historical Library of Swarthmore College.

59. Poskett argues that the materiality of the mind and the materiality of letters went hand in hand. Poskett, "Phrenology."

60. Graceanna Lewis to Esther Lewis, October 7, 1842, Lewis-Fussell Family Papers, RG5/087, Friends Historical Library of Swarthmore College.

61. Sallie Holley, letter, October 31, 1852, in *A Life for Liberty: Anti-Slavery and Other Letters of Sallie Holley*, ed. John White Chadwick (New York: G. P. Putnam's Sons, 1899), 292.

62. Susanna Strickland Moodie, letter to James Bird, November 17, 1827, in *Letters of a Lifetime*, ed. Carl Ballstadt, Elizabeth Hopkins, and Michael Peterman (Toronto: University of Toronto Press, 1985), 390.

63. Letitia Hargrave to Mrs. Dugald Mactavish, November 29, 1848, in *The Letters of Letitia Hargrave* (Toronto: Champlain Society, 1947), 310.

64. Lucy Chase, diary no. 1, October 19, 1841, to March 3, 1842, Chase Family Papers, American Antiquarian Society.

65. Mrs. L. Miles, *Phrenology and the Moral Influence of Phrenology: Arranged on 40 Cards, Illustrative of the System*, 6th ed. (London: Ackermann, 1835); Miles, *Phrenology*, 7th ed.

66. See interior of the "casket of knowledge." Miles, *Phrenology*. The language of "phrenological bijou" may have carried over from an earlier text: *A Phrenological Bijou, or Vademecum of Mental Science, Being a Brief Illustration of the Organs, as arranged on Mr. Akin's Palladian Bust, and Likewise His Set of Phrenological Models* (Bath: C. Duffield; London: Hearst, Chance, 1830).

67. Te Heesen, *World in a Box*, 7.

68. Miles, *Phrenology*, 7th ed., title page card. A similar image can be found on the title page of *Phrenological Bijou* (1830).

69. Mrs. L. Miles, *The Bello e Nuovo* (London: Ackermann, ca. 1840); "The Forsaken," *Monthly Magazine*, November 1837, 514.

70. Mrs. L. Miles, *Phrenology, and the Moral Influence of Phrenology: Arranged for General Study, and the Purposes of Education, from the First Published Works of Gall and Spurzheim, to the Latest Discoveries of the Present Period* (Philadelphia: Carey, Lea and Blanchard, 1835).

71. Edgar Allan Poe, *Southern Literary Messenger*, March 1836, 2:286–87. Italics in original.

72. "Literary Port Folio," *Atkinson's Casket*, February 1836, 108.

Chapter 7. Keeping Prussia's House in Order

1. For the particulars of this census reform, see Christine von Oertzen, "Machineries of Data Power: Manual Versus Mechanical Census Compilation in Nineteenth-Century Europe," in *Data Histories*, ed. Elena Aronova, Christine von Oertzen, and David Sepkoski, Osiris 32 (Chicago: University of Chicago Press, 2017), 129–50.

2. *Sizing* is a technical process in which the page is lined with a covering coat so that ink does not lute or smear into the linen or woodchip fibers. "Simple-sized" paper has been sized on only one side, because the other side is not used for writing on.

3. Dan Rosenberg, "Early Modern Information Overload," *Journal of the History of Ideas* 64, no. 4 (2003): 1–9.

4. Paul Edwards emphasizes this fact by calling attention to the friction that millions of tons of paper punch cards caused for data compilation in twentieth-century climatology. See Paul N. Edwards, *A Vast Machine: Computer Models, Climate Data, and the Politics of Global Warming* (Cambridge, MA: MIT Press, 2010).

5. See Carl von Clausewitz, *Vom Kriege: Hinterlassenes Werk, erläutert von Wilhelm von Scherff* (Berlin: F. Schneider, 1880).

6. On the concept of paper tools see first and foremost Ursula Klein, *Experiments, Models, Paper Tools: Cultures of Organic Chemistry in the Nineteenth Century* (Stanford, CA: Stanford University Press, 2002). The concept is discussed in detail in the introduction to this volume.

7. On the concept of sociomateriality see Bittel, Leong, and von Oertzen, this volume; Werrett, this volume; see also Wanda Orlikowski, "Sociomaterial Practices: Exploring Technologies at Work," *Organization Studies* 28, no. 9 (2009): 1435–48.

8. Francesca Bray, *Technology, Gender and History in Imperial China: Great Transformations Reconsidered* (London: Routledge, 2013), 5–6. On sociotechnological systems see Wiebe E. Bijker, Thomas P. Hughes, and Trevor Pinch, eds., *The Social Construction of Technological Systems: New Directions in the Sociology and History of Technology* (Cambridge, MA: MIT Press, 1989).

9. Eyferth, this volume; Jacob Eyferth, *Eating Rice from Bamboo Roots: The Social History of a Community of Handicraft Papermakers in Rural Sichuan, 1920–2000* (Cambridge, MA: Harvard University Press, 2009), 17–18.

10. Lorraine Daston, "The Immortal Archive: Nineteenth-Century Science Imagines the Future," in *Archiving Sciences*, ed. Lorraine Daston (Chicago: University of Chicago Press, 2017), 159–84; Staffan Müller-Wille, "Names and Numbers: 'Data' in Classical Natural History, 1758–1859," in Aronova, von Oertzen, and Sepkoski, *Data Histories*, 109–28 .

11. For nineteenth-century reminiscences, see Ernst Heilborn, *Die gute Stube: Berliner Geselligkeit im 19. Jahrhundert* (Rudolstadt, 1922); otherwise, Rebekka Habermas, *Frauen und Männer des Bürgertums: Eine Familiengeschichte, 1750–1850* (Göttingen: Vandenhoeck und Ruprecht, 2000).

12. Again, I am drawing inspiration from Bray, *Technology*, 3.

13. See Anke te Heesen, "The Notebook: A Paper-Technology," in Bruno Latour and Peter Weibel, *Making Things Public: Atmospheres of Democracy* (Cambridge, MA: MIT Press, 2003), 582–89; Anne Blair, *Too Much to Know: Managing Scholarly Information before the Modern Age* (New Haven, CT: Yale University Press, 2010); a useful overview on paper tools and technologies is provided by Isabelle Charmantier and Staffan Müller-Wille, "Carl Linnaeus's Botanical Paper Slips (1767–1773)," *Intellectual History Review* 24, no. 2 (2014): 215–38. See also Bittel, Leong, and von Oertzen, this volume.

14. A popular mass product throughout Europe for centuries, playing cards were also used to create library card catalogs, their uniform size allowing precise standardization, easy shuffling, and robust handling. Markus Krajewski, *Paper Machines: About Cards and Catalogs, 1548–1929* (Cambridge, MA: MIT Press, 2011), 33.

15. von Oertzen, *Machineries*.

16. Engel, "Aktenmässige Darstellung der Vorbereitungen zu den statistischen Aufnahmen im December 1867, insbesondere der Volkszählung im Preussischen Staate und

norddeutschen Bundesgebiete," *Zeitschrift des königlich preussischen statistischen Bureaus* 7 (1867): 303.

17. If people were unable to read or write, the entries were completed by enumerators when they collected the material. See counting letter face, 1871 census, I. HA, Rep. 77, Tit. 94, no. 132, vol. 1, 115, Secret State Archives Prussian Cultural Heritage (Geheimes Staatsarchiv Preußischer Kulturbesitz, hereafter GStA PK).

18. See Emil Blenck, *Zusammenstellung und Begutachtung der Erfahrungen, Bedenken und Abänderungsvorschläge, welche die Königlichen Regierungspräsidenten und der Magistrat zu Berlin in ihren Berichten über die Ausführung der Volkszählung vom 2. Dezember 1895 niedergelegt haben*, I. HA Rep. 77, Tit. 94, no. 159, vol. 2, 37, GStA PK.

19. See von Mayr, *Statistik und Gesellschaftslehre* (Tübingen: Mohr, 1914), 122. On the discussion of self-inscription and the evaluation of turnouts within Prussia, see Michael Schneider, *Wissensproduktion im Staat: Das Königlich-Preußische statistische Bureau, 1860–1914* (Frankfurt am Main: Campus, 2013), 236–37.

20. On the concept of "data doubles" see Dan Bouk, "The History and Political Economy of Personal Data over the Last Two Centuries in Three Acts," in Aronova, von Oertzen, and Sepkoski, *Data Histories*, 85–108. On the centrality of metadata in the history of data see Markus Krajewski, "Tell Data from Meta: Tracing the Origins of Big Data, Bibliometrics, and the OPAC," in Aronova, von Oertzen, and Sepkoski, *Data Histories*, 224–42.

21. Emil Blenck, *Das königlich Statistische Bureau im ersten Jahrhundert seines Bestehens, 1805–1905* (Berlin: Königliches Statistisches Bureau, 1905), 226.

22. Engel to the Prussian Ministry of the Interior, June 21, 1871, Kostenanschlag für die diesjährige Volkszählung, I. HA, Rep. 77, Tit. 94, no. 132, vol. 1, 44–47r, GStA PK.

23. Engel, "Die Kosten der Volkszählungen mit besonderer Rücksicht auf die im December 1870 im Preussischen Staate bevorstehende Zählung," *Zeitschrift des königlich preussischen statistischen Bureaus* 10, no. 1–2 (1870): 40.

24. Moving the heavy boxes within the bureau remained a process prone to friction despite the fact that an extra wing with a gas-driven mechanical elevator was added to the main building in 1875 to make room for more workers and to facilitate "maneuvering the comprehensive and heavy raw material." Blenck, *Das königlich statistische Bureau*, 227.

25. See Engel to the Ministry of the Interior, December 1, 1872, Bericht über den Fortgang der Volkszählungsarbeiten, I. HA, Rep. 77, Tit. 94, no. 132, vol. 2, 132, GStA PK.

26. Emil Blenck, "Die Berufs- und Gewerbezählung vom 14. Juni 1895 und die damit verbundene landwirtschaftliche Betriebszählung," *Zeitschrift des königlich preussischen statistischen Bureaus* 37 (1897): 302.

27. Verzeichnis sämmtlicher Hilfsarbeiter nach dem Stand vom 1. April 1890, I. HA, Rep. 77, Tit. 536, no. 30, vol. 1, n.p., GStA PK.

28. Emil Blenck to the Ministry of the Interior, August 18, 1898, Documentation regarding the housework of the wives of the bureau's secretaries Friedrich and Seidel, I. HA, Rep. 77, Tit. 536, no. 30, vol. 1, n.p., GStA PK.

29. Blenck to Ministry of the Interior, July 21, 1898, I. HA, Rep. 77, Tit. 536, no. 30, vol. 1, n.p., GStA PK. Hiring housemaids when abstraction work was done was a common practice; see Blenck to the Ministry of the Interior, May 8, 1899, I. HA, Rep. 77, Tit. 536, no. 30, vol. 2, n.p., GStA PK.

30. Verzeichnis sämmtlicher Hilfsarbeiter nach dem Stand vom 1. April 1890, I. HA, Rep. 77, Tit. 536, no. 30, vol. 1, n.p., GStA PK.

31. Workers' homes could be identified on the basis of two of the bureau's surviving registers from 1890 and 1898 as well as the address books of the City of Berlin. For this project, we used *Berliner Adressbücher* from the years 1890, 1897, and 1900, https://www.zlb.de/besondere-angebote/berliner-adressbuecher.html. A great many thanks to Laura Selle for locating the apartments and for producing the map.

32. Blenck to Ministry of the Interior, December 10, 1908, I. HA, Rep. 77, Tit. 536, no. 30, vol. 2, n.p., GStA PK. Whereas male relatives such as sons, nephews, and brothers-in-law also participated in the work, it was the housewife's orderliness that enabled men's participation.

33. Leong, this volume; Serrano, this volume; Yale, this volume.

34. Judith McGaw, "Reconceiving Technology: Why Feminine Technologies Matter," in *Gender and Archaeology*, ed. Rita P. Wright (Philadelphia: University of Pennsylvania Press, 1996), 52–75.

35. Blenck to Ministry of the Interior, December 10, 1908, I. HA, Rep. 77, Tit. 536, GStA PK.

36. Bray, *Technology, Gender and History*, 3–6.

37. See Harry Collins, *Artificial Experts: Social Knowledge and Intelligent Machines* (Cambridge, MA: MIT Press, 1992), 63–71.

38. Petersilie (one of Blenck's speakers) to Ministry of the Interior, January 27, 1909, I. HA, Rep. 77, Tit. 536, no. 30, vol. 2, n.p., GStA PK.

39. Director Blenck to ministry of the interior, August 18, 1898, re: die den Ehefrauen Friedrich und Seidel 1897/8 übertragenen Hausarbeiten, I. HA, Rep. 77, Tit. 536, no. 30, vol. 1, n.p., GStA PK.

40. The rich documentation exists because Lasswitz was denounced by a dismissed coworker for receiving a pension in addition to his high earnings from census work. As in other such cases, the census bureau had to justify its policy by providing detailed accounts of Lasswitz's private as well as financial circumstances: how much he earned and when, what kinds and how many cards his household received, and who assisted him. Director Blenck to Ministry of the Interior, May 8, 1899, I. HA, Rep. 77, Tit. 536, no. 30, vol. 2, n.p., GStA PK.

41. Lasswitz is listed under this address in the *Berliner Adressbuch des Jahres 1897* (Berlin: Scherl, 1898), 169.

42. *Berliner Adressbuch des Jahres 1897*, 169.

43. The following is based on an extensive report from the statistical bureau to the Ministry of the Interior, May 8, 1899, I. HA, Rep. 77, Tit. 536, no. 30, vol. 2, GStA PK.

44. Documentation of Payments to Worker Max Lasswitz (Geldempfangs-Nachweisung für den Hülfsarbeiter Max Lasswitz), May 8, 1899, I. HA, Rep. 77, Tit. 536, no. 30, vol. 2, n.p., GStA PK.

45. Blenck to Ministry of the Interior, May 13, 1895, Petition to subsidize the bureau's officials' efforts to support the daily worker Paul Benschkowski suffering from a stroke, I. HA, Rep. 77, Tit. 536, no. 30, vol. 1, n.p., GStA PK.

46. Helene Hoyer, Petition to the Statistical Bureau, October 28, 1895, and decline from the Prussian Ministry of the Interior, November 11, 1895, I. HA, Rep. 77, Tit. 536, no. 30, vol. 1, n.p., GStA PK.

47. Documentation of delivery and pickup of census material to the Lasswitz home between February 1896 and February 1897. Report from the statistical bureau to the Ministry of the Interior, May 8, 1899, I. HA, Rep. 77, Tit. 536, no. 30, vol. 2, n.p., GStA PK.

48. Hermann Schwabe, "Bericht der städtischen Volkszählungs-Commission über die Ausführung der Zählung nebst Anlage 1–12," in *Die Resultate der Berliner Volkszählung vom 3. December 1867* (Berlin, 1869). On Schwabe's efforts to document Berlin's housing conditions, see also Christine von Oertzen, "Datafication and Spatial Visualization Nineteenth-Century Statistics," in *Historical Studies in the Natural Sciences* 48, no. 5 (2018): 568–80.

49. See Erik Grimmer-Solem, *The Rise of Historical Economics and Social Reform in Germany, 1864–1894* (Oxford: Oxford University Press, 2003).

50. Ernst Engel, *Das Rechnungsbuch der Hausfrau und seine Bedeutung im Wirthschaftsleben des Staates* (Berlin: Leonhard Simion, 1882).

51. Engel, *Das Rechnungsbuch*, 7. Similarly, see Lorenz von Stein, *Die Frau auf dem Gebiete der Nationalökonomie* (Stuttgart: Cotta, 1875).

52. Gustav Schmoller, *Grundriß der allgemeinen Volkswirtschaftslehre*, part 1 (Leipzig: Duncker und Humblot, 1900), 249. See also Serrano, this volume; Szalay, this volume; and Werrett, this volume.

53. Blenck to the Prussian Ministry of the Interior, April 21, 1898, regarding the petition of veteran Wilhelm Mittelstädt, I. HA, Rep. 77, Tit. 536, no. 30, vol. 1, n.p., GStA PK.

Chapter 8. Tracing Paper, the Posture Sciences, and the Mapping of the Female Body

1. Lloyd T. Brown, "The Harvard Slouch: Four Out of Every Five Harvard Freshman Stand in a Faulty Posture," *New York Times Sunday Magazine*, March 18, 1917, 8.

2. "The Mosher-Lesley Schematograph," SC0011, box 11, folder 5, Clelia Mosher Papers, Stanford University Archives.

3. Bittel, Leong, and von Oertzen, this volume.

4. Lorraine Daston and Peter Galison, "The Image of Objectivity: The Talismanic Image," *Representations* 40 (1992): 81–128.

5. Bruno Latour, "Visualization and Cognition: Drawing Things Together," in *Knowledge and Society: Studies in the Sociology of Culture Past and Present*, vol. 6, ed. Henrika Kuklick and Elizabeth Long (Greenwich, CT: JAI, 1986), 16. Latour's analysis does not, however, take gender into account.

6. Clelia Duel Mosher, "Schematogram: A New Method of Graphically Recording Posture and Changes in the Contour of the Body," *School and Society* 1, no. 18 (May 1, 1915): 642–45.

7. For "physiological feminist," see Mary Roberts Coolidge, "Clelia Duel Mosher, the Scientific Feminist," *Research Quarterly* 12, no. 3 (1941): 642. See also Elizabeth Brownlee Griego, "A Part and Yet Apart: Clelia Duel Mosher and Professional Women at the Turn-of-the-Century" (PhD diss., University of California, Berkeley, 1983), 156.

8. Edward Clarke, *Sex in Education: or, A Fair Chance for Girls* (Boston: J. R. Osgood, 1873). For more on this debate and its effect on Mosher, see Griego, "A Part and Yet Apart," 152.

9. Rather than explicitly reproaching male researchers for relying on incomplete data and perpetuating pathological falsehoods about menstruation, Mosher took the somewhat less daring route and insisted that uneducated women presented the biggest problem.

10. Clelia D. Mosher, "Functional Periodicity in Women and Some Modifying Factors" (part 2), *California State Journal of Medicine* 9, no. 2 (1911): 56.

11. Mosher, "Functional Periodicity" (part 2), 56.

12. Clelia D. Mosher, "Functional Periodicity in Women and Some of the Modifying Factors" (part 1), *California State Journal of Medicine* 9, no. 1 (1911): 5.

13. Daniel M. Fox and James Terry, "Photography and the Self Image of American Physicians, 1880–1920," *Bulletin of the History of Medicine* 52 (1978): 450.

14. Anthropometry swept the United States after the Civil War. See Edward M. Hartwell, "A Preliminary Report on Anthropometry in the United States," *Publications of the American Statistical Association* 3, no. 24 (1893): 554–68.

15. Roberta J. Park, "Physiology and Anatomy Are Destiny!? Brains, Bodies and Exercise in Nineteenth Century American Thought," *Journal of Sport History* 18, no. 1 (1991): 31–63. See also John Kasson, *Houdini, Tarzan, and the Perfect Man: The White Male Body and the Challenge of Modernity in America* (New York: Hill and Wang, 2002).

16. For an example of the poster, with an explanation, see Beth Linker, "A Dangerous Curve: The Role of History in America's Scoliosis Screening Program," *American Journal of Public Health* 102, no. 4 (2012): 606–16.

17. Carla Bittel, *Mary Putnam Jacobi and the Politics of Medicine in Nineteenth-Century America* (Chapel Hill: University of North Carolina Press, 2009). See also Steven Peitzman, *A New and Untried Course: Woman's Medical College and Medical College of Pennsylvania, 1850–1998* (New Brunswick, NJ: Rutgers University Press, 2000); and Regina Morantz-Sanchez, *Sympathy and Science: Women Physicians in American Medicine* (Chapel Hill: University of North Carolina Press, 1985).

18. Clelia Mosher is perhaps most well known among today's scholars for her pathbreaking sex survey of Victorian-era women, research that was not discovered and made public until well after her death. See Mosher, *The Mosher Sex Survey: Sexual Attitudes of 45 Victorian Women*, ed. James MaHood and Kristine Wenburg (New York: Arno Press, 1980).

19. She worried about the "possible identification of the individual woman by anyone into whose hands the record may fall." See Mosher, "Schematogram," 642.

20. Walter G. Vincenti, "The Air-Propeller Tests of W. F. Durand and E. P. Lesley: A Case Study in Technological Methodology," *Technology and Culture* 20, no. 4 (1979): 712–51.

21. As Wendy Gamber demonstrates in her research on dressmaking, male designers in the late nineteenth century began to promote masculine systems of sewing patterns and drafting systems, undermining the expertise of female dressmakers who possessed specialized skills and knowledge in cutting garments that would conform to the buyer's body. Sewing patterns—a paper similar in feel and function to tracing papers used by draftsmen—were thus created by men, who possessed the scientific and engineering knowledge, and intended for the use of everyday women. Thus while sewing patterns would become a commonplace, feminine domestic technology, the process by which these tools became popular should be ultimately seen as deskilling a feminine domain of expertise. See Gamber, "Dressmaking," in *Gender and Technology: A Reader*, ed. Nina A. Lerman, Ruth Oldenziel, and Arwen P. Mohun (Baltimore: Johns Hopkins University Press, 2003), 238–66.

22. While there are no *inherently* masculine or feminine technologies, certain historians have traced the ways in which technologies are culturally assigned a gender and also presumed to be used by one sex or another. For examples of this, see Lerman, Oldenziel, and Mohun, *Gender and Technology*.

23. Lillian Curtis Drew who adopted Mosher's technique described having heard Mosher depict the benefits of the schematograph in this way. See Drew, *Individual Gymnastics: A Handbook of Corrective and Remedial Gymnastics* (Philadelphia: Lea and Febiger, 1926), 83.

24. Drew, *Individual Gymnastics*, 82.

25. Robert J. Cook, "Report of the Orthopædic Examination of 1393 Freshmen at Yale University," *Journal of Bone and Joint Surgery* 4, no. 2 (1922): 250.

26. Mosher, "Schematogram," 644.

27. Lloyd T. Brown, "A Combined Medical and Postural Examination of 746 Adults," *American Journal of Orthopedic Surgery* 15, no. 11 (1917): 775.

28. Mosher, "Schematogram," 643–44.

29. Lillian Curtis Drew, *Individual Gymnastics: A Handbook of Corrective and Remedial Gymnastics* (Philadelphia: Lea and Febiger, 1926), 95.

30. Joan Jacobs Brumberg, *The Body Project: An Intimate History of American Girls* (New York: Random House, 1997), 112–13.

31. "The Mosher-Lesley Schematograph," SC0011, box 11, folder 5, Clelia Mosher Papers, Stanford University Archives. See also Mosher's graphics in her respiration studies: "Hygiene and Physiology of Women," SC0011, box 3, Clelia Mosher Papers, Stanford University Archives. Other studies found the layering of schematographs also useful. See, in particular, Antioch College, *The Effects of Modern Shoes upon Proper Body Mechanics, 1924–1931* (Yellow Springs, OH: Antioch College, 1931), fig. 12, n.p.

32. Coolidge, "Clelia Duel Mosher," 643–44.

33. See figure 8.2 above. Goldthwait to Neilson, March 13, 1922, Office of the President, William Allan Neilson, 1917–1939, box 11, folder Joel E. Goldthwait, Smith College Archives.

34. William Allan Neilson, Northampton, MA, to Joel E. Goldthwait, Boston, MA, March 16, 1922, Office of the President, William Allan Neilson, 1917–1939, box 11, file Joel E. Goldthwait, Smith College Archives.

35. Neilson to Goldthwait, March 16, 1922.

36. Claire McRee, "The Debutante Slouch: Fashion and the Female Body in the United States, 1912–1935" (master's thesis, Bard College, 2015), 11.

37. Articles pertaining to posture health appear in a wide range of African American newspapers throughout the twentieth century. During this time, several prominent African American physicians and physical educators encouraged posture surveillance at home and at places of higher education. For a few examples, see E. Elliot Rawlins, "Keeping Fit: Postural Defects," *New York Amsterdam News*, February 15, 1928; "'Get Uplift in Your Bearing,' Say Posture Experts," *Philadelphia Tribune*, April 24, 1930; Algernon Jackson, "Afro Health Talk: Posture and Purpose," *Afro-American*, November 2, 1940; and "Watch Posture as Others Do," *Chicago Defender*, August 21, 1954, national edition, sec. Fashions and Home Service.

38. Mosher, "Schematogram," 642.

39. H. Ling Taylor (Sec. APL), New York to Members of the American Posture League, January 10, 1916, file "American Posture League, Report of Executive Committee," B MS c 81.4, Zabdiel Boylston Adams papers, Countway Library of Medicine, Harvard University.

40. "Fradd to Assume Post in Service Command: PT Director Will Aid with Rehabilitation," *Harvard Crimson*, June 16, 1944, http://www.thecrimson.com/article/1944/6/16/fradd-to-assume-post-in-service/. A graduate of Springfield College who before coming to Harvard oversaw physical reconditioning of injured veterans at Walter Reed Hospital, Fradd made public his disdain for the hyperintellectual types at Harvard, claiming that many students there were "by and large pitiful victims" of unbalanced living. See "Education in Action," *Journal of Education* 107, no. 6 (February 6, 1928): 157–59.

41. N. W. Fradd, "A New Method of Recording Posture," *Journal of Bone and Joint Surgery* 5, no. 4 (1923): 757.

42. Hubbard recounts Fradd's story and says it was M. C. Reed of Eastman Kodak Company. C. H. Hubbard, "The Silhouetteograph—Its Contribution to the Study of Posture and Body Mechanics," *Journal of Health and Physical Education* 5, no. 3 (1934): 49–67.

43. Hubbard, "Silhouetteograph," 49, 66–67.

44. Fradd's Harvard colleagues, Roger I. Lee and Lloyd T. Brown, discussed Fradd's view of how tracing methods introduced the problem of the "personal equation" in their article, "A New Chart for the Standardization of Body Mechanics," *Journal of Bone and Joint Surgery* 5, no. 5 (1923): 756.

45. Brown, "Combined Medical and Postural Examination of 746 Adults," 776.

46. Drew, *Individual Gymnastics*, 95.

47. To see how his posters and charts were adopted by other experts, see Armin Klein and Leah C. Thomas's report prepared for the US Department of Labor, *Posture and Physical Fitness* (Washington, DC: Government Printing Office, 1931), 6–11.

48. For examples of these charts and reprints, see Klein and Thomas, *Posture*.

49. "The Silhouettograph," *American Physical Education Review* 33, no. 9 (November 1928): 595.

50. "Individual Health and Hygiene: A List of Free and Inexpensive Teaching Materials," *American Journal of Nursing* 42, no. 2 (1942): 199–200.

51. Wellesley College was one of the first and most active all-women's college to adopt photography. See Charlotte G. MacEwana and Eugene C. Howea, "An Objective Method of Grading Posture," *Research Quarterly of the American Physical Education Association* 3, no. 3 (1932): 144–57; and Sarah Ligon, "A Crooked Science: The History of Posture Photos at Wellesley," *Counterpoint: Rational Discourse and Campus Life*, February 2002, 12–15. One physical educator in 1934 still recognized Mosher's device as "the most worth-while contribution [for] posture readings." Hubbard, "Silhouetteograph," 49.

52. Lee and Brown, "New Chart," 756.

53. Associated Press, "Photographing Nude Coeds Creates Furor at Wayne U.," *Washington Post*, October 4, 1937, 1.

54. "G.W. Taboos Nudity Except behind Screens, So Coeds Pose Only for Silhouette Photos," *Washington Post*, October 5, 1937, 1. Because of the bad press, however, schools such as American University stopped recording nude posture of its students entirely "on grounds of immorality" (1).

55. Ellen Kelly, "Taking Posture Pictures," *Health and Physical Education* 17, no. 8 (1946): 504–5, 464.

56. For contemporary examples of how good posture is unreflexively equated with good health, beauty, and material success, see Jane Brody, "Good Posture May Better Your Position," *New York Times*, December 28, 2015, D5; and Amy Cuddy, "Your iPhone Is Ruining Your Posture—and Your Mood," *New York Times*, December 13, 2015, SR3.

Chapter 9. A Letter Is a Paper House

My thanks to Carla Bittel, Elaine Leong, and Christine von Oertzen; the members of the "Working with Paper" working group; and the scholars and staff at the Max Planck Institute for the History of Science, especially Lorraine Daston, for their hospitality, support, and critical co-thinking and working with paper in Fall 2016. Thanks to the Huntington Library and The Andrew W. Mellon Fellowship of Scholars in Critical Bibliography for financial support of this research and (to the latter) for critical interrogation of the results.

1. Sarah Tarlow, *Ritual, Belief, and the Dead in Early Modern Britain and Ireland* (Cambridge: Cambridge University Press, 2011), 136.

2. By the early eighteenth century, families used more elaborate decorations on coffins to signal greater wealth and status. Ralph Houlbrooke, *Death, Religion and the Family in England, 1480–1750* (Oxford: Clarendon Press, 1998), 340.

3. Houlbrooke, *Death*, 285. Wine and biscuits were fashionable at London funerals, but these expensive treats may not have been popular in Essex.

4. Samuel Dale, "The Life of Mr. Ray by Mr. Dale of Braintree," in *Further Correspondence of John Ray*, ed. Robert W. T. Gunther (London: Ray Society, 1928), 7; Houlbrooke, *Death*, 285.

5. Two early entrants are Steven Shapin, "The House of Experiment in Seventeenth-Century England," *Isis* 79, no. 3 (1988): 373–404; and Deborah Harkness, "Managing an Experimental Household: The Dees of Mortlake and the Practice of Natural Philosophy," *Isis* 88, no. 2 (1997): 247–62.

6. On Ray's life and work, see Charles Raven, *John Ray, Naturalist: His Life and Works*, 2nd ed. (Cambridge: Cambridge University Press, 1950).

7. I touch on John and Margaret Ray's case in Yale, "The Book and the Archive in the History of Science," *Isis* 107, no. 1 (2016): 106–15. There, I write only briefly about Margaret's role in the passage of John's papers from home to posthumous publication. In the present essay, I delve more deeply into the ways in which women's labor provides the structure for scientific sociability, both in the home and on paper, and show how individuals constructed masculinity and femininity through paper transactions.

8. Harkness, "Managing an Experimental Household," 251–58.

9. Shapin, "House of Experiment"; and Shapin, *A Social History of Truth: Civility and Science in Seventeenth-Century England* (Chicago: University of Chicago Press, 1994), 370.

10. Margaret Pelling, "The Women of the Family? Speculations around Early Modern British Physicians," *Social History of Medicine* 8, no. 3 (1995): 383–401; Rob Iliffe and Frances Wilmoth, "Astronomy and the Domestic Sphere: Margaret Flamsteed and Caroline Herschel as Assistant Astronomers," in *Women, Science, and Medicine 1500–1700: Mothers and Sisters of the Royal Society*, ed. Lynette Hunter and Sarah Hutton (Stroud: Sutton, 1997), 235–65; Monika Mommertz, "The Invisible Economy of Science: A New Approach to the History of Gender and Astronomy at the Eighteenth-Century Berlin Academy of Sciences," in *Men, Women, and the Birthing of Modern Science*, ed. Judith P. Zinsser (DeKalb: Northern Illinois University Press, 2005), 159–78; Alix Cooper, "Homes and Households," in *The Cambridge History of Science*, vol. 3, *Early Modern Science*, ed. Katharine Park and Lorraine Daston (Cambridge: Cambridge University Press, 2008), 224–37; Mary Fissell, "Introduction: Women, Health, and Healing in Early Modern Europe," *Bulletin of the History of Medicine* 82, no. 1 (2008): 1–17; Elaine Leong, "Making Medicines in the Early Modern Household," *Bulletin of the History of Medicine* 82, no. 1 (2008): 145–68; Anna Marie Roos, "The Art of Science: A 'Rediscovery' of the Lister Copperplates," *Notes and Records of the Royal Society* 66, no. 1 (2012): 19–40; Alisha Rankin, *Panaceia's Daughters: Noblewomen as Healers in Early Modern Germany* (Chicago: University of Chicago Press, 2013); Amanda Herbert, *Female Alliances: Gender, Identity, and Friendship in Early Modern Britain* (New Haven, CT: Yale University Press, 2014), 78–116; Mary Terrall, "Masculine Knowledge, the Public Good, and the Scientific Household of Réaumur," in *Scientific Masculinities*, ed. Erika Lorraine Milam and Robert A. Nye, Osiris 30 (Chicago: University of Chicago Press, 2015),

182–201; Simon Werrett, *Thrifty Science*, forthcoming; Meghan Roberts, "Learned and Loving: Representing Women Astronomers in Enlightenment France," *Journal of Women's History* 29, no. 1 (2017): 14–37. On learned seventeenth-century women generally, see Carol Pal, *Republic of Women: Rethinking the Republic of Letters in the Seventeenth Century* (Cambridge: Cambridge University Press, 2012).

11. Elizabeth Yale, *Sociable Knowledge: Natural History and the Nation in Early Modern Britain* (Philadelphia: University of Pennsylvania Press, 2016). See also Iliffe and Wilmoth, "Astronomy and the Domestic Sphere," 250–51.

12. For a female servant assisting in natural philosophical work, see Vera Keller, "Scarlet Letters: Sir Theodore de Mayerne and the Early Stuart Color World in the Royal Society," in *Archival Afterlives: Life, Death, and Knowledge-Making in Early Modern British Scientific and Medical Archives*, ed. Vera Keller, Anna Marie Roos, and Elizabeth Yale (Leiden: Brill, 2018).

13. Mary Rhinelander McCarl, "Publishing the Works of Nicholas Culpeper, Astrological Herbalist and Translator of Latin Medical Works in Seventeenth-Century London," *Bulletin canadien d'histoire de la medicine* 13, no. 2 (1996): 239–41; Patrick Curry, "Culpeper, Nicholas (1616–1654)," *Oxford Dictionary of National Biography*.

14. Raven, *John Ray*, 213–14; *Further Correspondence*, 136; Dorothy Johnston, "The Life and Domestic Context of Francis Willughby," in *Virtuoso by Nature: The Scientific Worlds of Francis Willughby FRS (1635–1672)*, ed. Tim Birkhead (Leiden: Brill, 2016), 26–29; and Isabelle Charmantier, Dorothy Johnston, and Paul J. Smith, "The Legacies of Francis Willughby," in Birkhead, *Virtuoso by Nature*, 363–66.

15. Iliffe and Wilmoth, "Astronomy and the Domestic Sphere," 254–55.

16. Echoing Francesca Bray, *Technology, Gender and History in Imperial China: Great Transformations Reconsidered* (Abingdon, UK: Routledge, 2013), 5; and Maerker, this volume.

17. Derham, "The Life of Mr. Ray," in *Select Remains of the Learned John Ray* (London: George Scott, 1760), 38.

18. Raven, *John Ray*, 179.

19. Yale, *Sociable Knowledge*, 55–88; James Delbourgo, "Collecting Hans Sloane," in *From Books to Bezoars: Sir Hans Sloane and His Collections*, ed. Alison Walker, Arthur MacGregor, and Michael Hunter (London: British Library, 2012), 9–23.

20. Anthony Buxton, *Domestic Culture in Early Modern England* (Woodbridge: Boydell Press, 2015), 54. Buxton shows that gentlemen typically had houses of six or seven rooms and clerics four.

21. Raven, *John Ray*, 179–80.

22. On the evolution of hall and parlor as social spaces in the seventeenth-century Essex home, see Amanda Flather, *Gender and Space in Early Modern England* (Woodbridge: Boydell Press, 2007).

23. Buxton, *Domestic Culture*, 208–42.

24. Ray to Sloane, February 11, 1684, *Correspondence of John Ray*, ed. Edwin Lankester (1848), 138.

25. Ray to Sloane, November 17, 1697, *Correspondence of John Ray* (1848), 335.

26. Derham, "Life of Mr. Ray," 53.

27. "Junii mensis initio in hortulo nostro vesperi circumvolitans à *Maria* filiola capta est." John Ray, *Historia insectorum* (London: A. & J. Churchill, 1710), 317–18 (quote on 318). On Ray's insect research, see Brian Ogilvie, "Attending to Insects: Francis Willughby and

John Ray," *Notes and Records* 66, no. 4 (2012): 357–72, esp. 360–64 and 366–67; and Ogilvie, "Willughby on Insects," in Birkhead, *Virtuoso by Nature*, 339–45.

28. Raven, *John Ray*, 394–95.

29. Ray, *Historia insectorum*, 200; Raven, *John Ray*, 395.

30. On Dale and Allen, see H. R. French, "'Ingenious & Learned Gentlemen': Social Perceptions and Self-Fashioning among Parish Elites in Essex, 1680–1740," *Social History* 25, no. 1 (2000): 44–66.

31. Flather, *Gender and Space in Early Modern England*; Flather, "Space, Place, and Gender: The Sexual and Spatial Division of Labor in the Early Modern Household," *History and Theory* 52, no. 3 (2013): 344–60.

32. Flather, *Gender and Space in Early Modern England*, 96–110.

33. Flather, *Gender and Space in Early Modern England*, 108.

34. Flather, *Gender and Space in Early Modern England*, 103.

35. Ray to Aubrey, August 24, 1692, Bodleian Library (hereafter Bod) MS Aubrey 13, f. 176r.

36. See Flather, *Gender and Space in Early Modern England*, 100–110; Joan Thirsk, *Food in Early Modern England: Phases, Fads, Fashions 1500–1760* (London: Hambledon Continuum, 2007), 305–7, 310–11.

37. Ray to Aubrey, August 24, 1692, Bod MS Aubrey 13, f. 176r.

38. Natalie Zemon Davis, *Women on the Margins: Three Seventeenth-Century Lives* (Cambridge, MA: Harvard University Press, 1997), 148.

39. Raven, *John Ray*, 8, 200.

40. Yale, *Sociable Knowledge*, 83–87; Lindsay O'Neill, *The Opened Letter: Networking in the Early Modern British World* (Philadelphia: University of Pennsylvania Press, 2015), 128–35; Herbert, *Female Alliances*, 28, 34, 43.

41. Ray to Aubrey, August 24, 1692, Bod MS Aubrey 13, f. 176r.

42. See Kelsey Jackson Williams, *The Antiquary: John Aubrey's Historical Scholarship* (Oxford: Oxford University Press, 2016).

43. On women's market activity, see Flather, "Space, Place, and Gender," 354.

44. Ray to Lister, August 22, 1670; Ray to Lister, March 3, 1670/1; Ray to Lister, April 13, 1671, all in Lister, *The Correspondence of Dr. Martin Lister (1639–1712)*, ed. Anna Marie Roos (Leiden: Brill, 2015), 1:278–81, 331–34, 350–51.

45. Raven, *John Ray*, 111, 123, 163–67, 334–36. Ray and Willughby's scientific legacies were (and are) deeply entangled. Scientists and historians have debated who deserves more credit in their partnership. Raven judged the merit to Ray's side; this view has been recently reassessed in Birkhead, *Virtuoso by Nature.*

46. Lister to Ray, January 20, 1670/1; Lister to Ray, February 8, 1670/1; Lister to Ray, March 21, 1670, Lister, *Correspondence*, 1:307–8, 319–21, 343.

47. Lister to Ray, October 12, 1672, Lister, *Correspondence*, 1:477–78.

48. Ray to Sloane, May 25, 1692, British Library (hereafter BL) Sloane MS 4036, f. 123r.

49. Felicity Heal, *The Power of Gifts: Gift-Exchange in Early Modern England* (Oxford: Oxford University Press, 2014), 38, 61–62.

50. On food gifts and reciprocity, see Heal, *Power of Gifts*, 36.

51. For health advice as a gift, see Ray to Sloane, October 16, 1693, BL MS 4036, f. 156r-v.

52. Lisa Wynne Smith, "Sloane as Friend and Physician of the Family," in Walker, MacGregor, and Hunter, *From Books to Bezoars*, 50–51.

53. Sidney W. Mintz, *Sweetness and Power: The Place of Sugar in Modern History* (New York: Penguin, 1986), 92–93, 166–71.

54. Margaret Ray to Sloane, [1704/5], BL Sloane MS 4039, f. 430r.

55. This sequence of letters are Margaret's only known surviving correspondence. On the survival of early modern women's letters, see James Daybell, "Gendered Archival Practices and the Future Lives of Letters," in *Cultures of Correspondence in Early Modern Britain*, ed. Daybell and Andrew Gordon (Philadelphia: University of Pennsylvania Press, 2016), 210–36.

56. Margaret Ray to Sloane, [1704/5], BL Sloane MS 4039, f. 430r.

57. For an account of how Ray came to have these papers, as well as a brief discussion of their publication history as *Historia insectorum*, see Ogilvie, "Willughby on Insects," 339–45.

58. Raven, *John Ray*, 177.

59. Thomas Willughby to Sloane, March 27, [1705?], BL Sloane MS 4062, f. 24r.

60. Thomas Willughby to Sloane, October 2, [1706?], BL Sloane MS 4062, f. 20r.

61. Margaret Ray to Sloane, November 19, 1706, BL Sloane MS 4040, f. 255r.

62. Margaret Ray to Sloane, December 24, 1706, BL Sloane MS 4040, f. 279r; Samuel Smith to Sloane, December 28, 1706, BL Sloane MS 4040, f. 281r.

63. Margaret Ray to Sloane, November 19, 1706, BL MS Sloane 4040, f. 255r.

64. Margaret Ray to Sloane, November 19, 1706, BL MS Sloane 4040, f. 255r.

65. Dale to Sloane, October 8, 1705, BL Sloane MS 4040, f. 76r; Dale, "A Letter from Mr Samuel Dale to Dr Hans Sloane, R.S. Secr. Giving an Account of What Manuscripts Were Left by Mr John Ray, Together with Some Anatomical Observations Made at Padua by the Said Mr Ray," *Philosophical Transactions* 25, no. 307 (1706): 2282–303.

66. Dale to Sloane, October 8, 1705, BL Sloane MS 4040, f. 76r; Dale to Sloane, January 26, 1708, BL Sloane MS 4041, f. 278r.

67. Margaret Ray to Sloane, November 19, 1706, BL Sloane MS 4040, f. 255r.

68. Dale to Sloane, August 14, 1706, BL Sloane MS 4040, f. 206r.

69. On Plumier, see Brian J. Ford, "Scientific Illustration in the Eighteenth Century," in *The Cambridge History of Science*, vol. 4, *Eighteenth-Century Science*, ed. Roy Porter (Cambridge: Cambridge University Press, 2003), 566.

70. Petiver to Dale, September 20, 1706, BL Sloane MS 3335, f. 32v. Many thanks to Florence Hsia and Arnold Hunt for pointing me toward this reference.

71. *Bibliotheca Rayana: or, a Catalogue of the Library of Mr. John Ray* (London: Thomas Ballard, 1707/8), title page.

72. Dale to Sloane, August 14, 1706, BL Sloane MS 4040, f. 206r.

73. Dale to Sloane, October 23, 1706, BL Sloane MS 4040, f. 242r.

74. Dale to Sloane, March 5, 1706/7, BL Sloane MS 4040, f. 318r.

75. *Bibliotheca Rayana*, t.p., and conditions of sale statement. I have not determined whether Margaret Ray got her price.

76. On the emotional, moral, and neighborly aspects of economic obligation and credit, see Craig Muldrew, *The Economy of Obligation: The Culture of Credit and Social Relations in Early Modern England* (Basingstoke, UK: Macmillan, 1998), 123–30.

77. Thomas Willughby to Sloane, March 27, [1705?], BL Sloane MS 4062, f. 24r.

78. Thomas Willughby to Sloane, September 14, [1705?], BL Sloane MS 4062 f. 22r.

79. Charmantier, Johnston, and Smith, "Legacies of Francis Willughby," 366–68.

80. Cassandra Willughby, quoted in Alice T. Friedman, *House and Household in Eliza-*

bethan England: Wollaton Hall and the Willoughby Family (Chicago: University of Chicago Press, 1989), 162.

81. Robinson to Ray, August 29, 1684, *NHM* MS Ray, 98r; Ray to Peter Courthope, May 1662, in *Further Correspondence*, 29.

82. Ray, "The Preface," in *The Ornithology of Francis Willughby* (London: John Martyn, 1678); Derham, "Life of Mr. Ray," 22–38.

83. Johnston, "The Life and Domestic Context of Francis Willughby," in Birkhead, *Virtuoso by Nature*, 26–29.

84. Johnston, "Life and Domestic Context," 13–19, 26–27, and 28; Ray to Lister, March 3, 1670, *Further Correspondence*, 127. Willughby's descendants, Lord and Lady Middleton, continue to hold this view (Birkhead, *Virtuoso by Nature*, xii).

85. Mark Greengrass, Daisy Hildyard, Christopher D. Preston, and Paul J. Smith, "Science on the Move: Francis Willughby's Expeditions," in Birkhead, *Virtuoso by Nature*, 142–226; Rosemary O'Day, ed., *Cassandra Brydges, Duchess of Chandos, 1670–1735: Life and Letters* (Woodbridge: Boydell, 2007), 83n9.

86. William Derham, "To the Reader," in John Ray, *Philosophical Letters between the Late Learned Mr. Ray and Several of His Ingenious Correspondents, Natives and Foreigners*, ed. William Derham (London: William and John Innys, 1718).

87. My thanks to Lorraine Daston, whose comments on an earlier version of this paper prompted me to think through the acts of editing that made a letter "philosophical."

88. Derham, "To the Reader," in Ray, *Three Physico-Theological Discourses* (London: William Innys, 1713), xvii; Yale, "Book and the Archive," 107.

89. Derham, "To the Reader," in *Philosophical Letters*.

90. Derham, "The Life of Mr. Ray," in *Select Remains*, 37–38.

91. Terrall, "Masculine Knowledge," 195.

92. Michael Hunter and Jim Bennett, *The Image of Restoration Science: The Frontispiece to Thomas Sprat's History of the Royal Society (1667)* (London: Routledge, 1667).

93. In marketing their wares, publishers segmented the book-buying public by gender, creating divides not just between men's and women's craft but also between men's and women's craft and experimental practices. See Leong, *Recipes and Everyday Knowledge*, chapter 6.

Chapter 10. Family Notebooks, Mnemotechnics, and the Rational Education of Margaret Monro

1. Alexander Monro *Primus*, *The Professor's Daughter: Essay on Female Conduct*, ed. P. A. G. Monro (Edinburgh: Royal College of Physicians, [1739] 1996), 7, 9, 10.

2. Hereafter I refer to Alexander Monro *Primus* as "Primus" or "Monro's father."

3. Frances Yates, *The Art of Memory* (Chicago: University of Chicago Press, 1966); Walter Ong, *Ramus: Method, and the Decay of Dialogue* (Cambridge, MA: Harvard University Press, 1958); James Elkins, *The Domain of Images* (Ithaca, NY: Cornell University Press, 2001) extends Yates and Ong in relation to the mnemotechnics of words as images.

4. Primus and Margaret Monro, *An Essay on Female Conduct* (1739), Bound MS 6659 (hereafter MS 6659), National Library of Scotland (hereafter NLS). Margaret's edition built on Alexander Monro, *Essay on Female Conduct in Letters from a Father to His Daughter* (1738), Bound MS 6658 (hereafter MS 6658), NLS. An amalgamation of the two manuscripts was published in 1996 as Monro, *The Professor's Daughter: Essay on Female Conduct*. This publication includes material from both MS 6658 and MS 6659. When referring the shared content of the manuscripts, I cite the 1996 printed edition.

5. James Secord, "Knowledge in Transit," *Isis* 95, no. 4 (2004): 654–72.

6. See Yale, this volume.

7. Elizabeth Benger, *Memoirs of the Late Mrs. Elizabeth Hamilton* (London: Longman, 1818), 1:12–13.

8. Lindy Moore, "The Value of Feminine Culture: Community Involvement in the Provision of Schooling for Girls in Eighteenth-Century Scotland," in *Women in Eighteenth-Century Scotland: Intimate, Intellectual and Public Lives*, ed. Katie Barclay and Deborah Simonton (Aldershot, UK: Ashgate, 2013), 97–114; Rosalind Russell, "Women of the Scottish Enlightenment: Their Importance in the History of Scottish Education" (PhD diss., University of Glasgow, 1988).

9. Claire Grogan, *Politics and Genre in the Works of Elizabeth Hamilton, 1756–1816* (London: Routledge, 2016); Rosalind Carr, *Gender and Enlightenment Culture in Eighteenth-Century Scotland* (Edinburgh: Edinburgh University Press, 2014); Barclay and Simonton, *Women in Eighteenth-Century Scotland*; Katharine Glover, *Elite Women and Polite Society in Eighteenth-Century Scotland* (Woodbridge: Boydell, 2011); Emma Rothschild, *The Inner Life of Empires: An Eighteenth-Century History* (Princeton, NJ: Princeton University Press, 2011).

10. Elizabeth Hamilton, *Memoirs of the Modern Philosophers*, vol. 3 (Bath: Cruttwell, 1804), 99–100.

11. William Perry, *Royal Standard Dictionary* (Edinburgh: Willison, 1775), 156, a popular dictionary in Scotland, defined "girl" as "a young female child."

12. Glover, *Elite Women and Polite Society*.

13. The box of Lady Charlotte's educational manuscripts is held in the private collection of David Temperley, an antiquarian bookseller based in Edinburgh.

14. Anke te Heesen, *The World in a Box: The Story of an Eighteenth-Century Picture Encyclopedia* (Chicago: University of Chicago Press, 2002).

15. Monro, *Professor's Daughter*, 21 (evidence and argument), 31 (analogy), 48 (argument).

16. Examples of judgment are given throughout Monro, *Professor's Daughter*. Two illuminating instances occur on pages 31 and 43.

17. For written feminine culture, see Katharine Glover, "The Female Mind: Scottish Enlightenment Femininity and the World of Letters. A Case Study of the Women of the Fletcher of Saltoun Family in the Mid-eighteenth Century," *Journal of Scottish Historical Studies* 25, no. 1 (2005): 1–20.

18. Karen O'Brien, *Women and Enlightenment in Eighteenth-Century Britain* (Cambridge: Cambridge University Press, 2009); Michèle Cohen, *Fashioning Masculinity: National Identity and Language in the Eighteenth Century* (London: Routledge, 2002); Nancy J. Hirschmann, *The Subject of Liberty: Toward a Feminist Theory of Freedom* (Princeton, NJ: Princeton University Press, 2003).

19. Primus's empirical (a posteriori) interpretation of the sensory origins of ideas were most likely formed when he studied with Hermann Boerhaave in Leiden and William Cheselden in London. Alexander Monro, "Life of Dr. Ar. Monro, Sr. in His own Handwriting," ed. H. D. Erlaum, *University of Edinburgh Journal* 17 (1954): 81–82; Geoffrey M. Sill, "Neurology and the Novel: Alexander Monro *primus* and *secundus*, Robinson Crusoe, and the Problem of Sensibility," *Literature and Medicine* 16, no. 2 (1997): 250–65.

20. Primus's position was slightly different from what feminist historians rightly have called *civilized femininity* and *virtuous femininity*. See Carr, *Gender and Enlightenment Culture*, for the former and O'Brien, *Women and Enlightenment*, for the latter.

21. Monro, "Life of Dr Monro Sr," 81–82. Rex Earl Wright-St. Clair, *Doctors Monro: A Medical Saga* (London: Wellcome Historical Medical Library), 27–49.

22. Monro, *Professor's Daughter*, 17; Myra Reynolds, *The Learned Lady in England, 1650–1760* (New York: Houghton Mifflin, 1920).

23. Charles Rollin, *The Ancient History of the Egyptians, Carthaginians, Assyrians, Babylonians, Medes and Persians, Macedonians, and Grecians* (London: Knapton, 1738). Primus read books aloud to Margaret and her brothers. Monro, *Professor's Daughter*, 16. For middle-class family reading, see Abigail Williams, *The Social Life of Books: Reading Together in the Eighteenth-Century Home* (New Haven, CT: Yale University Press, 2017).

24. Monro, *Professor's Daughter*, addresses bookkeeping on pages 9, 10, 68, 106–7, 112, and 121. Matthew Daniel Eddy, "The Child Writer: Graphic Literacy and the Scottish Educational System, 1700–1820," *History of Education* 45, no. 6 (2016): 695–718. For the role of women in family bookkeeping in England, see Amanda Vickery, *The Gentleman's Daughter: Women's Lives in Georgian England* (New Haven, CT: Yale University Press, 2003).

25. Monro, *Professor's Daughter*, 17.

26. Margaret's inclusion in her brother's conversations is mentioned in Monro, *Professor's Daughter*, 136.

27. Monro, "Life of Dr Monro Sr," 94.

28. Matthew Daniel Eddy, "The Nature of Notebooks: How Enlightenment Schoolchildren Transformed the *Tabula Rasa*," *Journal of British Studies* 57, no. 2 (2018): 275–307.

29. *Proceedings of Latin Reading Club with Entries from Dec. 1740 to 20 March 1741*, MS 100, Bound MS, University of Otago Special Collections (hereafter UOSC). The role played by Primus in educating his sons is summarized in Wright-St. Clair, *Doctors Monro*, 50–58.

30. For Donald, Primus wrote *A Treatise on the Anatomical Encheireses or Manual Part of Anatomy* (1747), M169, Bound MS, UOSC. For Alexander *Secundus* he wrote *Commentary on Monro's Anatomy of the Bones* (1750), M163, Bound MS, UOSC.

31. Archibald Campbell (Marquis of Argyl), *Instructions to a Son, Containing Rules of Conduct in Publick and Private Life* (Glasgow: Foulis, 1743); David Hume, *Advice to a Daughter in a Letter* (Edinburgh: Reid, 1771); John Gregory, *A Father's Legacy to His Daughters* (Edinburgh: Creech, 1774). These texts were based on manuscripts read by the Campbell, Hume, and Gregory households.

32. Monro, *Professor's Daughter*, 7–8.

33. For an excellent example of a manuscript copybook, see Jean Greig, *Copybook Belonging to Jean Greig* (1763), dep. 190, box 3, Bound MS, Greig Papers, NLS.

34. William Duncan, professor at Marishal College, Aberdeen, expresses this common view in his *Elements of Logick* (London: Dodsley, 1752).

35. See for example the headings made by the student James Dunbar in *A Volume Completed by James Dunbar in 1710 Containing Arithmetic, Introduction to Algebra* (1710), Acc 5706/11, Bound MS, NLS.

36. Primus's views on the malleability of some vices is noteworthy, as it seems similar to the argument that Adam Smith would make in *The Theory of Moral Sentiments* (1759) about the mind's ability to positively redirect and repurpose potentially destructive emotions.

37. Monro, *Professor's Daughter*, 30, 45, 167, 170, 172–73, 174, 178.

38. Matthew Daniel Eddy, "The Interactive Notebook: How Students Learned to Keep Notes during the Scottish Enlightenment," *Book History* 19, no. 1 (2016): 87–131.

39. Adam Ferguson, for instance, used "parts" in *Institutes of Moral Philosophy* (Edinburgh: Kincaid and Bell, 1769).

40. *Oxford English Dictionary*, "range," eighteenth-century usages: to categorize (II.11); to lay or set down in a line (II.9.e); to order in a line (II.9.a).

41. William Perry, *The Man of Business, and Gentleman's Assistant* (Edinburgh: Perry, 1777), 210. See also William Gordon, *The Universal Accountant and Complete Merchant*, vol. 1 (Edinburgh: Gordon and Donaldson, 1763), 166.

42. Monro, *Professor's Daughter*, 10, 68.

43. Matthew Daniel Eddy, "The Shape of Knowledge: Children and the Visual Culture of Literacy and Numeracy," *Science in Context* 26, no. 2 (2013): 215–45. The classic work on the history of gestalt pictures is Rudolf Arnheim, *Art and Visual Perception: A Psychology of the Creative Eye* (Berkeley: University of California Press, 1954).

44. Eddy, "Shape of Knowledge."

45. Monro, *Professor's Daughter*, 18.

46. There were over twenty different kinds of notebooks that could potentially be involved in one master bookkeeping ledger. Perry, *Man of Business*, 209–12, and John Mair, *Book-Keeping Moderniz'd* (Edinburgh, 1773), ix, 408–9, 491–92.

47. Monro, "Life of Dr Monro Sr," 93, praises her wise "Management" skills, which would have included overseeing household accounts.

48. Barbara Montgomery, *Account Book Begun by Barbara Montgomery* (1787), MS.14098A, Bound MS, NLS. Montgomery married Alexander Walker of Bowland in 1811.

49. Jonathan Gibson, "Significant Space in Manuscript Letters," *Seventeenth Century* 12, no. 1 (1997): 1–10.

50. See Bittel, this volume.

51. Two examples where Primus suggested a break occur in MS 6658, ff. 17, 29.

52. For Margaret's underscored main headings, see MS 6659 ff. 5, 54, 207, 276, 289, 308. For Primus's main headings, see MS 6658, ff. 3, 14, 84, 125, 113, 144.

53. MS 6659, ff. iv recto, iv verso, and v recto.

54. Margaret's subheadings for Part II, "Of the General Conduct of Life," occur across MS 6659. ff. 54–207. Primus's subheadings for Part II, "Of the General Conduct of Life," occur across MS 6658, ff. 14–83.

55. Margaret MS 6659, ff. 5–8; Primus, MS 6658, f. 2.

56. Margaret MS 6659, ff. 38–40; Primus MS 6658, f. 10.

Chapter 11. Papier-Mâché Anatomical Models

1. See for instance Elizabeth Stephens, *Anatomy as Spectacle: Public Exhibitions of the Body from 1700 to the Present* (Liverpool: Liverpool University Press, 2011); Ann B. Shteir and Bernard V. Lightman, eds., *Figuring It Out: Science, Gender, and Visual Culture* (Hanover, NH: Dartmouth College Press, 2006).

2. Ludmilla Jordanova, *Sexual Visions: Images of Gender in Science and Medicine between the Eighteenth and Twentieth Centuries* (Madison: University of Wisconsin Press, 1989). See especially chapter 3, "Body Images and Sex Roles."

3. See, e.g., Mario Bucci, *Anatomia Come Arte* (Florence: Edizioni d'Arte Il Fiorino, 1969).

4. Francesca Bray, *Technology, Gender and History in Imperial China: Great Transformations Reconsidered* (Abingdon, UK: Routledge, 2013), 5.

5. Wanda Orlikowski, "Sociomaterial Practices: Exploring Technologies at Work," *Organization Studies* 28, no. 9 (2009): 1438. See also Bittel, Leong, and von Oertzen, this volume, and Werrett, this volume.

6. For an in-depth interrogation of the materiality of models, see Elizabeth Hallam, *Anatomy Museum: Death and the Body Displayed* (London: Reaktion, 2016).

7. Lucia Dacome, "Women, Wax and Anatomy in the 'Century of Things,'" *Renaissance Studies* 21, no. 4 (2007): 528–29. See also Dacome, *Malleable Anatomies: Models, Makers, and Material Culture in Eighteenth-Century Italy* (Oxford: Oxford University Press, 2017).

8. Anna Maerker, "Florentine Anatomical Models and the Challenge of Medical Authority in Late-Eighteenth-Century Vienna," *Studies in History and Philosophy of Science Part C: Studies in History and Philosophy of Biological and Biomedical Sciences* 43, no. 3 (2012): 730–40.

9. J. P. F. X. Fauken, *Entwurf zu einer Einrichtung der Heilkunde* (Göttingen: Johann Christian Dieterich, 1794), 24.

10. [Joseph Richter,] *Das Affenland oder der Doktor Fanfarone* [Vienna?, n.p., 1787], 38–39.

11. For models' materiality in relation to their claims to utility, see Anna Maerker, "'Wunderbare Vorrichtungen' oder 'nutzloses Spielzeug'? Debatten zum öffentlichen Nutzen der Visualisierung des Körperinneren," in *"Erkenne dich selbst!" Visuelle Gesundheitsaufklärung im 20. Jahrhundert*, ed. Sybilla Nikolow (Cologne: Böhlau, 2015), 133–43.

12. See especially Nicolas René Desgenettes, "Réflexions générales sur l'utilité de l'anatomie artificielle; et en particulier sur la collection de Florence, et la necessité d'en former de semblables en France," *Journal de Médecine, Chirurgie et Pharmacie* 94 (1793): 162–76, 233–52.

13. Archives nationales (Paris), F/17/2290: "Décret relatif à la formation d'un cabinet d'anatomie . . ." (10 ventôse an III); *Collection des lois et décrets approuvée et encouragée par le comité de salut public de la convention nationale* (Douai: Lagarde, n.d.), 8:178–79.

14. Percy and Laurent, "Muséum," in *Dictionnaire des sciences médicales* (Paris: Pankoucke, 1812), 35:11–41; Antoine Laurent Jessé Bayle, "Catalogue des préparations d'anatomie humaine du muséum de la faculté de médecine de Paris," in *Traité élémentaire d'anatomie*, 4th ed. (Paris: Deville Cavellin, 1833), 647–714.

15. On Galli see Dacome, "Women, Wax and Anatomy." On du Coudray and phantoms see, e.g., Nina Rattner Gelbart, *The King's Midwife: A History and Mystery of Madame du Coudray* (Berkeley: University of California Press, 1998), esp. 60–65; Ghislaine Lawrence, "An Obstetric Phantom," *Lancet* 358, no. 9296 (2001): 1916.

16. See Hallam, *Anatomy Museum*.

17. Jean-François Ameline, *Observations sur les pièces d'anatomie de M. le docteur Auzoux* (Caen: De Bonneserre, 1825), 3–4.

18. Jean-François Ameline, *Mémoire sur l'utilité des pièces d'anatomie artificielle chirurgicale* (Caen: A. Le Roy, 1820), 4–5.

19. See, e.g., George Dickinson, *English Papier-Mâché: Its Origin, Development and Decline* (London: Courier, 1925).

20. For paper technologies and paper tools see, e.g., Ursula Klein, *Experiments, Models, Paper Tools: Cultures of Organic Chemistry in the Nineteenth Century* (Stanford, CA: Stanford University Press, 2003); Charlotte Bigg, "Diagrams," in *A Companion to the History of Science*, ed. Bernard Lightman (Oxford: Wiley Blackwell, 2016), 557–71, esp. 561–62. In this chapter I use the term to encompass both textual and nontextual uses of paper objects in medical research and medical teaching.

21. Ameline, *Mémoire*, 16; italics in the original.

22. Ameline, *Mémoire*, 5.

23. Ameline, *Mémoire*, 10, 13.

24. *Rapport fait au Conseil royal d'instruction publique, sur les pièces d'anatomie artificielle de M. Ameline . . . et sur les demandes qu'il adresse à ce sujet au Conseil royal* (Caen: n.p., 1821); *Rapport fait à l'Académie royale de médecine sur les pièces d'anatomie artificielle de M. Ameline . . . : séance générale du 3 février* (Paris: n.p., 1824).

25. Margaret Olszewski, "Designer Nature: The Papier-Mâché Botanical Teaching Models of Dr Auzoux in Nineteenth-Century France, Great Britain and America" (PhD diss., University of Cambridge, 2009), 16.

26. "Rapport de la Société Médicale d'Emulation du 19 novembre," reprinted in Auzoux, *Notice sur les préparations artificielles de M. Auzoux* (Paris, 1825), 14.

27. Auzoux, *Notices*, 22. For earnings in nineteenth-century France see Christian Morrison and Wayne Snyder, "The Income Inequality of France in Historical Perspective," *European Review of Economic History* 4 (2000): 73.

28. See, e.g., John Harley Warner, "The Fielding H. Garrison Lecture: The Aesthetic Grounding of Modern Medicine," *Bulletin of the History of Medicine* 88, no. 1 (2014): 1–47.

29. See, e.g., *Précis analytique des travaux de l'Académie des science, des belles-lettres et des arts de Rouen, pendant l'année 1805* (Rouen: Periaux, 1807), 75.

30. Dr. F. Ratier, "Auzoux (le docteur)," in *Le Biographe universel: Revue générale biographique et nécrologique* . . . 9, no. 1, ed. E. Pascallet (Paris 1845), 15.

31. The drawings are on display at the Musée de l'Ecorché in Le Neubourg, France.

32. My description here follows the detailed analysis provided in Elizabet Nijhoff Asser, B. Reisland, B. J. W. Grob, and E. Goetz, "Lost Fingers, Scurfy Skin and Corroding Veins—Conservation of Anatomical Papier-Mâché Models by Dr Auzoux," *ICOM Committee for Conservation* 1 (2008): 287.

33. Ratier, "Auzoux," 16.

34. "Rapport fait à l'Académie royale de médecine par MM. Adelon, Ant. Dudois, Ribes . . . Séance du 10 mai 1831," reprinted in Louis Auzoux, *Leçons élémentaires d'anatomie et de physiologie humaine et comparée*, 1st ed. (Paris: Firmin Didot, 1839), 6.

35. Louis Auzoux, *Leçons élémentaires d'anatomie et de physiologie humaine et comparée*, 1st ed. (Paris: Firmin Didot, 1839); 2nd rev. ed. (Paris: Labé, 1858).

36. Ratier, "Auzoux," 21.

37. Ann La Berge, *Mission and Method: The Early Nineteenth-Century French Public Health Movement* (Cambridge: Cambridge University Press, 1992).

38. See, e.g., Thomas Le Roux, "Hygienists, Workers' Bodies and Machines in Nineteenth-Century France," *European Review of History* 20, no. 2 (2013): 255–70.

39. Auzoux, *Leçons*, 1st ed. (1839), 9, 10, 19, 127, 139, 213.

40. Auzoux, *Leçons*, 2nd ed. (1858), 131.

41. For contemporary French entrepreneurs pursuing similar goals see, e.g., R. Howard Bloch, *God's Plagiarist: Being an Account of the Fabulous Industry and Irregular Commerce of the Abbé Migne* (Chicago: University of Chicago Press, 1994).

42. "Examen par le professeur Berard de quelques ouvriers de l'atelier du Docteur Auzoux lors d'une exposition à Evreux," October 10, 1837, 242 AP 3 dossier 3, Archives Nationales, Paris.

43. *Medical and Surgical Reporter* 17 (1867): 496.

44. Anna Maerker, "Anatomizing the Trade: Designing and Marketing Anatomical Models as Medical Technologies, ca. 1700–1900," *Technology and Culture* 54, no. 3 (2013): 531–62.

45. For a British perspective see also George Knox, *Description of an Artificial Anatomical*

Figure: Constructed by the Chevalier Auzoux . . . Exhibited in 1832 before the King, in London (Madras: Church Mission, 1834), 273–74.

46. Antoine Clot, *Mémoires de A.-B. Clot Bey*, ed. Jacques Tagher (Cairo: Institut Français d'Archéologie Orientale, 1949); Khaled Fahmy, *All the Pasha's Men: Mehmed Ali, His Army and the Making of Modern Egypt* (Cambridge: Cambridge University Press, 1997), especially 148–51, chapter 5; James Heyworth-Dunne, *An Introduction to the History of Education in Modern Egypt* (London: Luzac, 1939), 122–33; LaVerne Kuhnke, *Lives at Risk: Public Health in Nineteenth-Century Egypt* (Berkeley: University of California Press, 1990); Amira el Azhary Sonbol, *The Creation of a Medical Profession in Egypt, 1800–1922* (Syracuse: Syracuse University Press, 1991).

47. Khaled Fahmy shows that European accounts of Egyptians' resistance to dissection was frequently exaggerated to justify the introduction of European approaches. Fahmy, "The Anatomy of Justice: Forensic Medicine and Criminal Law in Nineteenth-Century Egypt," *Islamic Law and Society* 6, no. 2 (1999): 224–71.

48. "Historico-statistical account of the present state of medicine in Egypt," report by Clot of December 27, 1837, quoted in John Bowring, *Report on Egypt and Candia, Parliamentary Papers, Reports from Commissioners* 11 (1840), 141; Clot Bey, *Aperçu général sur l'Égypte*, (Paris: Fortin Masson, 1840), 2:385.

49. Khaled Fahmy, "Women, Medicine and Power in Nineteenth-Century Egypt," in *Remaking Women: Feminism and Modernity in the Middle East*, ed. Lila Abu-Lughod (Princeton, NJ: Princeton University Press, 1998), 35–36. See also Bowring, *Report on Egypt and Candia*, 138–40; Baptistin Poujoulat, *Voyage dans l'Asie Mineure*, 2 vols (Paris: Ducollet, 1840), 2:517–20; Ehud R. Toledano, *Slavery and Abolition in the Ottoman Middle East* (Seattle: University of Washington Press, 1998); Mervat F. Hatem, "The Professionalization of Health and the Control of Women's Bodies as Modern Governmentalities in Nineteenth-Century Egypt," in *Women in the Ottoman Empire: Middle Eastern Women in the Early Modern Era*, ed. Madeline C. Zilfi (Leiden: Brill, 1997), 66–80.

50. "The Little Abyssinian Slave," *Chambers's Edinburgh Journal, New Series* 146, October 17, 1846, 252–53.

51. Poujoulat, *Voyage dans l'Asie Mineure*, 2:519.

52. *Jahrbücher für praktische Heilkunde* 1, no. 3 (1845): 502; "Faculté de Médecine feminine, au Caire, en Égypte," *Gazette médicale belge*, February 9, 1845, 25.

53. Victor Schoelcher, *L'Égypte en 1845* (Paris, 1846).

54. George Michael La Rue, "African Slave Women in Egypt, ca. 1820 to the Plague of 1834–35," in *Women and Slavery*, vol. 1, *Africa, the Indian Ocean World, and the Medieval North Atlantic*, ed. Gwyn Campbell, Suzanne Miers, and Joseph C. Miller (Athens: Ohio University Press, 2007), 1:175.

55. Hatem, "Professionalization of Health," 74.

56. Christopher Hoolihan lists Austin Flint, Gunning S. Bedford, C. W. Gleason, and Calvin Cutter, among others. Hoolihan, ed., *An Annotated Catalogue of the Edward C. Atwater Collection of American Popular Medicine and Health Reform* (Rochester, NY: University of Rochester Press, 2004), 2:200–201, 329.

57. Elizabeth Reis, ed., *American Sexual Histories*, 2nd ed. (Chichester, UK: Wiley-Blackwell, 2012), 167.

58. April R. Haynes, *Riotous Flesh: Women, Physiology, and the Solitary Vice in Nineteenth-Century America* (Chicago: University of Chicago Press, 2015), esp. chapter 4.

59. Quoted from Reis, *American Sexual Histories*, 168.

60. See, e.g., Frederick Hollick, *The Male Generative Organs in Health and Disease, from Infancy to Old Age . . .* (New York: T. W. Strong, 1851); Hollick, *The Origin of Life and Process of Reproduction in Plants and Animals . . .*, new ed. (Philadelphia: McKay, 1902 [1878]).

61. Hollick, *Male Generative Organs*, 417; Hollick, *Origin of Life*, vii.

62. Quoted in Nancy Isenberg, *Sex and Citizenship in Antebellum America* (Chapel Hill: University of North Carolina Press, 1998), 69; quoted in Hollick, *Outlines of Anatomy and Physiology* (Philadelphia: National Publishing Company, 1846), 1.

63. See, e.g., "Dr. Hollick's Lectures," *Pennsylvania Freeman*, March 12, 1846.

64. Quoted in Rickie Solinger, *Pregnancy and Power: A Short History of Reproductive Politics in America* (New York: New York University Press, 2005), 59; Hollick, *Male Generative Organs*, 422.

65. *New York Evening Post*, April 29, 1844, quoted in Hollick, *Outlines*, 2.

66. *Philadelphia Spirit of the Times* (December 4, [1844?]), quoted in Hollick, *Outlines*, 1.

67. Hollick, *Outlines*, n.p.

68. Haynes, *Riotous Flesh*. See also Isenberg, *Sex and Citizenship*; Reis, *American Sexual Histories*.

69. Paulina Wright Davis, "Letter from Mrs. Wright," *Water-Cure Journal*, June 1, 1846, 29. See also Isenberg, *Sex and Citizenship*, 166.

70. Series II, folder 1.7, Lectures on Anatomy, 18–20, Paulina Wright Davis Papers, Archives and Special Collections Library, Vassar College Libraries.

71. Series II, folder 1.7, Lectures on Anatomy, n.p. [33], Paulina Wright Davis Papers, Archives and Special Collections Library, Vassar College Libraries.

72. William Elder, "Mrs. Wright's Lectures to Ladies on Anatomy, Physiology, and Health," *Water-Cure Journal* 2, no. 1 (June 1, 1846): 11. For further praise of the "fine French Apparatus" and of Wright's "delicacy," see also "Lectures to Ladies," *Liberator*, March 2, 1849, 3.

Chapter 12. Women Who Worked with Documents to Rationalize Reproduction

1. This project would not have been possible without the skilled labor of Abby Balfour, Erin Burke, and Christy Mills. Thanks too to archivists Elizabeth Maurer at Miami University and Tom Rosenbaum at Rockefeller Archive Center. Joanna Radin, Monica Mercado, Antonio Barrera, David Robinson, Matthew Kato, and my fellow contributors to this book offered crucial critiques and suggestions.

2. On paper tools, see Ursula Klein, "Paper Tools in Experimental Cultures," *Studies in History and Philosophy of Science* 32, no. 2 (2001): 265–302.

3. Calvert Dedrick, "Personnel in Division of Statistical Research," December 19, 1939, folder "Div. of Stat. Research" Entry 210, "General Records Maintained by Calvert Dedrick, 1935–1942," box 217, Bureau of Census Records, RG 29, National Archives.

4. Emily R. Merchant, "Prediction and Control: Global Population, Population Science, and Population Politics in the Twentieth Century" (PhD diss., University of Michigan, 2015), chapter 5; Dan Bouk, *How Our Days Became Numbered: Risk and the Rise of the Statistical Individual* (Chicago: University of Chicago Press, 2015), chapter 7.

5. Michelle Murphy, *The Economization of Life* (Durham, NC: Duke University Press, 2017); Matthew Connelly, *Fatal Misconception: The Struggle to Control World Population* (Cambridge, MA: Belknap Press of Harvard University Press, 2008); Alison Bashford, *Global Population: History, Geopolitics, and Life on Earth* (New York: Columbia University

Press, 2014); Derek S. Hoff, *The State and the Stork: The Population Debate and Policy Making in US History* (Chicago: University of Chicago Press, 2012).

6. Ben Kafka, "Paperwork: The State of the Discipline," *Book History* 12 (2009): 340–53.

7. JoAnne Yates, *Control through Communication: The Rise of System in American Management* (Baltimore: Johns Hopkins University Press, 1989); Theodore Porter, *Trust in Numbers: The Pursuit of Objectivity in Science and Public Life* (Princeton, NJ: Princeton University Press, 1995).

8. Lisa Gitelman, *Paper Knowledge: Toward a Media History of Documents* (Durham, NC: Duke University Press, 2014).

9. Bruno Latour, *Science in Action: How to Follow Scientists and Engineers through Society* (Cambridge, MA: Harvard University Press, 1987), chapter 6.

10. See Margery W. Davies, *Woman's Place Is at the Typewriter: Office Work and Office Workers 1870–1930* (Philadelphia: Temple University Press, 1982); Sharon Hartman Strom, *Beyond the Typewriter: Gender, Class, and the Origins of Modern American Office Work, 1900–1930* (Urbana: University of Illinois Press, 1992); and Lisa M. Fine, *The Souls of the Skyscraper: Female Clerical Workers in Chicago, 1870–1930* (Philadelphia: Temple University Press, 1990). On the rise of gendered corporate spaces, see Angel Kwolek-Folland, *Engendering Business: Men and Women in the Corporate Office, 1870–1930* (Baltimore: Johns Hopkins University Press, 1994); Olivier Zunz, *Making America Corporate, 1870–1920* (Chicago: University of Chicago Press, 1990).

11. Marie Hicks, *Programmed Inequality: How Britain Discarded Women Technologists and Lost Its Edge in Computing* (Cambridge, MA: MIT Press, 2017); Jennifer S. Light, "When Computers Were Women," *Technology and Culture* 40, no. 3 (1999): 455–83; Lisa Nakamura, "Indigenous Circuits: Navajo Women and the Racialization of Early Electronic Manufacture," *American Quarterly* 66, no. 4 (2014): 919–41.

12. Margot Lee Shetterly, *Hidden Figures: The American Dream and the Untold Story of the Black Women Mathematicians Who Helped Win the Space Race* (New York: William Morrow, 2016).

13. Emily Marks to P. K. Whelpton, October 10, 1941, preceding "Form 2: Household Survey of Indianapolis. Sponsored by the Council of Social Agencies. Instructions to Canvassers," p. 3, box 7 or 9, Scripps Foundation Papers, Miami University Special Collections and University Archives (hereafter Scripps Papers). My thanks to Elizabeth Maurer, who shared digital reproductions of these files with me. The digital files did not always differentiate between materials from boxes 7 and 9, as noted in this case and others.

14. "First Private Household Survey Starts Here with 3 Schools Co-Operating," *Indianapolis Star*, March 23, 1941.

15. "First Private Household Survey"; P. K. Whelpton and Clyde V. Kiser, "Social and Psychological Factors Affecting Fertility: IV. Developing the Schedules, and Choosing the Type of Couples and the Area to Be Studied," *Milbank Memorial Fund Quarterly* 23, no. 4 (1945): 407.

16. "Form 2," 2.

17. "Form 2," 3.

18. Whelpton and Kiser, "Factors Affecting Fertility: IV," 400–404.

19. "P. K. Whelpton, 71, Researcher, Dies," *New York Times*, April 7, 1964, in Rockefeller Foundation Papers, RG 1.2, series 200, box 534, folder 4566, Rockefeller Archives (hereafter Rockefeller Foundation Papers).

20. Whelpton and Kiser, "Factors Affecting Fertility: IV," 390–91.

21. Whelpton and Kiser, "Factors Affecting Fertility: IV," 390–91.

22. Whelpton and Kiser, "Factors Affecting Fertility: IV," 391.

23. Ronald F. Freedman, Pascal K. Whelpton, and Arthur A. Campbell, "Family Planning in the U.S.," *Scientific American* 200, no. 4 (April 1959): 50–51.

24. *The Problems of a Changing Population: Report of the Committee on Population Problems to the National Resources Committee May 1938* (Washington, DC: US Government Printing Office, 1938), 7.

25. "P. K. Whelpton, 71, Researcher, Dies."

26. The Ford Foundation soon became the Population Council's chief funder. See Merchant, "Prediction and Control," 313.

27. RF 59004 on January 23, 1959, in box 534, folder 4564, Rockefeller Foundation Papers; RF 56086 on May 25, 1956, in box 533, folder 4555, Rockefeller Foundation Papers.

28. Ronald Freedman, "The Planned Family and American Population Growth," *Antioch Review*, March 1957, 39.

29. Merchant, "Prediction and Control," chapter 5.

30. "Form G-1 Indianapolis Study," box 7 or 9, Scripps Papers.

31. Whelpton and Kiser, "Factors Affecting Fertility: IV," 393.

32. Card 1 and Card 2 in Population Council Records, RG 1, series 1, box 22, folder 337–339, Rockefeller Archives (hereafter Population Council Records).

33. Merchant, "Prediction and Control," 383–84.

34. Scripps Foundation and Survey Research Center, "A Study of the Growth of American Families (Married Women)," February 15, 1955, question 44, p. 17, last updated June 21, 2016, accessed December 19, 2017, https://www.disc.wisc.edu/archive/gaf/index.html.

35. Scripps Foundation and Survey Research Center, "A Study of the Growth of American Families (Married Women)," February 15, 1955, question 59, p. 21.

36. Peter Stallybrass, "Printing and the Manuscript Revolution," in *Explorations in Communication and History*, ed. Barbie Zelizer (New York: Routledge, 2008), 112.

37. Mendelsohn, "Data as a State of Mind and the Problem of How It Came to Be in Early Modern Europe," paper read at Histories of Data and the Database conference at Huntington Library, CA, November 18, 2016.

38. Gitelman, *Paper Knowledge*, 25–26.

39. On significance of the categories that we count with, see Ian Hacking, "Biopower and the Avalanche of Printed Numbers," *Humanities in Society* 5, no. 3–4 (1982): 279–95. For an application of Hacking's ideas to the history of American social surveying, see Sarah E. Igo, *The Averaged American: Surveys, Citizens, and the Making of a Mass Public* (Cambridge, MA: Harvard University Press, 2007).

40. Emily Marks, "Additions to Filling Instructions for the Household Survey of Indianapolis," October 10, 1941, p. 6, box 7 or 9, Scripps Papers.

41. Marks, "Additions," 8.

42. If a "wife" refused to state her age, the canvasser could judge her overage independently, but then had to add a footnote to the card indicating "My estimate." Marks, "Additions," 2.

43. "Form 2," 7. On passing: "Determination of Sizes of Individual Native White Cohorts, 1917–1954," page 3, box 7 or 9, Scripps Papers.

44. Whelpton and Kiser, "Factors Affecting Fertility: IV," 404.

45. Marks, "Additions," 2.

46. Marks, "Additions," 2, 6, 8.

47. "Form F—Instructions Indianapolis Study, 1941" page 4, box 7 or 9, Scripps Papers.

48. Scripps Foundation, "A Study of the Growth of American Families, (Single Women)," February 15, 1955, questionnaire, box 7, folder 6, Scripps Papers.

49. Question 4 in Scripps Foundation, "A Study of the Growth of American Families, (Single Women)," February 15, 1955, questionnaire, box 7, folder 6, Scripps Papers.

50. Merchant also notes that the GAF was apparently more transparent—the Kinsey interviewers recorded sex histories using a memorized code. Merchant, "Prediction and Control," 378–80.

51. Scripps Papers, box 7, folders 10 and 11.

52. "First Private Household Survey."

53. P. K. Whelpton and Clyde V. Kiser, "Social and Psychological Factors Affecting Fertility: III. The Completeness and Accuracy of the Household Survey of Indianapolis," *Milbank Memorial Fund Quarterly* 23, no. 3 (1945): 256; Whelpton and Kiser, "Factors Affecting Fertility: IV," 407.

54. "First Private Household Survey."

55. Whelpton and Kiser, "Factors Affecting Fertility: IV," 407n21.

56. P. K. Whelpton and Clyde V. Kiser, "Social and Psychological Factors Affecting Fertility: V. The Sampling Plan, Selection, and the Representativeness of Couples in the Inflated Sample," *Milbank Memorial Fund Quarterly* 24, no. 1 (1946): 64.

57. "A Memorandum on Population Research at the University of Michigan," March 2, 1954, p. 7, box 22, folder 337–339, Population Council Records.

58. Igo, *Averaged American*, 214.

59. Whelpton and Kiser, "Factors Affecting Fertility: III," 256.

60. Based on calculations from table VII in Bureau of the Census, *Sixteenth Census of the United States: 1940. Population. The Labor Force (Sample Statistics). Wage and Salary Income in 1939* (Washington, DC: Government Printing Office, 1943), 7.

61. "Lucknow University," in folder "Staff: Whelpton, Pascal K. 1954," box 32, folder 470, Population Council Records.

62. Rockefeller Foundation, *Annual Report* (1952), 237.

63. See, for instance, "True Rate of Natural Increase. White Females in Each State of the United States, 1930. Births for Mothers of Unknown Age Distributed," oversize drawer 8-1, folder 3, Alfred J. Lotka Papers, Public Policy Papers, Department of Rare Books and Special Collections, Princeton University.

64. Joseph H. Willits to Warren Thompson, June 26, 1944, box 533, folder 4556, Rockefeller Foundation Papers.

65. Warren S. Thompson to Roger F. Evans, November 3, 1945, box 533, folder 4556, Rockefeller Foundation Papers.

66. "Statement of Transactions in Connection with Revolving Fund for Pacific Studies—Rockefeller Foundation, Scripps Foundation Miami University, Oxford Ohio, March 1, 1943, to June 30, 1944," in series 200, box 533, folder 4556, Rockefeller Foundation Papers.

67. Warren S. Thompson to Roger F. Evans, November 3, 1945.

68. Rockefeller Foundation officials extended the grant one year: J. H. Willits to Warren S. Thompson, December 7, 1945, in box 533, folder 4556, Rockefeller Foundation Papers. Whelpton's book acknowledges these eventual clerks: Shaffer and Brittain as supervisors, who oversaw Mrs. Rhelda Belch, Mrs. Jean Calhoun, Mrs. Marjorie Derr, Mrs. Maryalice Edington, Mrs. June Goldner, Miss Florence Schwegman, "and others who assisted them." Whelpton, *Cohort Fertility*, vi.

69. “Procedure for Correcting the Censuses of 1920–1950 by Five-Year Age Groups Native White Females,” box 7 or 9, Scripps Papers.

70. See folder labeled “Cohort Tables—NBR memos,” box 7 or 9, Scripps Papers.

71. See folder labeled “Cohort Tables—NBR memos,” box 7 or 9, Scripps Papers.

72. “Contents of Attached Tables” [1956], box 7 or 9, Scripps Papers.

73. “Contents of Attached Tables” [1956], box 7 or 9, Scripps Papers.

74. “Cohort Life Tables,” box 7 or 9, Scripps Papers.

75. Mills to Bouk, personal communication, May 22, 2017.

Afterword. Making and Using Paper in Late Imperial China

1. Gufosi stele, Jiajiang county, Yancheng township. Transcribed stele text in the possession of the author.

2. Werrett, this volume.

3. Tsien Tsuen-hsuin, *Paper and Printing,* part 1 of *Chemistry and Chemical Technology,* vol. 5 of *Science and Civilisation in China,* ed. Joseph Needham (Cambridge: Cambridge University Press, 1985), 38–40.

4. On the low cost of paper in China, see Kai-Wing Chow, *Publishing, Culture, and Power in Early Modern China* (Stanford, CA: Stanford University Press, 2005), 28–36.

5. As Wolfe shows, paper in early modern England was cheaper and more widely used than generally assumed.

6. Wang Juhua, ed., *Zhongguo gudai zaozhi gongcheng jishu shi* [The history of papermaking technology in ancient China] (Taiyuan: Shanxi Education Press, 2005), 318–23. See for example Samuel Wells Williams, *The Middle Kingdom: A Survey of the Geography, Government, Education, Social Life, Arts, Religion, etc. of the Chinese Empire and its Inhabitants* (New York: Wiley, 1876), 477–79.

7. Dorothy Ko, *The Social Life of Inkstones: Artisans and Scholars in Early Qing China* (Seattle: University of Washington Press, 2017).

8. Tsien, “Paper and Printing,” 114–15.

9. Aimee Lee, *Hanji Unfurled:One Journey into Korean Papermaking* (Ann Arbor, MI: Legacy Press, 2012), 95–117.

10. But see Leong’s observation that people in seventeenth-century Naples used devotional images of Francis Xavier to absorb and expel plague infections.

11. Gan Duan, *Jiajiang xian xiangtu zhilüe* [Abridged gazetteer of towns in Jiajiang County] (Jiajiang: Yitong Shuju, 1948), 17.

12. Hill Gates, “Money for the Gods,” *Modern China* 13, no. 3 (July 1987): 259–77.

13. Mu-Chou Poo, “Preparation for the Afterlife in Ancient China,” in *Mortality in Traditional Chinese Thought,* ed. Amy Olberding and Philip J. Ivanhoe (Albany: State University of New York Press, 2011), 13–36.

14. Tim Ingold, *The Perception of the Environment: Essays in Livelihood, Dwelling and Skill* (London: Routledge, 2000), 166.

15. François Sigaut, “Technology,” in *Companion Encyclopedia of Anthropology: Humanity, Culture, and Social Life,* ed. Tim Ingold (London: Routledge, 1994), 448.

16. Dorothy Ko, “Epilogue: Textiles, Technology, and Gender,” *East Asian Science, Technology, and Medicine* 36 (2012): 171.

17. Charlotte Furth, *A Flourishing Yin: Gender in China’s Medical History, 960–1665* (Berkeley: University of California Press, 1999).

18. Francesca Bray, *Technology, Gender and History in Imperial China: Great Transformations Reconsidered* (London: Routledge, 2013), 18.

19. Bray, *Technology,* 17–18.

20. Ko, *Social Life of Inkstones,* 201.

21. Adam Chau, "Script Fundamentalism: The Practice of Cherishing Written Characters (Lettered Paper 惜字紙) in the Age of Literati Decline and Commercial Revolution," in *Chinese and European Perspectives on the Study of Chinese Popular Religions,* ed. Philip Clart (Taipei: Boyang Wenhua, 2012), 130–67; Alexander Des Forges, "Burning with Reverence: The Economics and Aesthetics of Words in Qing (1644–1911) China," *PMLA* 121, no. 1 (2006): 139–55.

22. Chau, "Script Fundamentalism," 134–38. These texts contain lists of offenses against lettered paper, which can be read for information on how paper was used: people are punished in this life and the afterworld for "selling torn books to others, throwing lettered paper in the dirt, stepping or stomping on it, covering windows and walls with it, covering vats with it, using it to hold up paintings, to wipe desks and inkstones, to wipe dirt, to make a torch at night, to wrap things, to light pipes, cutting it with a knife or a pair of scissors, tearing it up while in a rage, using books as pillows, giving books to women to hold needles and threads, chewing paper into a paste and spitting it out, sticking it into cracks in the wall, burning it into ashes and throwing them on the ground."

23. Katherine Alexander, "Virtues of the Vernacular: Moral Reconstruction in Late Qing Jiangnan and the Revitalization of *Baojuan*" (PhD diss., University of Chicago, 2016).

24. These work songs (*haozi*) were described to me by local folklorists as "too dirty to be written down." Ordinarily, men carefully avoided offensive language when speaking to women in the village, since all covillagers were considered kin and an offense against them was an offense against community and ancestors.

25. Jacob Eyferth, *Eating Rice from Bamboo Roots: The Social History of a Community of Handicraft Papermakers in Rural Sichuan, 1920–2000* (Cambridge, MA: Harvard Asia Center, 2009), 42.

26. Karin Hausen, "Family and Role-Division: The Polarization of Sexual Stereotypes in the Nineteenth Century—An Aspect of the Dissociation of Work and Family Life," in *The German Family: Essays on the Social History of the Family in Nineteenth- and Twentieth-Century Germany,* ed. Richard Evans and W. R. Lee (London: Croom Helm, 1981), 51–83.

27. Susan Mann, "Household Handicrafts and State Policies in Qing Times," in *To Achieve Security and Wealth: The Qing Imperial State and the Economy, 1644–1911,* ed. Jane Kate Leonard and John R. Watts (Ithaca, NY: Cornell University East Asia Program, 1992), 87–88.

28. Zou Yiren, *Jiu Shanghai renkou bianqiande yanjiu* [Population studies of old Shanghai] (Shanghai: Shanghai Renmin Chubanshe, 1980), 104–5.

29. *Chengdu Echeng Wuhan shougongye diaocha* [Survey of the handcrafts of Chengdu, Echeng (Ezhou) and Wuhan] (Beijing: Caizheng jingji chubanshe, 1955), 12.

30. Hill Gates, *China's Motor: A Thousand Years of Petty Capitalism* (Ithaca, NY: Cornell University Press, 1996), 29–38.

SELECTED BIBLIOGRAPHY

This bibliography represents a list of secondary sources selected by authors. Additional readings and full references of primary sources can be found in the footnotes of individual chapters.

Agar, Jon. *The Government Machine: A Revolutionary History of the Computer*. Cambridge, MA: MIT Press, 2003.

Alexander, Katherine. "Virtues of the Vernacular: Moral Reconstruction in Late Qing Jiangnan and the Revitalization of *Baojuan*." PhD diss., University of Chicago, Department of East Asian Studies, 2016.

Apple, Rima D., Gregory J. Downey, and Stephen L. Vaughn. *Science in Print: Essays on the History of Science and the Culture of Print*. Madison: University of Wisconsin Press, 2012.

Arnold, David. *Everyday Technology: Machines and the Making of India's Modernity*. Chicago: University of Chicago Press, 2015.

Barrett, Timothy. "European Papermaking Techniques 1300–1800." *Paper through Time: Nondestructive Analysis of 14th- through 19th-Century Papers*. University of Iowa Libraries. July 14, 2014. Accessed January 8, 2018. http://paper.lib.uiowa.edu/european.php.

Becker, Peter, and William Clark, eds. *Little Tools of Knowledge: Historical Essays on Academic and Bureaucratic Practices*. Ann Arbor: University of Michigan Press, 2001.

Bijker, Wiebe E., Thomas P. Hughes, and Trevor Pinch, eds. *The Social Construction of Technological Systems: New Directions in the Sociology and History of Technology*. Cambridge, MA: MIT Press, 1989.

Bittel, Carla. *Mary Putnam Jacobi and the Politics of Medicine in Nineteenth-Century America*. Chapel Hill: University of North Carolina Press, 2009.

Bittel, Carla. "Woman, Know Thyself: Producing and Using Phrenological Knowledge in 19th-Century America." *Centaurus* 55, no. 2 (2013): 104–30.

Blair, Ann. "Note Taking as an Art of Transmission." *Critical Inquiry* 31, no. 1 (2004): 85–107.

Blair, Ann. *Too Much to Know: Managing Scholarly Information before the Modern Age*. New Haven, CT: Yale University Press, 2010.

Bouk, Dan. "The History and Political Economy of Personal Data over the Last Two Centuries in Three Acts." *Osiris* 32, no. 1 (2017): 85–106.

Bouk, Dan. *How Our Days Became Numbered: Risk and the Rise of the Statistical Individual*. Chicago: University of Chicago Press, 2015.

Bourdieu, Pierre. *Outline of a Theory of Practice*. Cambridge: Cambridge University Press, 1977.

Bray, Francesca. *Technology, Gender and History in Imperial China: Great Transformations Reconsidered*. London: Routledge, 2013.

Brayman, Heidi, Jesse M. Lander, and Zachary Lesser, eds. *The Book in History, the Book as History: New Intersections of the Material Text. Essays in Honor of David Scott Kastan*. New Haven, CT: Yale University Press, 2016.

Brayman Hackel, Heidi. *Reading Material in Early Modern England: Print, Gender, and Literacy*. Cambridge: Cambridge University Press, 2005.

Brokaw, Cynthia J. *Commerce in Culture: The Sibao Book Trade in Qing and Republican Periods*. Cambridge, MA: Harvard University Press, 2007.

Brown, Nancy Pollard. *Paperchase: The Dissemination of Catholic Texts in Elizabethan England*. Vol. 1 of *English Manuscript Studies 1100–1700*, edited by Peter Beal and Jeremy Griffiths. Oxford: B. Blackwell, 1989.

Brownlee Griego, Elizabeth. "A Part and Yet Apart: Clelia Duel Mosher and Professional Women at the Turn-of-the-Century." PhD diss., University of California, Berkeley, 1983.

Burke, Peter. *What Is the History of Knowledge?* Cambridge, UK: Polity Press, 2015.

Burke, Victoria E. "'The art of Numbering well': Late Seventeenth-Century Arithmetic Manuscripts Compiled by Quaker Girls." In *Material Readings of Early Modern Culture: Texts and Social Practices, 1580–1730*, edited by James Daybell and Peter Hinds, 246–65. Basingstoke: Palgrave Macmillan, 2010.

Burke, Victoria E. "Recent Studies in Commonplace Books." *English Literary Renaissance* 43, no. 1 (2013): 153–77.

Burns, Kathryn. *Into the Archive: Writing and Power in Colonial Peru*. Durham, NC: Duke University Press, 2010.

Butler, Judith. *Bodies That Matter: On the Discursive Limits of "Sex."* New York: Routledge, 1993.

Calhoun, Joshua. "The Word Made Flax: Cheap Bibles, Textual Corruption, and the Poetics of Paper." *PMLA* 126, no. 2 (2011): 327–44.

Canning, Kathleen. "The Body as Method? Reflections on the Place of the Body in Gender History." *Gender & History* 11, no. 3 (1999): 499–513.

Cavallo, Sandra. *Artisans of the Body in Early Modern Italy: Identities, Families and Masculinities*. Manchester: Manchester University Press, 2007.

Cavallo, Sandra. "The Motivations of Benefactors: An Overview of Approaches to the Study of Charity." In *Medicine and Charity before the Welfare State*, edited by Jonathan Barry and Colin Jones, 47–62. Studies in the Social History of Medicine. London: Routledge, 1991.

Cevolini, Alberto, ed. *Forgetting Machines: Knowledge Management Evolution in Early Modern Europe*. Leiden: Brill, 2016.

Charmantier, Isabelle, and Staffan Müller-Wille. "Carl Linnaeus's Botanical Paper Slips (1767–1773)." *Intellectual History Review* 24, no. 2 (2014): 215–38.

Charmantier, Isabelle, and Staffan Müller-Wille. "Worlds of Paper: An Introduction." *Early Science and Medicine* 19, no. 5 (2014): 379–97.

Coleman, D. C. *The British Paper Industry, 1495–1860: A Study in Industrial Growth*. Oxford: Clarendon Press, 1958.

Collins, Harry. *Artificial Experts: Social Knowledge and Intelligent Machines*. Cambridge, MA: MIT Press, 1992.

Cooper, Alix. "Homes and Households." In *The Cambridge History of Science*. Vol. 3, *Early Modern Science*, edited by Katharine Park and Lorraine Daston, 224–37. Cambridge: Cambridge University Press, 2008.

Cooper, Alix. "Picturing Nature: Gender and the Politics of Natural-Historical Description in Eighteenth-Century Gdansk/Danzig." *Journal for Eighteenth-Century Studies* 36, no. 4 (2013): 519–29.

Cooter, Roger. *The Cultural Meaning of Popular Science: Phrenology and the Organization of Consent in Nineteenth-Century Britain*. Cambridge: Cambridge University Press, 1984.

Cooter, Roger, and Stephen Pumfrey. "Separate Spheres and Public Places: Reflections on the History of Science Popularization and Science in Popular Culture." *History of Science* 32, no. 3 (1994): 237–67.

Cowan, Ruth Schwartz. "The Consumption Junction: A Proposal for Research Strategies in the Sociology of Technology." In *The Social Construction of Technological Systems: New Directions in the Sociology and History of Technology*, edited by Wiebe E. Bijker, Thomas P. Hughes, and Trevor Pinch, 261–80. Cambridge, MA: MIT Press, 1989.

Cowan, Ruth Schwartz. *More Work for Mother: The Ironies of Household Technology from the Open Hearth to the Microwave*. New York: Basic Books, 1983.

Dacome, Lucia. *Malleable Anatomies: Models, Makers, and Material Culture in Eighteenth-Century Italy*. Oxford: Oxford University Press, 2017.

Daston, Lorraine. "The History of Science and the History of Knowledge." *KNOW: A Journal on the Formation of Knowledge* 1, no. 1 (2017): 131–54.

Daston, Lorraine, ed. *Science in the Archives: Pasts, Presents, Futures*. Chicago: University of Chicago Press, 2017.

Daston, Lorraine. "Taking Note(s)." *Isis* 95, no. 3 (2004): 443–48.

Daston, Lorraine, and Peter Galison. "The Image of Objectivity." *Representations* 40, no. 1 (1992): 81–128.

Daston, Lorraine, and Peter Galison. *Objectivity*. Cambridge, MA: MIT Press, 2007.

Daybell, James. *The Material Letter in Early Modern England: Manuscript Letters and the Culture and Practices of Letter-Writing, 1512–1635*. London: Palgrave Macmillan, 2012.

de la Peña, Carolyn. "The History of Technology, the Resistance of Archives, and the Whiteness of Race." *Technology and Culture* 51, no. 4 (2010): 919–37.

Delbourgo, James, and Staffan-Müller Wille. "Introduction." *Isis* 103, no. 4 (2012): 710–15.

DiMeo, Michelle, and Sara Pennell, eds. *Reading and Writing Recipe Books, 1550–1800*. Manchester: University of Manchester Press, 2013.

Eddy, Matthew Daniel. "The Interactive Notebook: How Students Learned to Keep Notes during the Scottish Enlightenment." *Book History* 19, no. 1 (2016): 86–131.

Edgerton, David. *Shock of the Old: Technology and Global History since 1900*. London: Profile Books, 2006.

Epstein, Stephan R. "Craft Guilds in the Pre-modern Economy: A Discussion." *Economic History Review* 61, no. 1 (2008): 155–74.

Eyferth, Jacob. *Eating Rice from Bamboo Roots: The Social History of a Community of Handicraft Papermakers in Rural Sichuan, 1920–2000*. Cambridge, MA: Harvard Asia Center, 2009.

Fine, Lisa M. *The Souls of the Skyscraper: Female Clerical Workers in Chicago, 1870–1930*. Philadelphia: Temple University Press, 1990.

Fissell, Mary. "Introduction: Women, Health, and Healing in Early Modern Europe." *Bulletin of the History of Medicine* 82, no. 1 (2008): 1–17.

Flather, Amanda. *Gender and Space in Early Modern England*. Woodbridge: Boydell Press, 2007.

Frasca-Spada, Marina, and Nick Jardine, eds. *Books and the Sciences in History*. Cambridge: Cambridge University Press, 2000.

Frevert, Ute. *Mann und Weib, und Weib in Mann: Geschlechter-Differenzen in der Moderne*. Munich: Beck, 1995.

Friedman, Alice T. *House and Household in Elizabethan England: Wollaton Hall and the Willoughby Family*. Chicago: University of Chicago Press, 1989.

Frye, Susan. *Pens and Needles: Women's Textualities in Early Modern England*. Philadelphia: University of Pennsylvania Press, 2010.

Furth, Charlotte. *A Flourishing Yin: Gender in China's Medical History, 960–1665*. Berkeley: University of California Press, 1999.

Gamber, Wendy. "Industrial Genders: Constructing Boundaries." In *Gender and Technology: A Reader*, edited by Nina E. Lerman, Ruth Oldenziel, and Arwen P. Mohun, 123–52. Baltimore: Johns Hopkins University Press, 2003.

Garvey, Ellen Gruber. *Writing with Scissors: American Scrapbooks from the Civil War to the Harlem Renaissance*. New York: Oxford University Press, 2013.

Gates, Hill. "Money for the Gods." *Modern China* 13, no. 3 (1987): 259–77.

Gitelman, Lisa. *Paper Knowledge: Toward a Media History of Documents*. Durham, NC: Duke University Press, 2014.

Glover, Katharine. *Elite Women and Polite Society in Eighteenth-Century Scotland*. Woodbridge: Boydell, 2011.

Glover, Katharine. "The Female Mind: Scottish Enlightenment Femininity and the World of Letters." *Journal of Scottish Historical Studies* 25, no. 1 (2005): 1–20.

Greig, Hannah, Jane Hamlett, and Leonie Hannan, eds. *Gender and Material Culture in Britain since 1600*. London: Palgrave, 2016.

Gruber Garvey, Ellen. *Writing with Scissors: American Scrapbooks from the Civil War to the Harlem Renaissance*. New York: Oxford University Press, 2013.

Harkness, Deborah. "Managing an Experimental Household: The Dees of Mortlake and the Practice of Natural Philosophy." *Isis* 88, no. 2 (1997): 247–62.

Harvey, Karen. *The Little Republic: Masculinity and Domestic Authority in Eighteenth-Century Britain*. Oxford: Oxford University Press, 2012.

Harvey, Karen. *Reading Sex in the Eighteenth Century: Bodies and Gender in English Erotic Culture*. Cambridge: Cambridge University Press, 2004.

Hausen, Karin. "Family and Role-Division: The Polarization of Sexual Stereotypes in the Nineteenth Century—An Aspect of the Dissociation of Work and Family Life." In *The German Family: Essays on the Social History of the Family in Nineteenth- and Twentieth-Century Germany*, edited by Richard Evans and W. R. Lee, 51–83. London, 1981. Originally published as "Die Polarisierung der 'Geschlechtscharaktere'—Eine Spiegelung der Dissoziation von Erwerbs- und Familienleben." In *Sozialgeschichte der Familie in der Neuzeit Europas: Neue Forschungen*, edited by Werner Conze, 363–93. Stuttgart: Ernst Klett, 1976.

Hausen, Karin. *Geschlechtergeschichte als Gesellschaftsgeschichte*. Göttingen: Vandenhoeck & Ruprecht, 2012.

Heal, Felicity. *The Power of Gifts: Gift-Exchange in Early Modern England*. Oxford: Oxford University Press, 2014.

Heesen, Anke te. *The Newspaper Clipping: A Modern Paper Object*. Manchester: Manchester University Press, 2014.

Heesen, Anke te. "The Notebook: A Paper-Technology." In *Making Things Public: Atmospheres of Democracy*, edited by Bruno Latour and Peter Weibel, 582–89. Cambridge, MA: MIT Press, 2003.

Heesen, Anke te. *The World in a Box: The Story of an Eighteenth-Century Picture Encyclopedia*. Chicago: University of Chicago Press, 2002.

Herbert, Amanda E. *Female Alliances: Gender, Identity, and Friendship in Early Modern Britain*. New Haven, CT: Yale University Press, 2014.

Herzig, Rebecca M. "Does Google™ Have Gender? Technologies of Everyday Life in Affluent Industrial Societies." *Icon* 16 (2010): 92–97.

Hess, Volker, and J. Andrew Mendelsohn. "Case and Series: Medical Knowledge and Paper Technology, 1600–1900." *History of Science* 48, no. 3–4 (2010): 287–314.

Hess, Volker, and J. Andrew Mendelsohn. "Paper Technology und Wissensgeschichte." *NTM. Zeitschrift für Geschichte der Wissenschaften, Technik und Medizin* 21, no. 1 (2013): 1–10.

Hicks, Marie. *Programmed Inequality: How Britain Discarded Women Technologists and Lost Its Edge in Computing*. Cambridge, MA: MIT Press, 2017.

Hills, Richard Leslie. *Papermaking in Britain, 1488–1988: A Short History*. London: Athlone Press, 1988.

Hoffmann, Christoph. "Processes on Paper: Writing Procedures as Non-material Research Devices." *Science in Context* 26, no. 2 (2013): 279–303.

Hull, Matthew S. "The File: Agency, Authority, and Autography in an Islamabad Bureaucracy." *Language and Communication* 23, no. 3–4 (2003): 287–314.

Hunter, Dard. *Papermaking: The History and Technique of an Ancient Craft*. 2nd ed. London: Alfred A. Knopf, 1947. 3rd ed. New York: Alfred A. Knopf, 1974.

Iliffe, Rob, and Frances Wilmoth. "Astronomy and the Domestic Sphere: Margaret Flamsteed and Caroline Herschel as Assistant Astronomers." In *Women, Science, and Medicine 1500–1700: Mothers and Sisters of the Royal Society*, edited by Lynette Hunter and Sarah Hutton, 235–65. Stroud: Sutton, 1997.

Ingold, Tim. "Materials against Materiality." *Archaeological Dialogues* 14, no. 1 (2007): 1–16.

Ingold, Tim. *The Perception of the Environment: Essays in Livelihood, Dwelling, and Skill*. London: Routledge, 2000.

Jardine, Boris. "State of the Field: Paper Tools." *Studies in History and Philosophy of Science* 64 (2017): 53–63.

Jenkins, Penny. "India Proof Prints." *Paper Conservator* 14 (1990): 46–57.

Johns, Adrian. *The Nature of the Book: Print and Knowledge in the Making*. Chicago: University of Chicago Press, 1998.

Jordanova, Ludmilla. *Sexual Visions: Images of Gender in Science and Medicine between the Eighteenth and Twentieth Centuries*. Madison: University of Wisconsin Press, 1989.

Kafka, Ben. *The Demon of Writing: Powers and Failures of Paperwork*. New York: Zone Books, 2012.

Kafka, Ben. "Paperwork: The State of the Discipline." *Book History* 12, no. 1 (2009): 340–53.

Klein, Ursula. *Experiments, Models, Paper Tools: Cultures of Organic Chemistry in the Nineteenth Century*. Stanford, CA: Stanford University Press, 2002.

Klein, Ursula. "Paper Tools in Experimental Cultures." *Studies in History and Philosophy of Science* 32, no. 2 (2001): 265–302.

Klein, Ursula, and Emma C. Spary, eds. *Materials and Expertise in Early Modern Europe: Between Market and Laboratory*. Chicago: University of Chicago Press, 2010.

Kline, Ronald, and Trevor Pinch. "Users as Agents of Technological Change: The Social Construction of the Automobile in the Rural United States." *Technology and Culture* 37, no. 4 (1996): 763–95.

Knorr-Cetina, Karin. "Objectual Practice." In *The Practice Turn in Contemporary Theory*, edited by Theodore R. Schatzki, Karin Knorr-Cetina, and Eike von Savigny, 175–88. London: Routledge, 2001.

Ko, Dorothy. *The Social Life of Inkstones: Artisans and Scholars in Early Qing China*. Seattle: University of Washington Press, 2017.

Kohlstedt, Sally Gregory, ed. *History of Women in the Sciences: Readings from* Isis. Chicago: University of Chicago Press, 1999.

Kohlstedt, Sally Gregory, and Helen E. Longino, eds. *Women, Gender, and Science: New Directions*. Osiris 12. Chicago: University of Chicago Press, 1997.

Kraemer, Fabian. "Ulisse Aldrovandi's *Pandechion Epistemonicon* and the Use of Paper Technology in Renaissance Natural History." *Early Science and Medicine* 19, no. 5 (2014): 398–423.

Kraemer, Fabian, and Helmut Zedelmaier. "Instruments of Invention in Renaissance Europe: The Cases of Conrad Gesner and Ulisse Aldrovandi." *Intellectual History Review* 24, no. 3 (2014): 321–41.

Krajewski, Markus, and Peter Krapp. *Paper Machines: About Cards & Catalogs, 1548–1929*. Cambridge, MA: MIT Press, 2011.

Krill, John. *English Artists' Paper: Renaissance to Regency*. New Castle, DE: Oak Knoll Press, 2002.

Kurlansky, Mark. *Paper: Paging through History*. New York: Norton, 2017.

Latour, Bruno. *The Making of Law: An Ethnography of the Conseil d'Etat*. Cambridge, UK: Polity Press, 2009.

Latour, Bruno. *Science in Action: How to Follow Scientists and Engineers through Society*. Cambridge, MA: Harvard University Press: 1987. Reprinted 1988.

Latour, Bruno. "Visualization and Cognition: Drawing Things Together." In *Knowledge and Society: Studies in the Sociology of Culture Past and Present*, vol. 6, ed. Henrika Kuklick and Elizabeth Long, 1–40. Greenwich, CT: JAI Press, 1986.

Leong, Elaine. "Collecting Knowledge for the Family: Recipes, Gender and Practical Knowledge in the Early Modern Household." *Centaurus* 55, no. 2 (2013): 81–103.

Leong, Elaine. "'Herbals she peruseth': Reading Medicine in Early Modern England." *Renaissance Studies* 28, no. 4 (2014): 556–78.

Leong, Elaine. *Recipes and Everyday Knowledge: Medicine, Science, and the Household in Early Modern England*. Chicago: University of Chicago Press, 2018.

Lerman, Nina E. "Categories of Difference, Categories of Power: Bringing Gender and Race to the History of Technology." *Technology and Culture* 51, no. 4 (2010): 893–918.

Lerman, Nina E., Ruth Oldenziel, and Arwen P. Mohun, eds. *Gender and Technology: A Reader*. Baltimore: Johns Hopkins University Press, 2003.

Light, Jennifer S. "When Computers Were Women." *Technology and Culture* 40, no. 3 (1999): 455–83.

Lightman, Bernard V. *Victorian Popularizers of Science: Designing Nature for New Audiences*. Chicago: University of Chicago Press, 2007.

Long, Pamela O. *Artisans/Practitioners and the Rise of the New Sciences, 1400–1600*. Corvallis: Oregon State University Press, 2011.

Love, Harold. *The Culture and Commerce of Texts: Scribal Publication in Seventeenth-Century England*. Amherst: University of Massachusetts Press, 1998.

Mann, Susan. "Household Handicrafts and State Policies in Qing Times." In *To Achieve Wealth and Security: The Qing Imperial State and the Economy*, edited by Jane Kate Leonard and John R. Watts, 75–96. Ithaca, NY: Cornell University Press, 1992.

Marotti, Arthur. *Manuscript, Print, and the English Renaissance Lyric*. Ithaca, NY: Cornell University Press, 1995.

McGaw, Judith A. *Most Wonderful Machine: Mechanization and Social Change in Berkshire Paper Making, 1801–1885*. Princeton: Princeton University Press, 1987.

McGaw, Judith A. "Why Feminine Technologies Matter." In *Gender and Technology: A Reader*, edited by Nina E. Lerman, Ruth Oldenziel, and Arwen P. Mohun, 13–36. Baltimore: Johns Hopkins University Press, 2003.

McKenzie, Donald Francis, Peter D. McDonald, and Michael Felix Suarez, eds. *Making Meaning: "Printers of the Mind" and Other Essays*. Amherst: University of Massachusetts Press, 2002.

McKitterick, David. *Print, Manuscript and the Search for Order, 1450–1830*. Cambridge, UK: Cambridge University Press, 2003.

Milam, Erika Lorraine, and Robert A. Nye, eds. In *Scientific Masculinities*. Osiris 30. Chicago: University of Chicago Press, 2015.

Mommertz, Monika. "The Invisible Economy of Science: A New Approach to the History of Gender and Astronomy at the Eighteenth-Century Berlin Academy of Sciences." In *Men, Women, and the Birthing of Modern Science*, edited by Judith P. Zinsser, 159–78. DeKalb: Northern Illinois University Press, 2005.

Morantz-Sanchez, Regina Markell. *Sympathy and Science: Women Physicians in American Medicine*. New York: Oxford University Press, 1985.

Müller, Lothar. *Weiße Magie: Die Epoche des Papiers*. Munich: Hanser, 2012.

Nakamura, Lisa. "Indigenous Circuits: Navajo Women and the Racialization of Early Electronic Manufacture." *American Quarterly* 66, no. 4 (2014): 919–41.

Nowotny, Helga, and Karin Hausen, eds. *Wie männlich ist die Wissenschaft?* Frankfurt am Main: Suhrkamp, 1986.

Oertzen, Christine. von "Machineries of Data Power: Manual versus Mechanical Census Compilation in Nineteenth-Century Europe." In *Data Histories*, ed. Elena Aronova, Christine von Oertzen, and David Sepkoski, 129–50. Osiris 32. Chicago: University of Chicago Press, 2017.

Oertzen, Christine von, Maria Rentetzi, and Elizabeth Siegel Watkins. "Finding Science in Surprising Places: Gender and the Geography of Scientific Knowledge. Introduction to 'Beyond the Academy: Histories of Gender and Knowledge.'" *Centaurus* 55, no. 2 (2013): 73–80.

Ogborn, Miles. *Indian Ink: Script and Print in the Making of the English East India Company*. Chicago: University of Chicago Press, 2007.

Oldenziel, Ruth. "Why Masculine Technologies Matter." In *Gender and Technology: A Reader*, edited by Nina E. Lerman, Ruth Oldenziel, and Arwen P. Mohun, 37–71. Baltimore: Johns Hopkins University Press, 2003.

Opitz, Donald L., Staffan Bergwik, and Brigitte Van Tiggelen, eds. *Domesticity in the Making of Modern Science*. Basingstoke, Hampshire: Palgrave Macmillan, 2016.

Orlikowski, Wanda. "Sociomaterial Practices: Exploring Technology at Work." *Organization Studies* 28, no. 9 (2007): 1435–48.

Orlikowski, Wanda J. "The Sociomateriality of Organisational Life: Considering Technology in Management Research." *Cambridge Journal of Economics* 34, no. 1 (2009): 125–41.

Orlikowski, Wanda J., and Susan V. Scott. "Sociomateriality: Challenging the Separation of Technology, Work and Organization." *Academy of Management Annals* 2, no. 1 (2008): 433–74.

Oudshoorn, Nelly, and Trevor Pinch, eds. *How Users Matter: The Co-Construction of Users and Technologies*. Cambridge, MA: MIT Press, 2003.

Pal, Carol. *Republic of Women: Rethinking the Republic of Letters in the Seventeenth Century*. Cambridge: Cambridge University Press, 2012.

Park, Katharine. *Secrets of Women: Gender, Generation, and the Origins of Human Dissection*. New York: Zone Books, 2006.

Pelling, Margaret. "The Women of the Family? Speculations around Early Modern British Physicians." *Social History of Medicine* 8, no. 3 (1995): 383–401.

Plotnick, Rachel. "At the Interface: The Case of the Electric Push Button, 1880–1923." *Technology and Culture* 53, no. 4 (2012): 815–45.

Porter, Theodore. *Trust in Numbers: The Pursuit of Objectivity in Science and Public Life*. Princeton, NJ: Princeton University Press, 1995.

Powell, Margaret K., and Joseph Roach. "Big Hair." *Eighteenth-Century Studies* 38, no. 1 (2004): 79–99.

Prescott, Heather Munro. *Student Bodies: The Influence of Student Health Services in American Society and Medicine*. Ann Arbor: University of Michigan Press, 2007.

Rabier, Christelle. "Introduction: The Crafting of Medicine in the Early Industrial Age." *Technology and Culture* 54, no. 3 (2013): 437–59.

Raman, Bhavani. *Document Raj: Writing and Scribes in Early Colonial South India*. Chicago: University of Chicago Press, 2012.

Rankin, Alisha. *Panaceia's Daughters: Noblewomen as Healers in Early Modern Germany*. Chicago: University of Chicago Press, 2013.

Rauser, Amelia. "Hair, Authenticity, and the Self-Made Macaroni." *Eighteenth-Century Studies* 38, no. 1 (2004): 101–17.

Roberts, Lissa. "Practicing *Oeconomy* during the Second Half of the Long Eighteenth Century: An Introduction." *History and Technology* 30, no. 3 (2014): 133–48.

Roberts, Lissa, Simon Schaffer, and Peter Dear, eds. *The Mindful Hand: Inquiry and Invention from the Late Renaissance to Early Industrialization*. Amsterdam: Royal Netherlands Academy of Arts and Sciences, 2007.

Roberts, Lissa, and Simon Werrett, eds. *Compound Histories: Materials, Governance and Production, 1760–1840*. Leiden: Brill, 2017. Online version. Accessed January 2, 2018. http://booksandjournals.brillonline.com/content/books/9789004325562.

Roberts, Meghan. "Learned and Loving: Representing Women Astronomers in Enlightenment France." *Journal of Women's History* 29, no. 1 (2017): 14–37.

Rockman, Seth. "Introduction." Special section: "Forum: The Paper Technologies of Capitalism." *Technology and Culture* 58, no. 2 (2017): 487–505.

Rosenband, Leonard N. *Papermaking in Eighteenth-Century France: Management, Labor, and Revolution at the Montgolfier Mill, 1761–1805*. Baltimore: Johns Hopkins University Press, 2000.

Rosenberg, Daniel. "Early Modern Information Overload." *Journal of the History of Ideas* 64, no. 1 (2003): 1–9.

Rossiter, Margaret W. *Women Scientists in America: Before Affirmative Action, 1940–1972*. Baltimore: Johns Hopkins University Press, 1995.

Rossiter, Margaret W. *Women Scientists in America: Forging a New World since 1972*. Baltimore: Johns Hopkins University Press, 2012.

Rossiter, Margaret W. *Women Scientists in America: Struggles and Strategies to 1940*. Baltimore: Johns Hopkins University Press, 1982.

Rusnock, Andrea. "Quantifying Infant Mortality in England and France, 1750–1800." In *Body Counts: Medical Quantification in Historical and Sociological Perspective*, edited by Gérard Jorland, Annick Opinel, and Geroge Weisz, 65–86. London: McGill-Queen's University Press, 2005.

Russett, Cynthia Eagle. *Sexual Science: The Victorian Construction of Womanhood*. Cambridge, MA: Harvard University Press, 1989.

Salzberg, Rosa. "Masculine Republics: Establishing Authority in the Early Modern Venetian Printshop." In *Governing Masculinities in the Early Modern Period: Regulating Selves and Others*, edited by Susan Broomhall and Jacqueline van Gent, 47–65. Farnham, UK: Ashgate, 2011.

Schiebinger, Londa L. *The Mind Has No Sex? Women in the Origins of Modern Science*. Cambridge, MA: Harvard University Press, 1989.

Schiebinger, Londa L. *Nature's Body: Gender in the Making of Modern Science*. Boston: Beacon Press, 1993.

Schmidt-Bachem, Heinz. *Aus Papier: Eine Kultur- und Wirtschaftsgeschichte der Papier verarbeitenden Industrie in Deutschland*. Berlin: de Gruyter, 2011.

Scott, Joan W. "Gender: A Useful Category of Historical Analysis." *American Historical Review* 91, no. 5 (1986): 1053–75.

Secord, Anne. "Corresponding Interests: Artists and Gentlemen in Nineteenth-Century Natural History." *British Journal for the History of Science* 27, no. 4 (1994): 383–408.

Secord, James A. "Knowledge in Transit." *Isis* 95, no. 4 (2004): 654–72.

Secord, James A. *Victorian Sensation: The Extraordinary Publication, Reception, and Secret Authorship of* Vestiges of the Natural History of Creation. Chicago: University of Chicago Press, 2003.

Secord, James A. *Visions of Science: Books and Readers at the Dawn of the Victorian Age*. Chicago: University of Chicago Press, 2015.

Serrano, Elena. "Chemistry in the City: The Scientific Role of Female Societies in Eighteenth-Century Madrid." *Ambix* 60, no. 2 (2013): 146–52.

Shapin, Steven. "The House of Experiment in Seventeenth-Century England." *Isis* 79, no. 3 (1988): 373–404.

Shapin, Steven. "The Invisible Technician." *American Scientist* 77, no. 6 (1989): 554–63.

Shteir, Ann B., and Bernard V. Lightman, eds. *Figuring It Out: Science, Gender, and Visual Culture*. Hanover, NH: Dartmouth College Press, 2006.

Sigaut, François. "Technology." In *Companion Encyclopedia of Anthropology: Humanity, Culture, and Social Life*, edited by Tim Ingold, 420–59. London: Routledge, 1994.

Smith, Pamela H., Amy R. W. Meyers, and Harold J. Cook, eds. *Ways of Making and Knowing: The Material Culture of Empirical Knowledge*. Ann Arbor: University of Michigan Press, 2014.

Soll, Jacob. *The Information Master: Jean-Baptiste Colbert's Secret State Intelligence System*. Ann Arbor: University of Michigan Press, 2009.

Stallybrass, Peter. "Printing and the Manuscript Revolution." In *Explorations in Communication and History*, edited by Barbie Zelizer, 111–18. London: Routledge, 2008.

Stallybrass, Peter, Roger Chartier, J. Franklin Mowery, and Heather Wolfe. "Hamlet's Tables and the Technologies of Writing in Renaissance England." *Shakespeare Quarterly* 55, no. 4 (2005): 379–419.

Stoler, Ann Laura. *Along the Archival Grain: Epistemic Anxieties and Colonial Common Sense*. Princeton, NJ: Princeton University Press, 2010.

Terrall, Mary. "Masculine Knowledge, the Public Good, and the Scientific Household of Réaumur." In *Scientific Masculinities*, edited by Erika Lorraine Milam and Robert A. Nye, 182–201. Osiris 30. Chicago: University of Chicago Press, 2015.

Thiel, Viktor. "Papiererzeugung und Papierhandel vornehmlich in den deutschen Landen von den ältesten Zeiten bis zum Beginn des 19. Jahrhunderts." *Archivalische Zeitschrift* 41 (1932): 147–50.

Tsien, Tsuen-hsuin. *Paper and Printing*. Part 1 of *Chemistry and Chemical Technology*, vol. 5 of *Science and Civilisation in China*, edited by Joseph Needham. Cambridge: Cambridge University Press, 1985.

Verbrugge, Martha H. *Able-Bodied Womanhood: Personal Health and Social Change in Nineteenth-Century Boston*. New York: Oxford University Press, 1988.

Verbrugge, Martha H. *Active Bodies: A History of Women's Physical Education in Twentieth-Century America*. Oxford University Press, 2012.

Vismann, Cornelia. *Files: Law and Media Technology*. Stanford, CA: Stanford University Press, 2008.

Walsh, Judith. "The Use of Paper Splitting in Old Master Print Restorations." *Print Quarterly* 17, no. 4 (2000): 383–90.

Walsham, Alexandra. "The Social History of the Archive: Record-Keeping in Early Modern Europe." *Past and Present* 230, suppl. 11 (2016): 9–48.

Wang, Juhua, ed. *Zhongguo gudai zaozhi gongcheng jishu shi* [The history of papermaking technology in ancient China]. Taiyuan: Shanxi Education Press, 2005.

Werrett, Simon. "Household Oeconomy and Chemical Inquiry." In *Compound Histories: Materials, Governance and Production, 1760–1840*, edited by Lissa Roberts and Simon Werrett, 35–56. Leiden: Brill, 2017. Online version. Accessed January 2, 2018, http://booksandjournals.brillonline.com/content/books/9789004325562.

Werrett, Simon. "Matter and Facts: Material Culture in the History of Science." In *Material Evidence: Learning from Archaeological Practice*, edited by Robert Chapman and Alison Wylie, 339–52. New York: Routledge, 2014.

Werrett, Simon. *Thrifty Science: Making the Most of Materials in the History of Experiment*. Chicago: University of Chicago Press, 2019.

Wiggins, Alison. *Bess of Hardwick's Letters: Language, Materiality, and Early Modern Epistolary Culture*. New York: Routledge, 2016.

Wolfe, Heather. "'Neatly Sealed, with Silk, and Spanish Wax or Otherwise': The Practice of Letter-Locking with Silk Floss in Early Modern England." In *In the Prayse of Writing: Early Modern Manuscript Studies*, edited by S. P. Cerasano and Steven W. May, 169–89. London: British Library, 2012.

Woudhuysen, H. R. *Sir Philip Sidney and the Circulation of Manuscripts, 1558–1640*. Oxford: Clarendon Press, 1996.

Yale, Elizabeth. "The History of Archives: The State of the Discipline." *Book History* 18, no. 1 (2015): 332–59.

Yale, Elizabeth. *Sociable Knowledge: Natural History and the Nation in Early Modern Britain*. Philadelphia: University of Pennsylvania Press, 2016.

Yates, JoAnne. *Control through Communication: The Rise of System in American Management*. Baltimore: Johns Hopkins University Press, 1989. Reprinted 1993.

Yates, JoAnne. *Structuring the Information Age: Life Insurance and Technology in the Twentieth Century*. Baltimore: Johns Hopkins University Press, 2005.

Yeo, Richard. *Notebooks, English Virtuosi and Early Modern Science*. Chicago: University of Chicago Press, 2014.

CONTRIBUTORS

Carla Bittel is associate professor of history at Loyola Marymount University in Los Angeles. Her research focuses on gender and the politics of knowledge. She has written on the history of women's health, women physicians, and the role of science in medicine. Bittel is the author of *Mary Putnam Jacobi and the Politics of Medicine in Nineteenth-Century America* (North Carolina, 2009). Her current work examines gender, race, and phrenology in the American context.

Dan Bouk researches the history of bureaucracies, quantification, and other modern things shrouded in cloaks of boringness. His first book, *How Our Days Became Numbered: Risk and the Rise of the Statistical Individual* (Chicago, 2015), explores the spread into ordinary Americans' lives of the US life insurance industry's methods for quantifying people, discriminating by race, justifying inequality, and thinking statistically. His recent writings put today's political and economic values of personal data in a wider historical context and explore the ways social scientists working with corporations and governments have rationalized the life cycle. He is currently writing a narrative history of the US census of 1940, and sharing his discoveries at censusstories.us.

Matthew Daniel Eddy is Durham University's chair in the history of philosophy and science. He researches the rise of modern science and its interface with the everyday life of the mind. He is especially interested in the historical factors that shaped how scientists experienced and represented the environment, cognition, and evolution. He has also published on a variety of themes relevant to the history of gender, religion, and childhood.

Jacob Eyferth is associate professor of modern Chinese history at the University of Chicago. He has published widely on social and cultural history of twentieth-century China, focusing especially on rural China, the history of work, technology, gender, and everyday life. His ethnographic study *Eating Rice from Bamboo Roots: The Social History of Community of Handicraft Papermakers in Rural Sichuan, 1920–2000* (Harvard, 2009) won the 2011 Joseph Levenson Prize. Currently he is working on a monograph tentatively titled "Cotton, Gender, and Revolution in Twentieth-Century China," using cloth and clothing as a lens through which to analyze how the monumental changes of the twentieth century—revolution, collectivization, industrialization, etc.—transformed the lives of rural women.

Elaine Leong holds a Wellcome Trust University Award at the History Department of University College London. From 2012 to 2019 she led the Minerva Research Group "Reading and Writing Nature in Early Modern Europe" at the Max Planck Institute for the History of Science, Berlin. She has published on histories of health, vernacular knowledge, and gender in early modern Europe. She is the author of *Recipes and Everyday Knowledge: Medicine, Science, and the Household in Early Modern England* (Chicago, 2018) and coeditor of *Secrets and Knowledge: Medicine, Science and Commerce 1500–1800* (Ashgate, 2011) and "Testing Drugs and Trying Cures in Medieval and Early Modern Europe," *Bulletin of the History of Medicine* (2017).

Beth Linker is an associate professor at the University of Pennsylvania in the Department of the History and Sociology of Science. Her research and teaching interests include the history of science and medicine, the body, health policy, and disability. She is the author of *War's Waste: Rehabilitation in World War I America* (Chicago, 2011). Linker is also the coeditor of *Civil Disabilities: Citizenship, Membership, and Belonging* (Pennsylvania, 2014). In her current book project, "Slouch: The Rise and Fall of the American Posture Sciences," she explores the rise of noncommunicable epidemics in the twentieth century, using posture as a case study.

Anna Maerker is senior lecturer in history of medicine at King's College London. Her publications on the material culture of medicine and science and on public history include *Model Experts: Wax Anatomies and Enlightenment in Florence and Vienna, 1775–1815* (Manchester, 2011), and, with Simon Sleight and Adam Sutcliffe, *History, Memory and Public Life: The Past in the Present* (Routledge, 2018).

Christine von Oertzen is a senior research scholar at the Max Planck Institute for the History of Science, Berlin. Her monographs *The Pleasure of a Surplus Income* (Berghahn, 2007) and *Science, Gender, Internationalism* (Palgrave Macmillan, 2014) explore gender politics in society and in academia. In her recent publications, she extends her interest in gender to the material culture of collecting, compiling, and visualizing data, from manual and machine processing of nineteenth-century census data to at-home observation of infants. She is coeditor, with Maria Rentetzi and Elizabeth Watkins, of the 2013 *Centaurus* special issue, titled *Beyond the Academy: Histories of Gender and Knowledge*; and, with Elena Aronova and David Sepkoski, of the 2017 Osiris volume *Data Histories.*

Elena Serrano is a research scholar at the Max Planck Institute for the History of Science, Berlin. She is interested in how knowledge travels and becomes embedded in societies, paying special attention to material culture, longue durée history of artifacts, and communities of practitioners beyond the academy. She earned her PhD at the Universitat Autònoma de Barcelona, Spain, with a prize-winning dissertation, which she is currently turning into her first monograph, titled "Women Improvers: Gendering Useful Knowledge in Spanish Enlightenment Spain (1753–1808)." Her research has been funded by the Chemical Heritage Foundation, the Spanish Ministry of Culture, and the Thyssen Foundation.

Gabriella Szalay is a PhD candidate at the Georg August University in Göttingen. Her dissertation, "Materializing the Past: The History of Art and Natural History in Germany, 1750–1800," examines how a network of German-speaking connoisseurs went about the task of writing an as yet unwritten history of the mechanical arts. In the fall of 2018 she began a two-year curatorial fellowship at the Harvard Art Museums.

Simon Werrett is a senior lecturer in the Department of Science and Technology Studies at University College London, where he teaches the history of early modern science. Before coming to UCL he taught in the Department of History at the University of Washington. His publications include *Fireworks: Pyrotechnic Arts and Sciences in European History* (Chicago, 2010); *Compound Histories: Materials, Governance, and Production, 1760–1840,* with Lissa Roberts (Brill, 2017); and *Thrifty Science: Making the Most of Materials in the History of Experiment* (Chicago, 2019).

Heather Wolfe is curator of manuscripts at the Folger Shakespeare Library. She has a PhD in English literature from the University of Cambridge and an MLIS from the University of California, Los Angeles. Her current projects include Shakespeare Documented, an online repository of images, transcriptions, and descriptions of all known references and allusions to Shakespeare, his family, and his writings in his lifetime; Early Modern Manuscripts Online, a transcription and crowdsourcing project that contributes to an online corpus of images and transcriptions of early modern manuscripts; and Before "Farm to Table," a Mellon-funded multidisciplinary research project that explores early modern foodways in Britain and the Atlantic. She also is working on early modern writing paper and Shakespeare's coat of arms, and has published widely on manuscript culture in early modern England.

Elizabeth Yale teaches in the Department of History and at the Center for the Book at the University of Iowa. Her book, *Sociable Knowledge: Natural History and the Nation in Early Modern Britain,* was published by University of Pennsylvania Press in 2016 in the Material Texts series. She is the author of essays published with *Isis, Book History,* and *Studies in History and Philosophy of Biological and Biomedical Sciences.* With Vera Keller and Anna Marie Roos, she edited *Archival Afterlives: Life, Death, and Knowledge-Making in Early Modern British Scientific and Medical Archives* (Brill, 2018). Her current book project explores gender, archives, and information management in the sciences in early modern Britain.

INDEX